Lecture Notes in Electrical Engineering

Volume 541

Lecture Notes in Electrical Engineering (LNEE) is a book series which reports the latest research and developments in Electrical Engineering, namely:

- Communication, Networks, and Information Theory
- Computer Engineering
- Signal, Image, Speech and Information Processing
- Circuits and Systems
- Bioengineering
- Engineering

The audience for the books in LNEE consists of advanced level students, researchers, and industry professionals working at the forefront of their fields. Much like Springer's other Lecture Notes series, LNEE will be distributed through Springer's print and electronic publishing channels.

More information about this series at http://www.springer.com/series/7818

Yidong Gu · Ming Gao · Guangheng Zhao
Editors

Proceedings of the Tiangong-2 Remote Sensing Application Conference

Technology, Method and Application

Springer

Editors
Yidong Gu
Technology and Engineering Center
for Space Utilization, CAS
Beijing, China

Guangheng Zhao
Technology and Engineering Center
for Space Utilization, CAS
Beijing, China

Ming Gao
Technology and Engineering Center
for Space Utilization, CAS
Beijing, China

ISSN 1876-1100 ISSN 1876-1119 (electronic)
Lecture Notes in Electrical Engineering
ISBN 978-981-13-3500-6 ISBN 978-981-13-3501-3 (eBook)
https://doi.org/10.1007/978-981-13-3501-3

Library of Congress Control Number: 2018962984

This Springer imprint is published by the registered company Springer Nature Singapore Pte Ltd.
The registered company address is: 152 Beach Road, #21-01/04 Gateway East, Singapore 189721, Singapore

Organization

Expert Committee

Shancong Zhang	Technology and Engineering Center for Space Utilization (CSU), Chinese Academy of Sciences (CAS)
Congmin Lv	CSU, CAS
Xuzhi Li	CSU, CAS
Jun Wei	Shanghai Institute of Technical Physics, CAS
Yunhua Zhang	National Space Science Center, CAS
Shurong Wang	Changchun Institute of Optics, Fine Mechanics and Physics, CAS

Advisory Committee

Fang Shen	East China Normal University
Fuxiang Huang	National Satellite Meteorological Center
Guangjian Yan	Beijing Normal University
Ji Zhou	University of Electronic Science and Technology of China
Jihua Meng	Institute of Remote Sensing and Digital Earth, CAS
Jingjuan Liao	Institute of Remote Sensing and Digital Earth, CAS
Jinsong Chong	Institute of Electronics, CAS
Minzheng Duan	Institute of Atmospheric Physics, CAS
Pifu Cong	National Marine Environmental Monitoring Center
Qianguo Xing	Yantai Institute of Coastal Zone Research, CAS
Shudong Wang	Institute of Remote Sensing and Digital Earth, CAS
Wenquan Zhu	Beijing Normal University
Xiaofeng Yang	Institute of Remote Sensing and Digital Earth, CAS

Xiaoxian Huang	Shanghai Institute of Technical Physics, CAS
Xin Li	Anhui Institute of Optics and Fine Mechanics, CAS
Yong Zhang	National Satellite Meteorological Center
Yuchu Qin	Institute of Remote Sensing and Digital Earth, CAS
Yutian Fu	Shanghai Institute of Technical Physics, CAS
Zengxiang Zhang	Institute of Remote Sensing and Digital Earth, CAS
Zhengqiang Li	Institute of Remote Sensing and Digital Earth, CAS
Bo Zhong	Institute of Remote Sensing and Digital Earth, CAS
Dongdong Shao	Beijing Normal University
Hongyan Ren	Institute of Geographic Sciences and Natural Resources Research, CAS
Jian Chen	Nanjing University of Information Science and Technology
Jinyang Du	Institute of Remote Sensing and Digital Earth, CAS
Lingmei Jiang	Beijing Normal University
Liwei Li	Institute of Remote Sensing and Digital Earth, CAS
Lixin Dong	National Satellite Meteorological Center
Runhe Shi	East China Normal University
Shilin Tang	South China Sea Institute of Oceanology, CAS
Shuying Jin	Wuhan University
Xiaohuan Xi	Institute of Remote Sensing and Digital Earth, CAS
Yonghua Sun	Capital Normal University
Wei Chen	Institute of Remote Sensing and Digital Earth, CAS

Compiling Committee

Shengyang Li	CSU, CAS
Lingli Mu	CSU, CAS
Bangyong Qin	CSU, CAS
Kang Liu	CSU, CAS
Chunying Wu	CSU, CAS
Wanfeng Zhang	CSU, CAS
Zhiwen Liu	CSU, CAS
Hong Tan	CSU, CAS
Xue Wan	CSU, CAS

Contents

Earth Observation Payloads and Data Applications of Tiangong-2 Space Laboratory

Ming Gao(✉)

Key Laboratory of Space Utilization, Technology and Engineering Center for Space Utilization, Chinese Academy of Sciences, Beijing 100094, China
mgao@csu.ac.cn

Abstract. As China's first Space Laboratory, Tiangong-2 was designed to carry on a number of space scientific experiments and application tests. For the Earth observation, Tiangong-2 is equipped with Wide-band Imaging Spectrometer (WIS), Interferometric Imaging Radar Altimeter (InIRA) and Multi-band Ultraviolet Edge Imaging Spectrometer (MUEIS). Up to now, these earth observation payloads have acquired a large amount of data, which has been released and applied in many fields. In this paper, the working mechanism, technical characters and advantages of these payloads are introduced, and the data acquirement, processing and service are summarized. With the data accumulation and application promotion, a series of application results have emerged. Some typical application achievements are presented as well. In the future, the earth observation payloads will continue providing data, and more and more applications are prospected.

Keywords: Tiangong-2 · Earth observation payload
Wide-band Imaging Spectrometer · Interferometric Imaging Radar Altimeter
Multi-band Ultraviolet Edge Imaging Spectrometer · Data application

1 Introduction

Tiangong-2, successfully launched on September 15, 2016, is the first space laboratory in China [1]. It was designed to carry on a number of space scientific experiments and application tests, which represents the development direction of China's space science frontier and strategic high-technology [2]. The Schematic diagram of Tiangong-2 is shown as Fig. 1.

Tiangong-2 Space Laboratory carried on more tasks than those of Tiangong-1. Tiangong-2 boarded more than ten space scientific experiments and application tests, covering earth observation and space earth science, space astronomy, microgravity physics, microgravity fluid physics and space material science, space life science and space environment, etc. [4].

Y. Gu et al. (Eds.): *Proceedings of the Tiangong-2 Remote Sensing Application Conference*, LNEE 541, pp. 1–13, 2019.
https://doi.org/10.1007/978-981-13-3501-3_1

Fig. 1. Schematic diagram of Tiangong-2 Space Laboratory [3]

2 Earth Observation Payloads

The earth observation payloads on Tiangong-2 includes Wide-band Imaging Spectrometer (WIS), Interferometric Imaging Radar Altimeter (InIRA) and Multi-band Ultraviolet Edge Imaging Spectrometer (MUEIS) based on advanced technology [2]. These earth observation payloads have been playing an important role in such fields as land, ocean and atmosphere and so on [5].

2.1 Wide-Band Imaging Spectrometer

WIS is a new generation moderate resolution optical payload, with wide-band and wide field of view [4]. WIS contains 14 programmable visible and near-infrared spectral channels (0.40 μm ~ 1.04 μm), 2 short wavelength infrared channels (1.232 μm ~ 1.654 μm) and 2 thermal infrared channels (8.125 μm ~ 9.275 μm). The ground resolution of each band in nadir point is 100 m, 200 m and 400 m respectively. WIS adopted push-broom and multi-module fields of view splicing imaging technology, with 300 km swath and 42° field of view [4]. WIS data are suitable for large-scale ground object detection, inland lake extraction, land and atmospheric monitoring, ocean and coastal water color monitoring, and water temperature observation, etc. [4]. The schematic diagram of WIS is shown as Fig. 2.

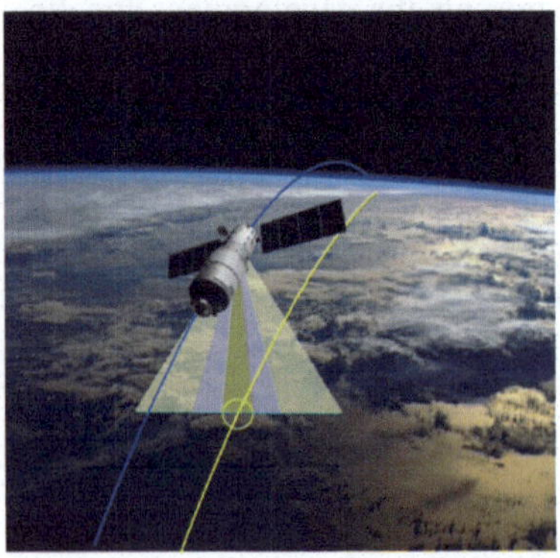

Fig. 2. Schematic diagram of on-orbit observation for WIS

2.2 Interferometric Imaging Radar Altimeter

InIRA uses small incidence angle to acquire high-precision ocean-land interference data, which can obtain echo signals of sea surface or terrestrial surface with high-coherence, based on receivers with dual-antenna and dual-channel [4]. It can extract the ocean interference phase map and three-dimensional ocean morphology. In the land applications, it also can obtain the three-dimensional topographic information. Figure 3 is the schematic diagram of on-orbit observation for InIRA Fig. 3. Schematic diagram of on-orbit observation for InIRA [1].

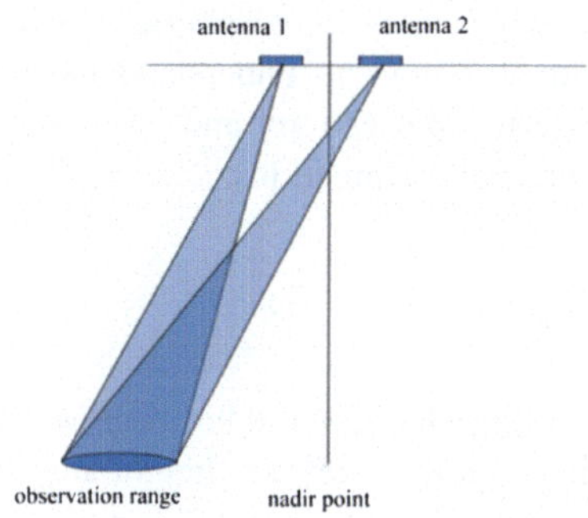

Fig. 3. Schematic diagram of on-orbit observation for InIRA [1]

The spatial resolution of InIRA in ocean interference model is 10 km × 10 km, with elevation precision is less than 8.2 cm (when orbital accuracy is less than 3 cm), and that in land model is 200 m × 200 m, with elevation precision is less than 10 m. The spatial resolution of the two-dimensional image is 100 m × 100 m.

InIRA data plays an important role in research of the global marine and land dynamic environment, including sea level altitude measurement, sea surface wind speed measurement, ocean currents monitoring, etc. [1].

2.3 Multi-band Ultraviolet Edge Imaging Spectrometer

MUEIS consists of Multi-Azimuth UV Imager (MAVI) and Front-azimuth Broadband Hyperspectrometer (FABH) [4].

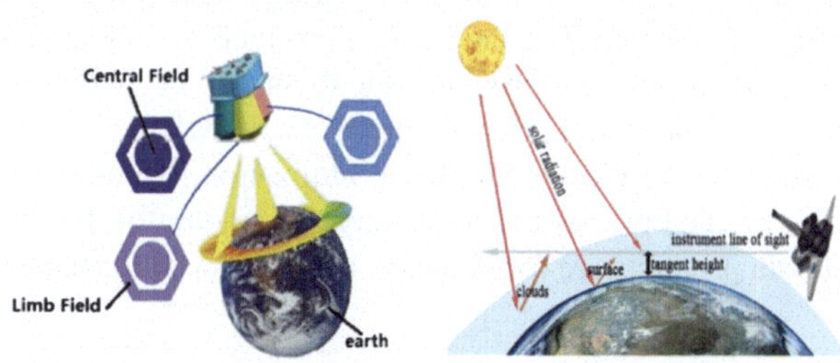

Fig. 4. Schematic diagram of on-orbit observation for MUEIS [6]

MAVI is used to detect radiation characteristics of the global mesosphere atmosphere with large field of view. FABH can provide precise spectral detection of atmosphere with over 120 channels. The schematic diagram is shown in Fig. 4.

MAVI has two fields of view with three channels (265 nm, 295 nm and 360 nm). One is central field used to detect the radiation brightness of whole atmosphere within 10 degrees of the nadir point, and the other one is limb field used to observe radiation brightness of the earth in six-azimuth with the field view about 140°. FABH with spectrum range of 290 nm ~ 1000 nm is used to detect the change of atmospheric precise spectrum change at different altitude from 10 km ~ 60 km [6].

MUEIS data is able to extract the vertical structure and three-dimensional dynamic distribution of atmospheric trace gases, atmospheric density, ozone distribution and aerosol, which can help human beings to understand the interaction between mesosphere and lower thermosphere, solar activity and space weather, as well as the relationship between earth weather and climate [4].

3 On-Orbit Operation

Since Tiangong-2 was successfully launched, it has been working for more than 2 years in a near-circular orbit with 340 km ~ 450 km height and 42° ~ 43° inclination [7]. The observation coverage range is between north 42° and south 42° latitude, shown in Fig. 5.

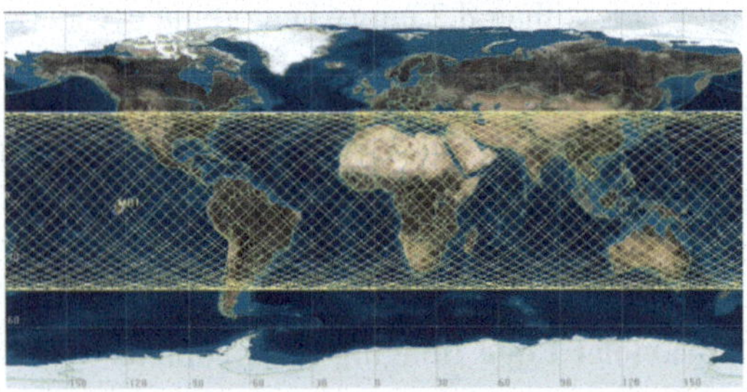

Fig. 5. The observation coverage range in Tiangong-2

At present, a large amount of high-quality image data have been obtained by Tiangong-2. Among them, the WIS standard data products are about 8.98 TB covering 119.1 million km^2. The InIRA data products are about 1.4 TB, with a global coverage of 33.97 million km^2. The standard data products of MUEIS are about 3.98 TB. Technology and Engineering Center for Space Utilization, Chinese Academy of Sciences (CSU) carries on data processing and generates amount of data products, shown in Table 1. All of data products are available on the website (http://www.msadc.cn/en).

Table 1. Standard data products of each earth observation payloads

Product Level	WIS	InIRA	MUEIS
Level 1	Radiometric correction products	Multiple-visual replica image products	Preprocessing products
Level 2	Geometric correction products	Geometric correction products	Primary atmospheric products
Level 3	Geometric correction products with high precision	Ortho-rectification products	None
Level 4	Ortho-rectification products	Three-dimensional elevation products	None
Level 5	General thematic products	None	None
Product format	Geo TIFF and XML	DAT/Geo TIFF and XML	HDF 5

4 Data Applications

With the data accumulation, plenty of data has been utilized in the fields of land and resource investigation, ocean and coastal zone monitoring, lake and farmland extraction, environment evaluation, etc. A series of application results have emerged, which showed great application potential.

4.1 The Applications of WIS

WIS data has been widely used in coastal zone change monitoring, lake extraction and change monitoring, land cover classification and change detection, environment monitoring and evaluation, crop identification, etc.

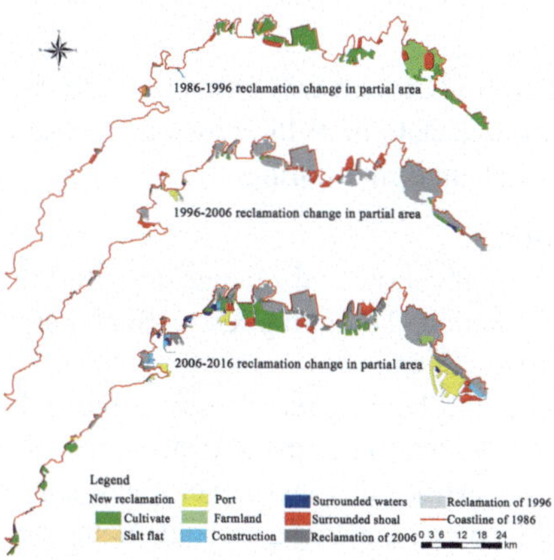

Fig. 6. Distribution of reclamation area in the Liaodong Bay

Coastal Zone Change Monitoring

Based on WIS data and other data, the researchers have carried out coastal zone change monitoring of Laizhou Bay, land change monitoring of Liaodong Bay (Fig. 6) [8], and Pearl River Estuary in the last 30 years.

The above applications reveal the spatial-temporal relationship between the coastal zone and the reclamation area in the Liaodong Bay. The result shows that it has the largest new reclamation area (266.55km2) in Linghai City of Liaoning Province in the past 30 years. The second largest one (226.13 km^2) is Dayi County of Liaoning Province [8].

Lake Extraction

The near-infrared data of WIS has also been used to extract water bodies and monitor change in Qinghai Lake, Poyang Lake, and Cuoqin County (Fig. 7).

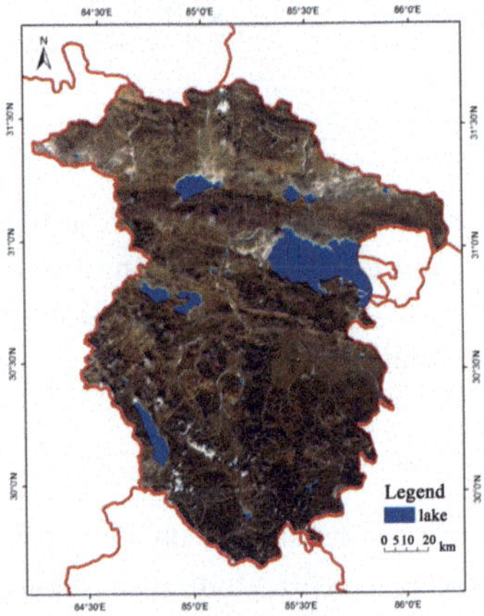

Fig. 7. Lake extraction in Cuoqin County, China

In Cuoqin, about 1254 km^2 of lake was extracted, and the identification accuracy was 93.74%. The near-infrared data of WIS shows the broad application prospects in water body extraction and change monitoring.

Land Cover Classification

In terms of land cover classification and change monitoring, the data of WIS was utilized in northwestern Mexico, Dongying City, Shandong Province (Fig. 8), and Nanjing City, Jiangsu Province, etc.

Using the data of WIS, the land cover classification of Dongying City has been carried out with accuracy of 84.5%. The spatial distribution of seven typical land cover in Dongying City was well extracted. Land cover classification results show that the near-infrared data of WIS contains a large amount of information suitable for high-precision land use classification.

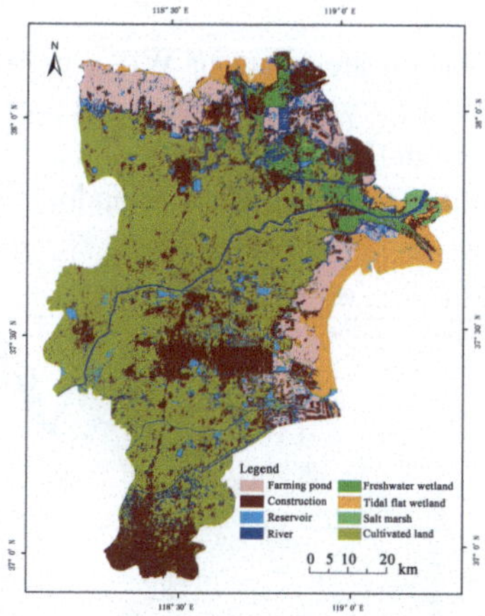

Fig. 8. Distribution of land cover types in Dongying, China

Ecological Environment Monitoring and Evaluation

At present, WIS data played an important role in wetland change monitoring in the Yellow River Delta and Panjin City, Liaoning Province [9]. Also the governments have used the data to identify algae in Taihu Lake and Chaohu Lake, and mangroves in the Beibu Gulf, Guangxi Province.

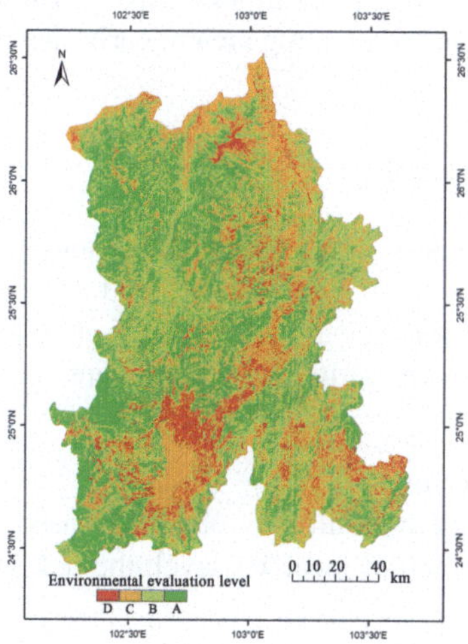

Fig. 9. Environment evaluation for Kunming City, China

As shown in Fig. 9, from A (green) to D (red), the environmental condition is reduced. In this application, area A is account for 62.75% of the total area of Kunming, which has a good ecological environment quality [10]. The monitoring and evaluation results reflect the ecological environment, which can provide decision support for the government to protect and restore ecological environment scientifically.

Crop Identification
In crop identification, the near-infrared data of WIS can reflect the planting structure and area, which provide support for yield evaluation. At present, the data had been used in cotton planting extraction in Hami, Xinjiang [11], winter wheat planting area extraction in Binzhou City and Dongying City, Shandong Province.

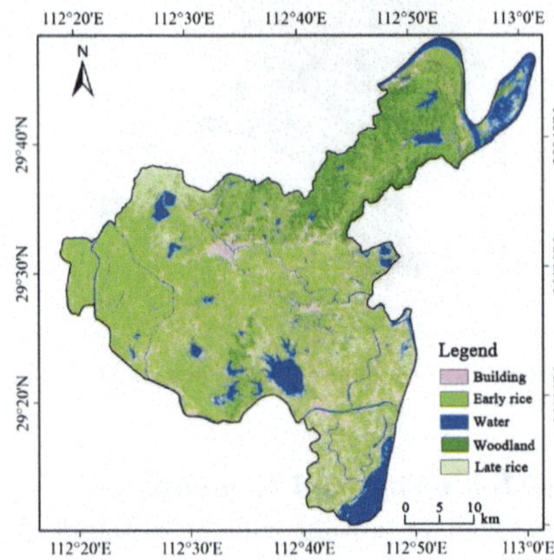

Fig. 10. Rice planting extraction of Huarong County, China

The spatial distribution of rice planting in Huarong, Hunan Province [12] is shown in Fig. 10. In this application, the planting area of early rice and later rice was extracted with high accuracy.

4.2 The Applications of InIRA

The data obtained by InIRA has been used in a large number of application researches on sea level altitude measurement, sea surface wind speed measurement, sea surface wavelength and wave inversion, ocean vortex detection, lake level inversion and so on. This paper takes the sea level altitude measurement and sea surface wind speed measurement as examples.

Sea level Altitude Measurement
Sea level altitude measurement result was obtained based on InIRA data. Figure 11 shows the InIRA elevation map of local sea level, the red line is shooting trajectory of 5° incident angle.

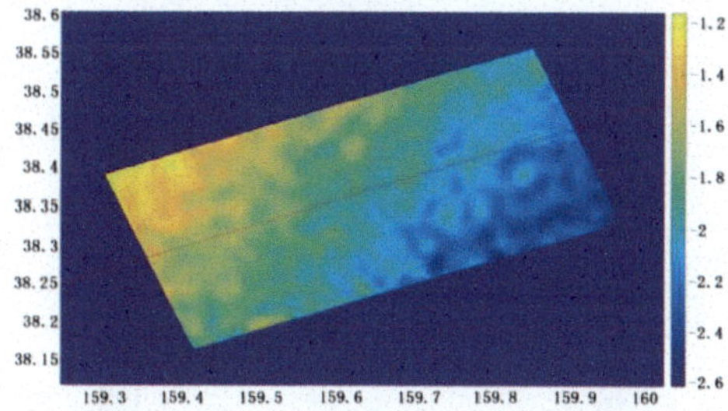

Fig. 11. InIRA elevation map of the local sea level in the Northwest Pacific

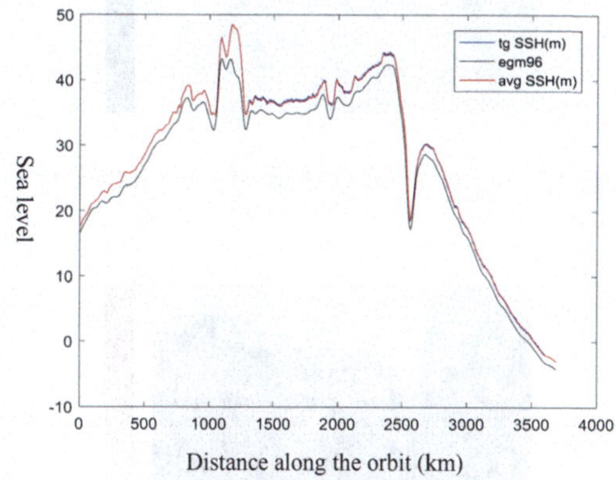

Fig. 12. Sea level altitude along the rail at 5° incident angle in the western Pacific

Figure 12 is the sea level altitude obtained by inversion of the position along the 5° incident angle. The blue line is the absolute elevation of the sea level derived from the InIRA data. The red line is the average of elevation in the corresponding area, and black line is the geoid. As shown in Fig. 12, the sea level altitude derived from InIRA data is consistent with the trend of the geoid.

Sea surface Wind Speed Measurement

Using the InIRA data obtained on September 22, 2016, the sea surface wind speed inversion in the North Atlantic Ocean is shown in Fig. 13.

Figure 14 showed the wind speed reanalysis result by US National Centers for Environmental Prediction. Comparing of Fig. 13 with Fig. 14, the measurement accuracy of InIRA is less than 2 m/s. In terms of sea surface wind speed measurement, InIRA data can provide effective support for ocean dynamics research.

4.3 The Applications of MUEIS

MUEIS data can be used to obtain radiation brightness image and atmospheric parameter information with high spatial, high time and high vertical resolution.

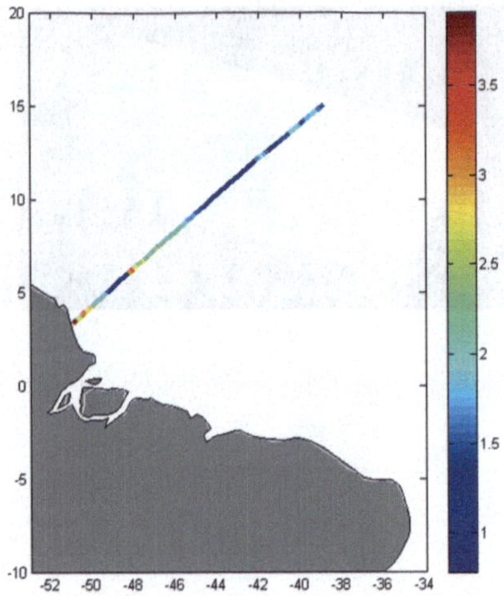

Fig. 13. Inversion of wind speed using InIRA data

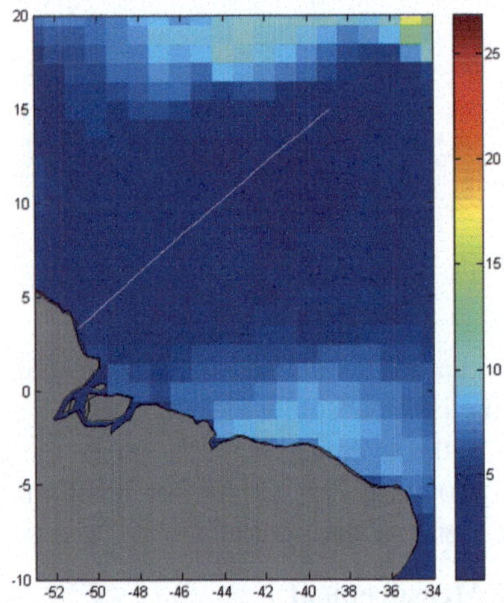

Fig. 14. Wind speed reanalysis result by NCEP (2016.09.22)

Furthermore, it can extract the vertical structure and three-dimensional dynamic distribution of atmospheric trace gases, atmospheric density, ozone and aerosol.

As shown in Fig. 15, radiation brightness image obtained by MAVI illustrates the radiance stratification phenomenon obviously, which can be used in stratosphere ozone inversion, ozone distribution change monitoring and ozone product accuracy evaluation of similar payloads.

Figure 16 is pseudo color image acquired by FABH. Vertical direction represents different atmosphere from 10 km to 60 km, and horizontal direction represents wavelength (290 nm ~ 1000 nm). As the height increasing, the gradual change and layered phenomenon can be found in Figs. 16 and 17.

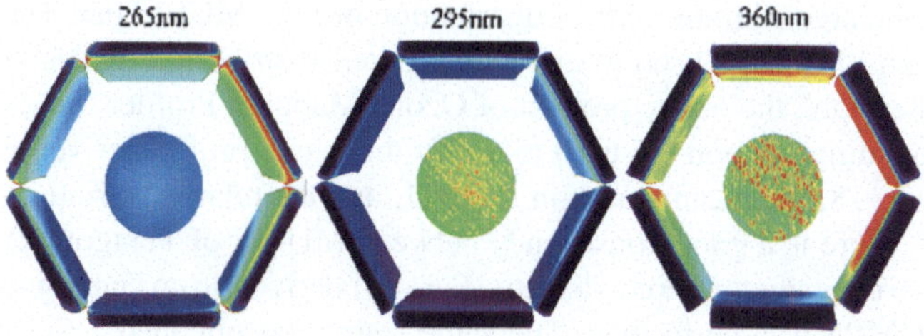

Fig. 15. Radiation brightness image obtained by MAVI on April 11, 2017

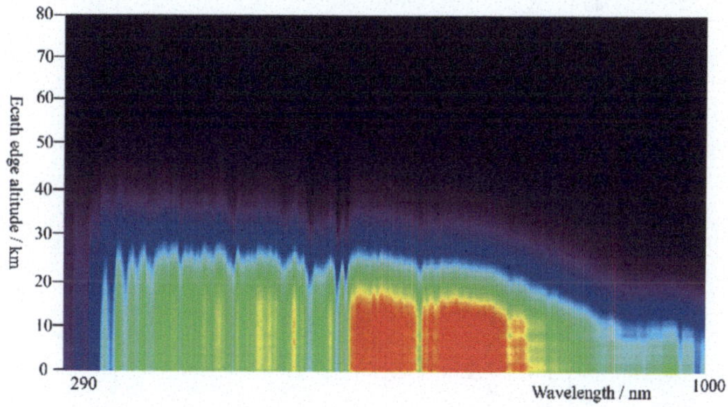

Fig. 16. Atmosphere spectral image obtained by FABH on April 11, 2017

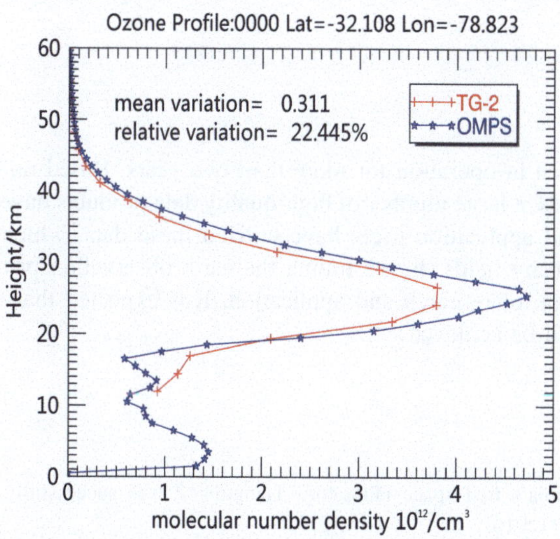

Fig. 17. Comparison between ozone profiles derived from Tiangong-2 and OMPS (2016.12.22)

Based on the algorithm of vertical distribution profile, MUEIS data (vertical spatial resolution: 3 km) has been used to extract the ozone distribution of Pacific Ocean and Beijing. Furthermore, the ozone product of Ozone Mapping Profiler Suite (OMPS) and the data of sounding balloon are used to verify the ozone profile derived from MUEIS data. In the Fig. 18, the mean variation is 0.311, and the relative variation is 22.445%. That is to say, there is a good consistency between MUEIS of Tiangong-2 and OMPS. Further comparison among ozone density (Fig. 18) derived fromTiangong-2, sounding balloon and OMPS was performed. The same result was obtained.

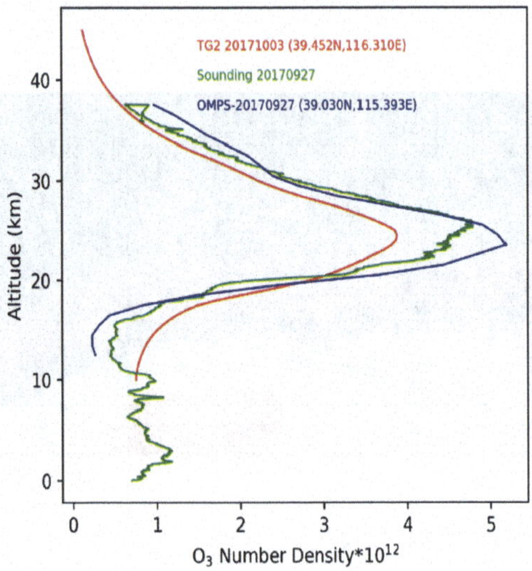

Fig. 18. Comparison of ozone density derived from Tiangong-2, sounding balloon and OMPS (2017.10.03)

5 Conclusion

Tiangong-2 has been in operation for more than two years. Based on advanced earth observation payloads, a large number of high-quality data products have been obtained. Many scientists and application users have utilized these data, which has played an important role in many fields. In the future, the earth observation payloads will continue providing data for research and application. It is expected that more and more valuable results will be achieved.

References

1. Jinghua, W.: China's first space laboratory Tiangong 2 was successfully launched. China Aerospace (10), 9 (2016)
2. Qiong, L.: Black Technology in "Tiangong No. 2". China Economic Report (10), 111–114 (2016)

3. Ji, Y.: The Tiangong 2 is the "most busy" space laboratory. China Aerospace (11), 11–12 (2016)
4. Manned Space Space Application Data Promotion Service Platform. http://www.msadc.cn/rwjstgeh/. Accessed 21 Oct 2018
5. China Manned Space Official Website. http://www.cmse.gov.cn/art/2017/12/11/art_19_32017.html. Accessed 21 Oct 2018
6. Zhiwen, L., Shengyang, L.: Tiangong 2 UV forward spectrometer atmospheric edge height calculation. Optics Precis. Eng. **06**, 1517–1523 (2018)
7. The official website of the Space Application Engineering and Technology Center of the Chinese Academy of Sciences. http://www.csu.cas.cn/gcdt/fzdt/201809/t20180926_5099083.html. Accessed 21 Oct 2018
8. Kang, L.I.U., Bang-yong, Q.I.N., Ling-li, M.U., Sheng-yang, L.I.: Monitoring and analysing of reclamation change based on multi-data remote sensing images: a case study of partial area of Liaoning Province. MARINE Environ. Sci. **36**(06), 911–917 (2017)
9. Jing, X., Dan, M.: Wetland monitoring in Panjin City based on TianGong-2 remote sensing data
10. Kang, L., Bangyong, Q., Shengyang, L.: The application of the Tiangong-2 WIS data in the ecological environment evaluation: a case study of Kunming. In: Proceeding of International Workshop on Environment and Geoscience, June 15–17, Hangzhou, China (2018)
11. Xiaojun, S., Kangyu, F., Jie, W., Wenchao, Q., Xiaolu, L.: Identification of Cotton by Using Random Forest with Tiangong-2 Data
12. Zhuang, Z., Shengyang, L., Yuyang, S., et al.: Early and Late Rice Identification from Tiangong-2 Wide Band Images Based on CNN//Symposium on Electrical, Electronics and Information Engineering, IEEE International (2018)

Design, Performance and In-Orbit Evaluation Results of Tiangong-2 Wide-Band Imaging Spectrometer

Jun Wei[1,4(✉)], Lei Ding[1,4], Xianqiang He[2], Yuyu Tang[1,4], Xin Feng[1,4], and Mingsen Lin[3]

[1] Shanghai Institute of Technical Physics, Chinese Academy of Sciences, Shanghai 200083, China
weijun@mail.sitp.ac.cn
[2] State Key Laboratory of Satellite Ocean Environment Dynamics, Second Institute of Oceanography, SOA, Hangzhou 310012, China
[3] National Satellite Ocean Application Service, SOA, Beijing 100081, China
[4] Key Laboratory of Infrared System Detection and Imaging Technology, Chinese Academy of Sciences, Shanghai 200083, China

Abstract. The system design, performance and in-orbit evaluations of a push-broom wide-band imaging spectrometer (WIS) on Tiangong-2 space laboratory is introduced. The WIS has 100 m spatial resolution in ocean color bands (FOV is around 42°), which is the highest compared to other in-orbit ocean color sensors (GOCI, MODIS, VIIRS, and OLCI). Besides, WIS includes visible and near infrared (VNIR), shortwave infrared (SWIR) and thermal infrared (TIR) channels and has excellent performance in ocean color monitoring, and SNR in 0.4–0.8 μm is more than 400 at typical oceanic radiance. Average SNR at SWIR is more than 550 at typical atmosphere radiance. Moreover, the instrument's average NEΔT at TIR is lower than 20 mK, which is one of the best. The in-orbit tests reveal that the ocean color components retrieved by WIS are highly consistent with the results from the typical ocean color satellite sensors (MODIS, VIIRS and GOCI), but have finer structures due to higher spatial resolution.

Keywords: Tiangong-2 · Wide-band imaging spectrometer
Spectral channel programming · InGaAs detector assembly
HgCdTe detector Dewar assembly

1 Introduction

The WIS is an ocean color remote sensor on Tiangong-2 space laboratory launched on Sep.15, 2016. It has fourteen programmable VNIR channels (V1–V14, 0.40–1.04 μm), two SWIR channels (S1: 1.232–1.252 μm and S2: 1.630–1.654 μm) and two TIR channels (T1: 8.125–8.825 μm and T2: 8.925–9.275 μm), and its spatial resolutions at nadir are 100 m, 200 m and 400 m for the VNIR, SWIR and TIR, respectively. Combining push-broom and multiple module FOV stitch technologies, WIS achieves wide swatch of 300 km.

Y. Gu et al. (Eds.): *Proceedings of the Tiangong-2 Remote Sensing Application Conference*, LNEE 541, pp. 14–27, 2019.
https://doi.org/10.1007/978-981-13-3501-3_2

To meet high performance monitoring over ocean color and temperature of water and sea ice, WIS' radiation has high resolution at each channel, the average signal-to-noise ratio (SNR) in band 0.4–0.8 μm is more than 400 at typical oceanic radiance, average SNR at SWIR is more than 550 at typical atmosphere radiance, and average Noise-equivalent temperature difference (NE$\varDelta$T) in TIR is better than 20 mK at 300 K blackbody, which is one of the best among similar ocean sensors in-orbit. Full field of view (FOV) and full aperture on-board calibrator is designed using internal reference radiation source with limited resource. The calibration unit is composed of pointing mirror, surface blackbody and halogen lamp with Pr-Nd glass cover of specific absorption peak. The calibrator rotates the mirror to point to the lamp and blackbody in turn to do the calibration in orbit. In this paper, the details of the design, performance tests and in-orbit evaluations of the WIS are introduced.

2 System

Like MERIS and OLCI of ESA, the WIS is a push-broom remote imaging system. Compared to the optical-mechanical scanning method, the push-broom system has the advantage of improving the system SNR, which is very important for weak ocean color signal detection.

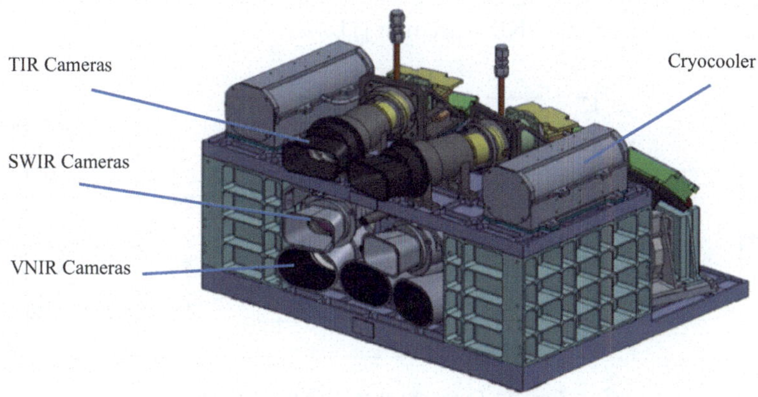

Fig. 1. Layout of the WIS optical head without calibrator

The system consists of an optical head and three control boxes. The optical head, mainly composed of VNIR imaging module, SWIR imaging module, TIR imaging module and calibrator, is the core part of WIS. As shown in the Fig. 1, the three modules are stacked in three layers from bottom to top, and each module is composed of 2 or 3 fan-mounted camera sets with a small FOV overlap.

2.1 VNIR Imaging Module

VNIR imaging module is designed for ocean color and atmospheric correction. Three VNIR cameras are installed fan-shaped to achieve the view of total 42 ° field (about 0.5° field overlap). Each camera is composed of VNIR lens, Spectrograph, CCD array and video electronic unit (Fig. 2). One distinguished characteristic of WIS is the programmability of its VNIR spectral channels in their width and central wavelength. The programming spectral range is 400 ~ 1040 nm with 2.5 nm step. The targets on the earth are first imaged onto the plane of the slit by VNIR lens, and the Spectrograph unit disperses the light through the slit and forms its dispersed image on the CCD array. The 2D array corresponds to the spatial extension of the slit and spectral dispersing dimension.

Seven pieces of lenses construct a transmission optics for VNIR lens. The focal length is 120 mm with F number of 3.2, and optical MTF is 0.85 or higher. The exit pupil of the lens matches the entrance pupil of the spectrograph unit. By integrating reflection and transmission optical design, small spectral bending and spatial distortion are achieved. The linear dispersion rate of the spectrograph is 111 nm/mm, which meets the requirements of system spectral resolution (2.5 nm). The selected CCD (CCD55-30) is a thinned back-illuminated, frame-transfer, non-inverted mode operation CCD from E2 V corp. it's made of 1243 × 576 detector elements, each detector pixel is a 22.5 μm square, corresponding to one pixel in the image. Its dynamic range is better than 96 dB at 1 MHz readout frequency. Mid-band coated is applied to achieve high quantum efficiency in VNIR spectrum [1].

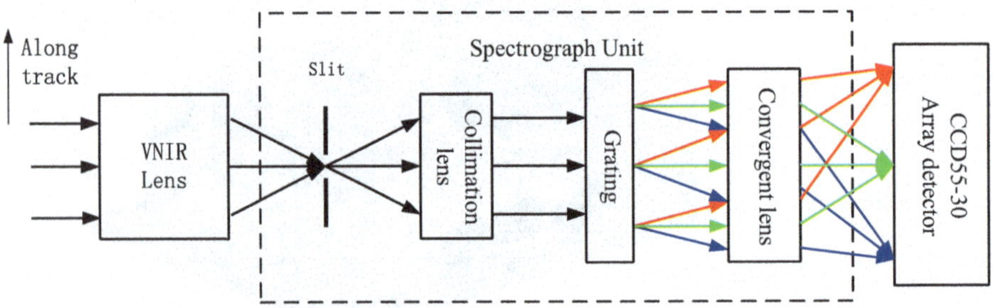

Fig. 2. Diagram of the VNIR camera

2.2 SWIR Imaging Module

SWIR channels aim to monitor cloud-snow and sea-land boundaries as well as to run atmospheric correction of turbid water. To achieve 42° FOV SWIR imaging, two SWIR cameras (FOV 21.5°) are mounted fan-shaped. Each camera is composed of SWIR lens, double band 800 (spatial pixel) *2 (spectral channel) InGaAs detector assembly and signal processing circuit unit. Five pieces of lenses develop the transmission optics of SWIR lens, its focal length is 50 mm with F number of 2.0, and optical MTF is better than 0.85. The SWIR dual-channel InGaAs detector assembly is independently developed by Shanghai Institute of Technical Physics (SITP), Chinese

Academy of Sciences. As shown in Fig. 3. It consists of window, cover plate, filter plate, optical stop, InGaAs focal plane array (FPA), transition electrode plate, thermoelectric cryocooler and metal tube shell. The InGaAs FPA consists of an 800 × 2 responsive elements and readout circuit, which can measure the spectral response of two detection channels by narrow bandpass filter. Responsive element uses a thinned back-illuminated structure, and its size is 25 μm square. Interface of the two channels are mirrored with 1.2 mm distance. Selectable working temperature of 15 °C or 5 °C is guaranteed by way of semiconductor refrigeration. The dark current is less than 5 pA, dynamic range is greater than 80 dB at 1 MHz read frequency, non-uniformity is 3% or less, and there is no blind responsive element.

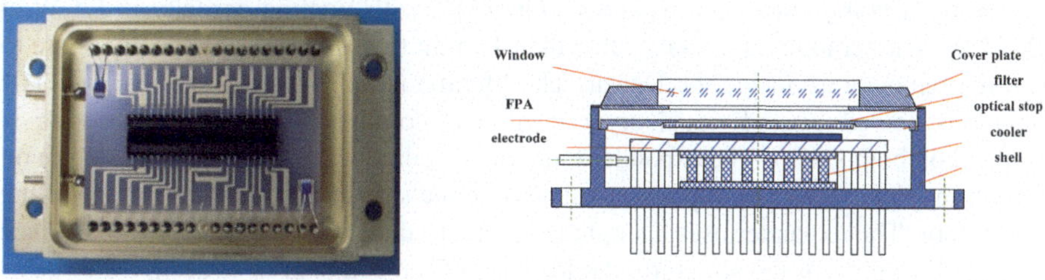

Fig. 3. The photo and schematic of SWIR dual-channel InGaAs detector assembly

2.3 TIR Imaging Module

TIR channels focus on temperature monitoring on sea ice and water. Two TIR cameras with same FOV (21.5°) are combined fan-shaped. Each camera is composed of TIR lens, dual-channel 400 (spatial pixel) × 2 (spectral channel) HgCdTe detector Dewar assembly, Stirling cryocooler and signal processing unit. The Stirling cryocooler can keep 80 K working temperature, and it's independently developed by SITP. Four pieces of lenses form the transmission optics for TIR camera. Its focal length is 28 mm with F number of 1.5, and optical MTF is better than 0.6. The Dewar assembly provides a good low-temperature vacuum environment for the detector, which is

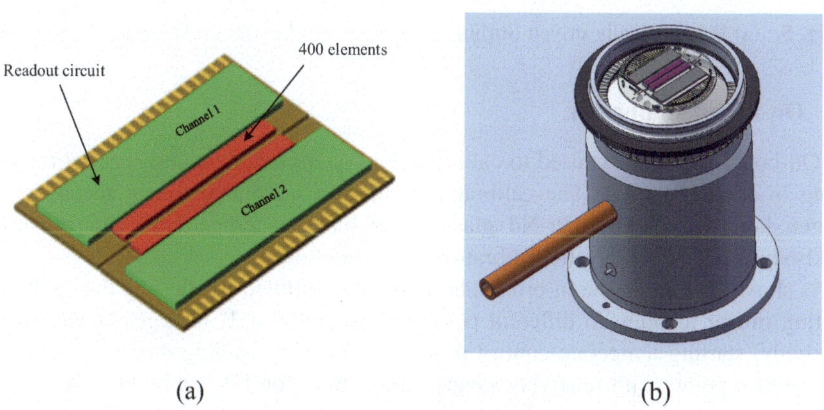

Fig. 4. (a) Dual-channel TIR FPA (b) Detector Dewar assembly

composed of infrared window, lead ring, shell, core, HgCdTe FPA, infrared filter, cold stop, and degassing agent. The cooling loss is less than 400 mW, and the temperature difference between the cold finger and the FPA is controlled to be under 2 K. As shown in Fig. 4, the FPA detector consists of two monolithically integrated responsive areas (400 elements) and a readout circuit based on capacitive feedback transimpedance amplifier (CTIA). The infrared filter of two spectral channels is installed on two responsive areas by bridge. The responsive element adopts back thinned structure. The size of element is 28 μm square. The D-star(D*) is greater than 8×10^{10} $cmHz^{1/2}W^{-1}$ at 80 K, the non-uniformity is less than 6%, and the responsive area of the detector has no blind element.

Developed by SITP, the TIR camera uses pneumatically driven Stirling cryocooler consisting of compressor and expander unit (Fig. 5). Its cooling capacity is 1.5 W at 80 K FPA temperature, and it takes less than 15 min to drop from room temperature to working temperature inside the Dewar. The lifetime is more than 20,000 h. Each TIR camera is cooled by individual cryocooler. The compressor of each cryocooler is arranged on both sides of the expander unit, and installed with the thermal insulation on the bottom plate. Two heat pipes are installed on the side and the bottom surface of the compressor. The expander uses a single heat pipe to dissipate heat. In order to enhance the overall rigidity of the structure, the location of the compressor is designed to be as low as possible.

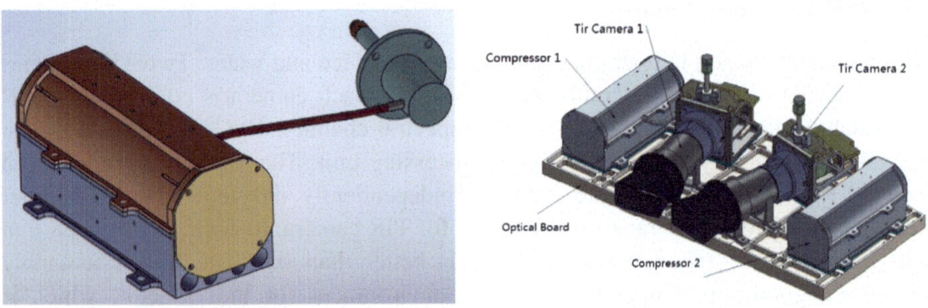

(a) (b)

Fig. 5. (a) Pneumatically driven Stirling cryocooler. (b) Layout of TIR imaging module

2.4 On-Board Calibrator

The On-board calibrator is used to calibrate the system radiation and spectral response during in-orbit operation. The calibrator includes a calibrated mirror mechanism, a halogen lamp box with a Pr-Nd glass cover plate, a temperature control surface blackbody and a VNIR/SWIR calibration detectors unit.

As shown in Fig. 6, the in-orbit calibration was fulfilled by moving the calibration pointing mirror to point to different positions. At position 1, it targets at the surface blackbody, starting temperate control process and during cooling down, select two to three control points with relatively steady temperature for TIR calibration. At position 2, the halogen lamp box with Pr-Nd glass cover plate is targeted. By opening and closing the six halogen lamp sources with various combinations, we can get three radiation levels for VNIR and SWIR calibration. At the same time, the spectral absorption peak generated from the same lamp sources passing through the Pr-Nd glass is utilized for spectrum calibration using envelop detection algorithm. During normal ground observation, the pointing mirror is at position 3, ensuring that the imaging field is not obstructed. The VNIR and SWIR bands are each equipped with a set of full-spectrum calibration detectors for monitoring the long-term stability of the lamp source.

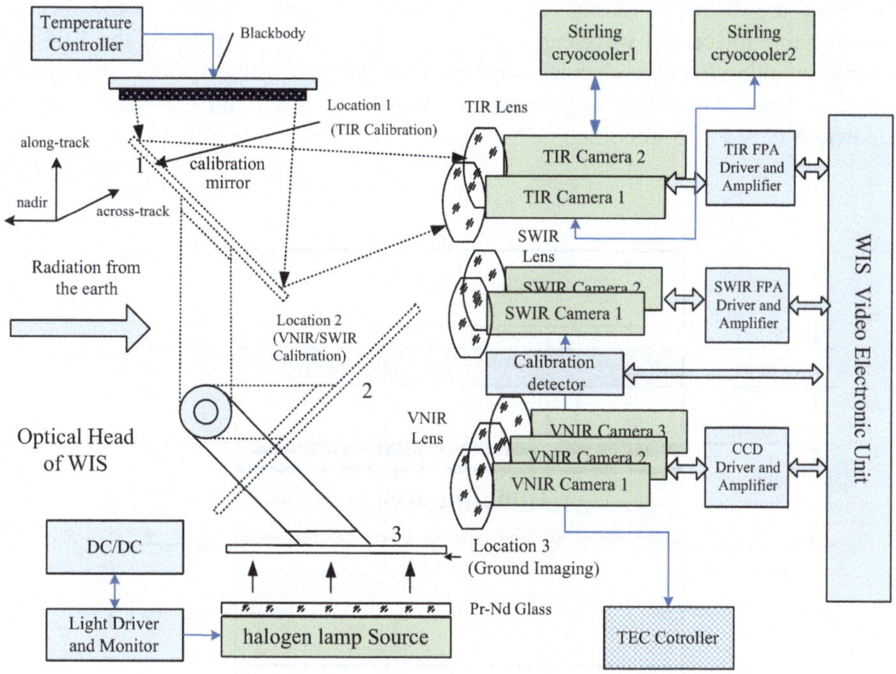

Fig. 6. Diagram of Optical head of the WIS

2.5 VNIR in-Orbit Spectral Channel Programming

The 14 VNIR spectrum channels of WIS support in-orbit programming of spectral center wavelength and bandwidth, which has significant advantages including improving the efficiency of the instrument when the data transmission rate is limited, carrying out in-orbit spectrum research for specific target; correcting the spectral properties of channels when they alter in orbit [2].

(1) **Implementation**

The target is imaged on the slit of the spectrograph unit through the lens (Fig. 2). The light passing through the slit is dispersed into 2D spectral and spatial images by the spectrograph and cast to the area-array CCD.

As shown in Fig. 7, the detector pixel lattice of size 1024 × 576 is read from the CCD array. There are 1024 effective pixels on each line in the horizontal spatial cross-orbit dimension while there are 576 lines along the vertical spectral dimension. In spectral dimension, one CCD line corresponds to 2.5 nm spectral sampling interval. The effective spectral dimension readout area covers a total of 256 rows containing 0.4–1.04 μm band, and the total width is 640 nm (256 × 2.5 nm). The remaining spectral lines are all dark pixels, and the fixed 20 lines in the shadow work as correction channel which is used for smear correction in the frame transfer process. The remaining dark lines are discarded during reading. In the figure, line 1 represents the spectral sampling line for central wavelength of 0.4 μm, and line 256 represents the 1.04 μm spectral sampling line. If a channel starts from line M and continues to accumulate and merge N lines, its central wavelength is (M + N/2) × 2.5 nm + 400 nm, and its bandwidth is N × 2.5 nm.

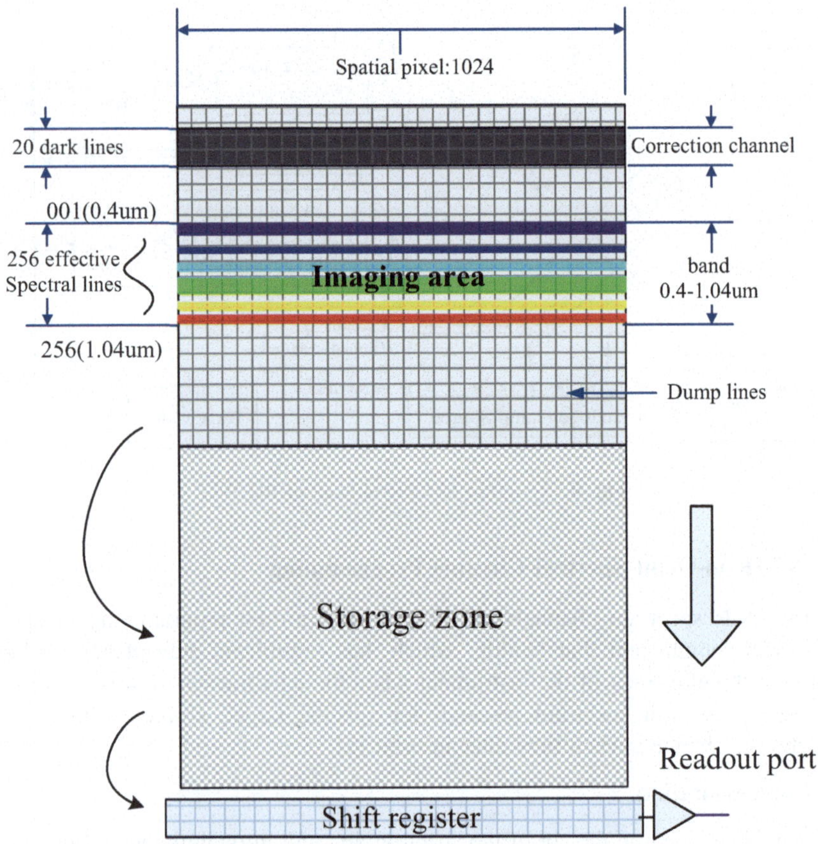

Fig. 7. Effective area read by CCD array

The merging of spectrum can be divided into two types: on-chip charge accumulation and off-chip AD data accumulation, and the selection on the type to be used should be determined according to spectral energy calibration. Under unsaturation condition, internal charge accumulation mode is highly recommended to improve SNR. While saturated, external AD data accumulation mode should be used to improve the dynamic range. The CCD selected has the function of line accumulation and line dumping, by readout timing control, which can greatly improve the system's frame frequency.

(2) **Sub-channel and full frame readout mode**

VNIR camera has two reading modes, sub-channel readout and full frame readout, which is selected through ground instruction injection. Sub-channel reading is usually selected when imaging and observing ground object, and the full frame method is mainly used for the spectrum calibration in orbit.

Sub-channel readout utilizes channel programming table to control readout timing. 14 channel image and 1 correction channel are selected from 256 spectrum lines. Furthermore, spectral line location (which decides the center wavelength) and bandwidth of all channels are also determined by the channel programming table, and the bandwidth is a multiple of 2.5 nm.

Full frame readout method reads out 256 effective spectral lines line by line. To make output digital interface fully compatible with sub-channel readout method, 15 lines are read on completion of each frame transfer, corresponding to the downlink data of fifteen 2.5 nm-bandwidth channels. There are 576 lines of pixels on CCD array in total, and pixel data of 570 lines are readout in 38 times. Middle area contains complete set of 256 effective spectral lines, and the readout of start and end areas are used for full frame correction.

3 Performance

WIS has been tested in detail before launching. Below introduce the results of sensitivity, dynamic range, MTF, spectrum and stray light tests closely related to the application of ocean color.

3.1 Sensitivity, Dynamic Range and MTF

The dynamic range and sensitivity of VNIR, SWIR and TIR channels were tested by indoor integrating sphere calibration and infrared radiation calibration experiments. The results are shown in Table 1.

FPGA is used to control the timing sequence to realize the programmability of each spectral channel, which reasonably balances the contradiction between large dynamic range and highly sensitive signal readout. In the Table 1, V1 ~ V14 is the default spectral configuration of the system, and the sensitivity value is under normal integration time. The VNIR channel integration time can be cut half and dynamic range be double of the value in the table on instructions. With MTF being tested by test card method, VNIR channels are more than 0.30 on average, SWIR channels are more than

Table 1. Dynamic range and sensitivity of each channel [3]

Channel[a]	Center wavelength	Bandwidth	Typical Radiance[b]	SNR or NEΔT (K)	Dynamic range (albedo)
V1	413	403–423	7.86	390	38.9%
V2	443	433–453	7.02	515	54.2%
V3	490	480–500	5.31	557	41.8%
V4	520	510–530	4.58	556	41.7%
V5	565	555–575	3.39	555	38.7%
V6	620	610–630	2.19	467	38.5%
V7	665	655–675	1.60	399	40.1%
V8	682.5	677.5–687.5	1.45	273	42.7%
V9	750	740–760	0.93	302	46.4%
V10	820	810–830	0.59	203	53.9%
V11	865	845–885	0.62	255	51.3%
V12	905	895–915	0.47	138	73.7%
V13	940	930–950	0.36	100	110.0%
V14	980	970–990	0.39	59	275.3%
S1	1.242	1.232–1.252	0.54	510	110%
S2	1.642	1.630–1.654	0.73	678	110%
T1	8.475	8.125–8.825	0.95(330 K)	0.014 K	100 ~ 340 K
T2	9.100	8.925–9.275	0.98(330 K)	0.026 K	100 ~ 340 K

[a]Channels V1 ~ V14 are in nm; S1 ~ S2 and T1 ~ T2 are in μm.
[b]Unit of spectral Radiance values is mW/ (cm^2 × μm × sr).

0.45 on average, and TIR channels are more than 0.30 on average. MTF of the viewing field at the edge of the channel slightly decreases by about 20%.

3.2 Spectrum Tests

Laboratory monochromator collimation method was used for spectrum calibration before launching. The input light source and grating of the monochromator are adjustable according to the test spectrum band. The VNIR and SWIR channels use halogen tungsten lamp, and TIR channels use infrared globar as the monochromator light source. The light passes through collimator and expands. As long as SNR can be guaranteed, the slit width of the monochromator should be adjusted to minimum, so that the output spectral width is far less than the width of the tested spectral channel.

The monochromator carries out wavelength scan and collects response value of spectral channel. Input light and monochromator form light source system, whose spectral radiance curves are measured by a standard spectral radiometer, and the outputs are used for normalization and correction of test result of each channel. Each camera consists of 1024, 800 or 400 pixels along cross orbit dimension. Although it is reliable to test each pixel, it is too time-consuming to be practical. Theoretical analysis and sample test show that these pixels are little discrete and thus in the test, 6 samples uniformly distributed in the FOV are selected for spectral scanning, and the spectral response of the other pixels are obtained by interpolation. Figures 8 and 10 show the spectral response curves of 6 instantaneous FOV in VNIR, SWIR and TIR channels. There are slight differences in the central wavelength and bandwidth of different FOV in the same channel, but each spectral characteristic meets the requirements of the design specification of the instrument [4].

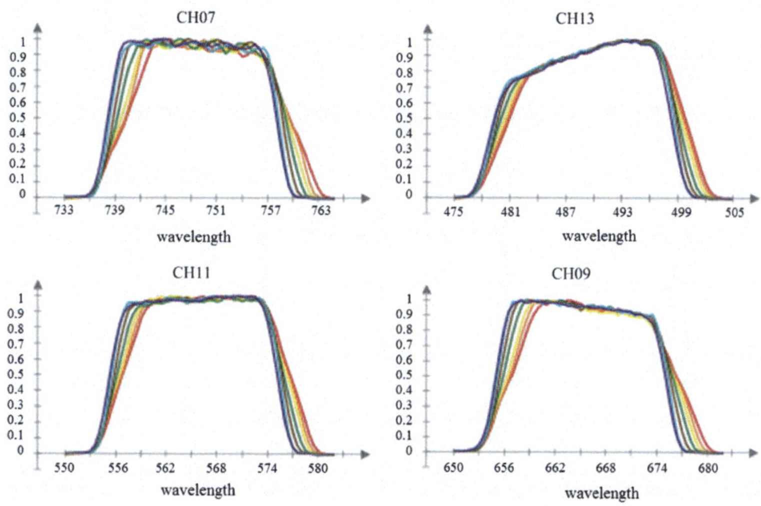

Fig. 8. VNIR channel spectral response

Setting the VNIR readout to full frame mode, step length of monochromator to 0.25 nm and scan, the center wavelength of each spectral line is shown in Fig. 9. Spectral dimension pixel No. and wavelength are well linear correlated. Calculated by data at line 464 (401.14 nm) and 210 (1034.78 nm), line spectral bandwidth averages at 2.495 nm per line. The full frame spectrum calibration result is the basis for the channel programming table configuration in the sub-channel mode.

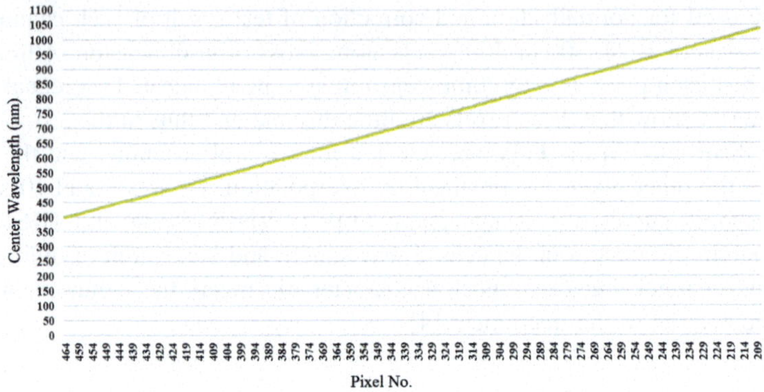

Fig. 9. VNIR camera full frame pixel center wavelength

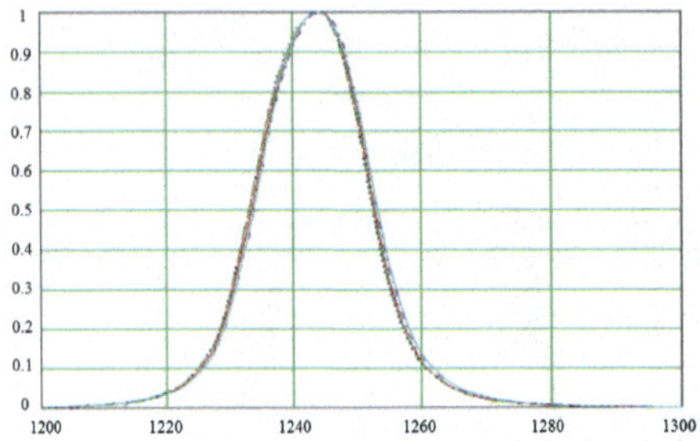

(a) S1 Channel spectral response

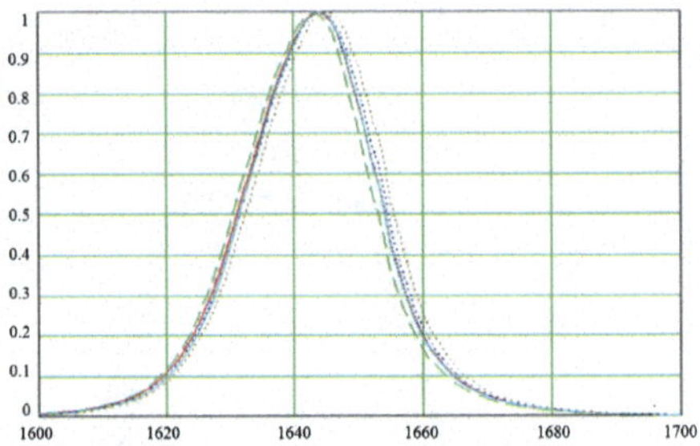

(b) S2 Channel spectral response

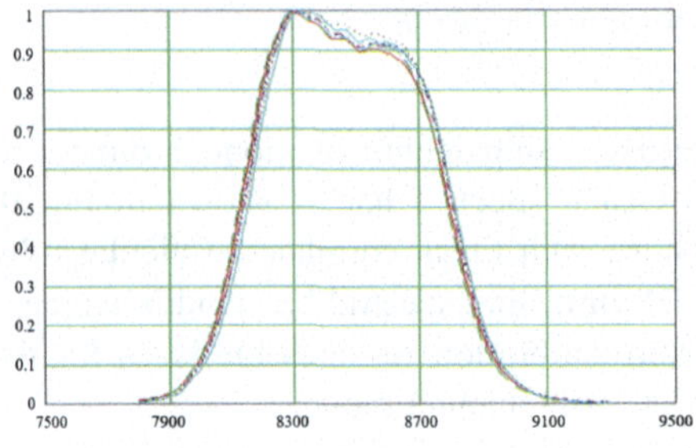

(c) T1 Channel spectral response

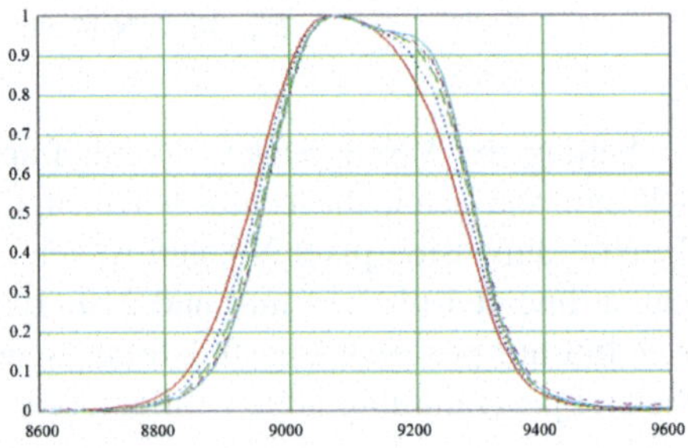

(d) T2 Channel spectral response

Fig. 10. Spectral response of SWIR and TIR channels (x-axis is pixel No., y-axis is Norm. response)

3.3 Stray Light Testing

Stray light is the radiance that enters directly into the effective FOV of the responsive element other than those coming from the target. It is required that the stray light coefficients in 0.5, 0.6 and 0.8 μm spectrums should be less than 3%.

The integrating sphere with blackbody cavity on the back of the sphere is used to record the VNIR camera's response value V_{black}, and after covering the cavity with white diffuse reflection plate, the response value V_{normal} is recorded. The stray light coefficient $\eta = V_{black}/V_{normal}$ is calculated.

Table 2 shows the stray light coefficient of the VNIR cameras measured by experiments, which meets the requirements.

Table 2. Stray light coefficient of test channels

Channel	Bandwidth (nm)	Stray light coefficient (%)
CH3	480–500	1.92
CH7	655–675	1.51
CH10	810–830	1.51
CH13	930–950	1.85

4 In-Orbit Operation and Preliminary Evaluation Results

On September 15, 2016, WIS was successfully launched along with Tiangong-2. Since starting up, the instrument has worked normally, and the images obtained are clear and with plentiful layers. Figure 11 are the first spectral images of VNIR, SWIR and TIR channels after WIS entering into orbit. Ground imaging and in-orbit tests are carried out afterwards, and in-orbit radiation and spectrum calibration test are also implemented on a regular basis. The results show that ground resolution, swath, dynamic MTF, sensitivity parameters meet the requirements of task, and plenty of images of ocean color and water temperature with high quality are obtained.

The retrieval performance of WIS for ocean color and temperature is evaluated by using image data acquired during the in-orbit operation. The results of in-orbit testing and application show that the precision of WIS output reaches the target. In the monitoring of offshore suspended solid concentration, it is in good agreement with GOCI retrieval results in terms of magnitude and spatial distribution. The correlation coefficient of both is 0.96 with an average relative deviation of 18.6%. Compared to GOCI's 500 m spatial resolution, its 100 m spatial resolution can significantly improve the resolution of suspended solid spatial structure and monitor suspended solid concentration in rivers and large or medium-sized lakes. In monitoring of chlorophyll concentration in continental shelf and ocean water, it is consistent with the retrieval results of MODIS and VIIRS, with a correlation coefficient up to 0.89. In addition, the spatial structure of its chlorophyll concentration distribution retrieval is better refined, and it can distinguish small-scale vortices and frontal structures. The retrieval accuracy of the TIR channel to sea surface temperature is better than 1 K, and the average NEΔT of the two channels is better than 20 mK (300 K blackbody) [5].

(a) VNIR color composite image

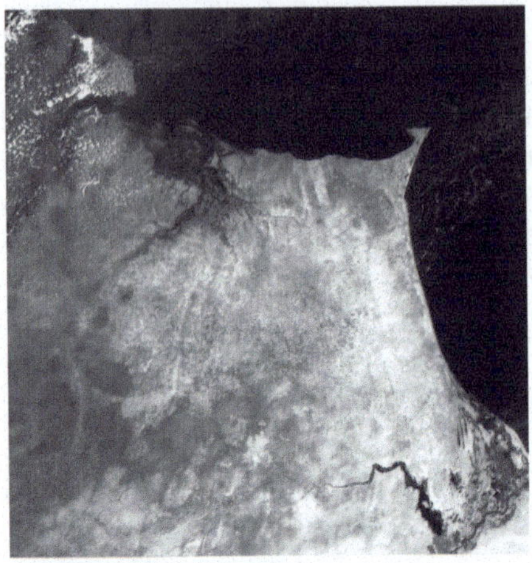

(b) SWIR channel S1 image

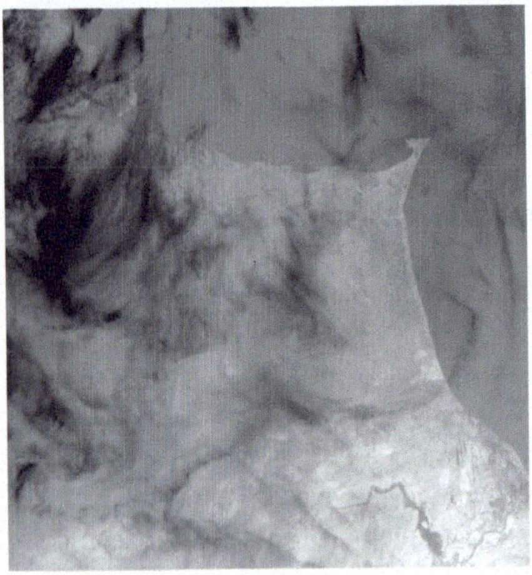

(c) TIR T1 channel image

(d) TIR T2 channel image

Fig. 11. First images after launching (Dakar, Africa)

In addition, the WIS can clearly invert the land aerosol products with fine resolution. The accuracy is comparable to that of MODIS products, and it has good consistency and meets the requirements. When the atmospheric images are retrieved by instruments, the information advantage of high spatial resolution is fully utilized. Combined with more detailed surface and atmospheric parameter information, the fine distribution of local water vapor can be retrieved.

5 Conclusion

This paper introduces the system design, main performance, in-orbit operation and major evaluation results of WIS. The ocean color components retrieved by WIS are highly consistent with the results from the typical ocean color remote sensors (MODIS, VIIRS and GOCI), but have finer structures due to higher spatial resolution. It is proved that the in-orbit function test, performance and application specifications of WIS meet the requirements. Its water color detection property lays solid basis for future data processing and application. The 100 m resolution water color remote sensing data provided by WIS has great potential for the ocean color detection in the coastal and inland waters.

Acknowledgement. Thanks to China Manned Space Engineering for providing space science and application data products of Tiangong-2.

References

1. Cui, Y.: Study of large field push-broom imaging spectrometer using CCD detector. Graduate school of Chinese academy of sciences (Shanghai institute of technical physics) (2014)
2. Wei, J.: Research on push-broom wide field imaging spectrometer technology. Graduate School of Chinese Academy of Sciences (Shanghai Institute of Technical Physics) (2009)
3. Jun, W., Xianqiang, H., Mingsen, L., Lei, D.: Moderate-resolution ocean color and temperature imager on Chinese Tian-Gong II space lab. In: 2017 42nd International Conference on Infrared, Millimeter, and Terahertz Waves (IRMMW-THz) (2017)
4. Wang, H.: Research on in-orbit spectral calibration and polarization correction technology of visible/near-infrared push-broom imaging spectrometer. In: Graduate School of Chinese Academy of Sciences (Shanghai Institute of Technical Physics) (2016)
5. He, Xianqiang, Bai, Yan, Wei, Jun, Ding, Jing, Shanmugam, Palanisamy, Wang, Difeng, Song, Qingjun, Huang, Xiaoxian: Ocean color retrieval from MWI onboard the Tiangong-2 Space Lab: preliminary results. Opt. Express **25**(20), 23955–23973 (2017)

Research on on-Board Calibration System of Tiangong-2 Wide-Band Imaging Spectrometer

Yuyu Tang[1(✉)], Jun Wei[1], Xiaoxian Huang[1], Xin Feng[1], and Qingjun Song[2]

[1] Shanghai Institute of Technical Physics, Chinese Academy of Sciences, Shanghai 200083, China
tangyuyu@mail.sitp.ac.cn
[2] National Satellite Ocean Application Service, SOA, Beijing 100081, China

Abstract. With the increasing requirements for quantification of in-orbit data products, the research of high-precision calibration system is of vital importance. Based on the research background of Tiangong-2 (TG-2) Wide-band imaging spectrometer payload, a complete scheme of full detection band radiometric and VNIR spectral calibration is proposed in this paper. A full FOV and aperture calibrator with internal reference radiation sources is designed, which consists of pointing mirror, halogen lamp with Pr-Nd glass and surface blackbody, and calibrates by rotating the mirror to point to the lamp and blackbody. The results of in-orbit calibration show that the scheme is reasonable and the calibration accuracy meets the requirements.

Keywords: Tiangong-2 · Imaging Spectrometer · Spectral calibration Radiometric calibration · Halogen lamp · Surface blackbody

1 Introduction

TG-2 "Wide-band Imaging Spectrometer" (WIS) is a push-broom space-to-ground optical remote sensor with high spectral range in the visible near-infrared (VNIR) spectrum and multi-spectral detection in short-wave (SWIR) and thermal infrared (TIR) spectrum, which is mainly used for remote sensing of ocean color and temperature, as well as land, atmosphere and cirrus detection. With the increasing demand for remote sensing accuracy of oceans detection, the instrument needs to solve the problems of gain variation, performance degradation and spectral positioning deviation caused by long-term in-orbit operation of the detector, especially in the application of high spectrum resolution such as ocean water color detection. The spectral characteristics of the detection channel and the accuracy of data retrieval are more demanding, so in-orbit radiation and spectral calibration are necessary [1]. In response of this demand, WIS is equipped with an on-board calibrator, which can realize high-frequency, continuous real-time in-orbit data authenticity verification and calibration, and effectively improve the quality of data product inversion.

Y. Gu et al. (Eds.): *Proceedings of the Tiangong-2 Remote Sensing Application Conference*, LNEE 541, pp. 28–39, 2019.
https://doi.org/10.1007/978-981-13-3501-3_3

2 Calibration Scheme

The on-board calibration of WIS includes full-detection band radiometric calibration and VNIR spectral calibration. Calibration sources usually adopt artificial radiation sources such as standard lamps and blackbodies, or natural radiation sources such as the sun, the moon, and deep space. Due to the platform conditions, artificial radiation sources are chosen.

In this paper, a tungsten halogen lamp is used as radiation calibration source for VNIR and SWIR bands [2]. A temperature-changing extend blackbody is used as radiation calibration source for TIR. These calibration sources are characterized by high luminance, wide field of view and stable performance. The Pr-Nd glass with standard absorption peak is set in the light source optical path for VNIR spectrum calibration. The glass also serves as the light uniformity component of the light source.

Considering that the change of the performance of artificial radiation sources in-orbit will affect the calibration, the on-board calibrator uses the introduced standard unit detector to monitor the attenuation of the light source.

2.1 Calibration Methods

WIS includes three VNIR sub-modules, two SWIR sub-modules, and two TIR sub-modules, each of which has an independent optical system. TIR, SWIR and VNIR are arranged in a fan shape according to a certain angle on different installation planes. There are three subsystems installed in three layers from top to bottom, as shown in Fig. 1.

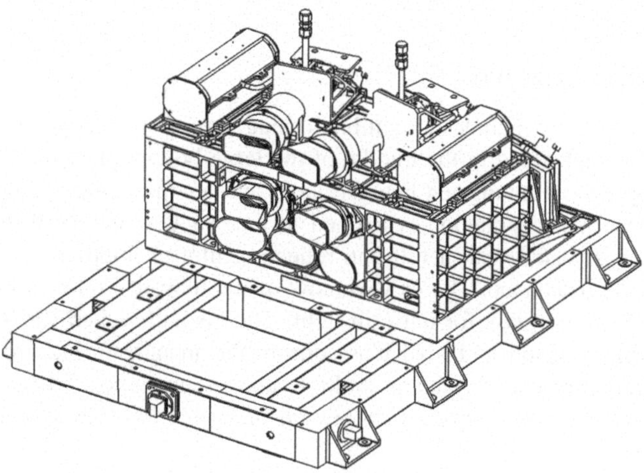

Fig. 1. WIS system model drawing

According to the characteristics of multi-module optical field of view synthesis, the calibration source is placed at the confluence of the sub-module field of view to ensure that the sub-module field of view is covered. To this end, the on-board calibration system introduces a pointing mirror rotation mechanism, that is, between the sub-module lens and the optical path of the field of view, the plane calibration is directed to the mirror. And the shift between on-board calibration mode and ground observation mode is completed by changing the direction of the optical path, thus realizing time-sharing observation between the calibration source and the Earth target of the optical system field of view. Figures 2 and 3 respectively show the working principle and the mold structure of WIS. The motor driving mechanism rotates to different positions through the calibration mirror with the crank to realize three working states: VNIR radiation and spectral calibration 、 SWIR and TIR radiometric calibration and ground observation.

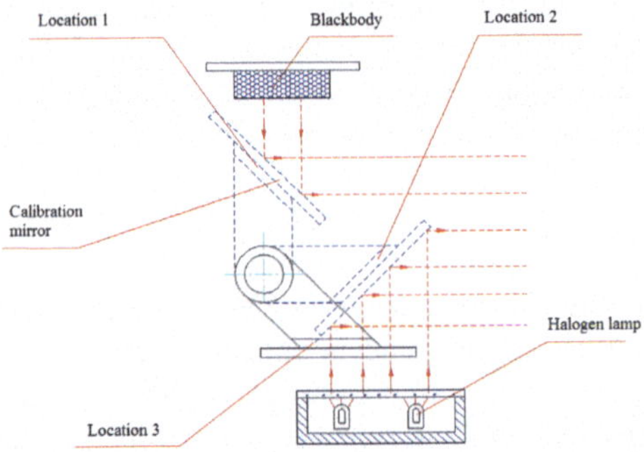

Fig. 2. The working principle diagram of on-board calibrator

2.2 Calibration Light Box

The light box assembly includes a light box casing, a tungsten halogen lamp and a Pr-Nd glass, and is embedded in the middle of the large bottom plate of WIS. The light source is composed of a tungsten halogen lamp and a converging collimating optical element. The thermal conduction is applied to heat light source, and the collimated optical element is plated with a full anti-reflection film to improve the stability and uniformity of the irradiance. The exit window of the tungsten halogen light screen is installed with bismuth glass for uniform light. The bismuth-doped (Pr-Nd) rare earth element process is added to Pr-Nd glass during the manufacturing process, and its spectral transmittance curve contains multiple absorption peaks. These characteristic peaks are used for on-orbit spectral calibration of visible light near-infrared modules (Figs. 4 and 5).

The power and color temperature of the tungsten halogen lamp determine the luminous flux of the lamp. WIS adopts imported tungsten halogen lamp with low driving voltage and high illumination efficiency. The power consumption is 10 W/12 V with the single luminous flux of 300 lm and the color temperature of 3000 K. The average life is greater than 300 h, and the size is Φ10 mm × 33 mm. Considering the typical input energy and dynamic range of a single channel of the instrument, it is

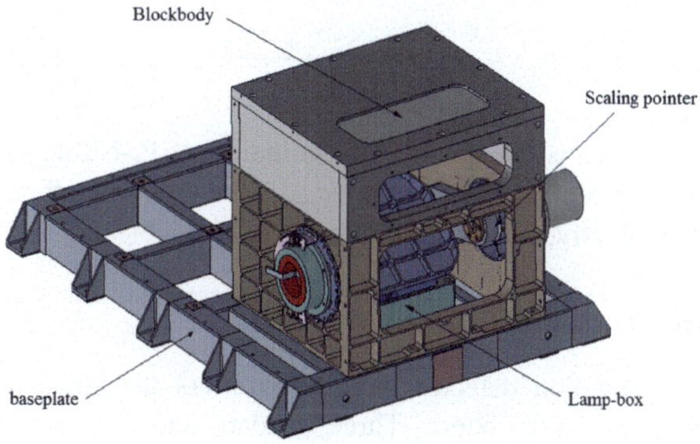

Fig. 3. The mold structure drawing of on-board sacler

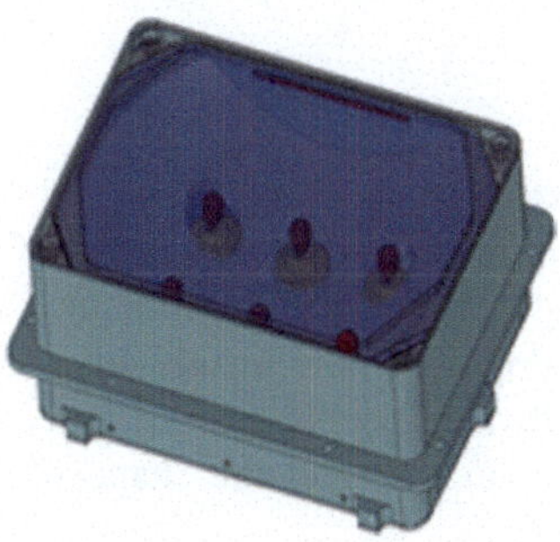

Fig. 4. The model drawing of light box

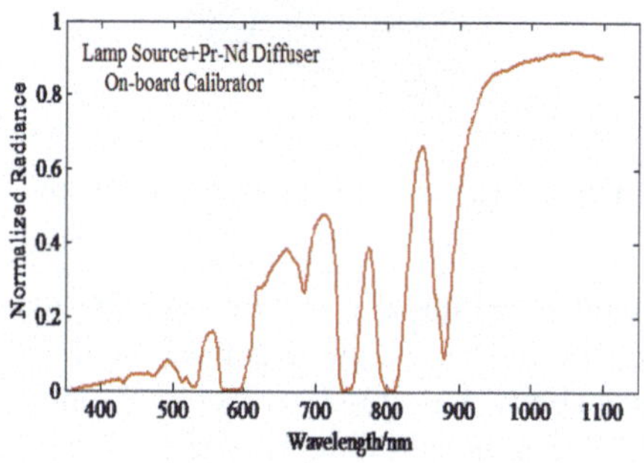

Fig. 5. The spectral radiance curve of the lamp

determined that the total number of tungsten halogen lamps is 6 and are evenly divided into 2 groups. The light source switch realizes the control of the two levels of radiance in the orbit calibration source. The radiance of the calibration lamp source is calculated as,

$$L_{cal} = \frac{\tau_{\mathrm{Pr-Nd}} c_{lamp} P_{lamp}}{\pi A_s} \quad (1)$$

Among which, $\tau_{\mathrm{Pr-Nd}}$ is the spectral transmittance of Pr-Nd diffuse glass, P_{lamp} is spectral power for tungsten halogen lamps, c_{lamp} is the emission efficiency for halogen lamp, and A_s is the effective area for diffusing screen.

2.3 Light Source Attenuation Detector Module

The light source attenuation detector module monitors the attenuation of the light source in the VNIR and SWIR bands. Three standard unit detectors with the uniform performance are configured for VNIR and SWIR respectively. Visible standard detector is C30807 of HAMAMATSU Company, the shortwave standard detector is C30641 of HAMAMATSU Company, and the full spectrum detector is not equipped with window filter.

2.4 Extend Blackbody

All-optical path calibration based on large-area blackbody is adopted for TIR radiation calibration. The blackbody is placed at the front end of the TIR light path, and its radiation source with different temperatures is obtained as the reference source for calibration. The blackbody adopts a whole metal body (copper) to open the V-shaped groove, and the material with the specific emissivity is more than 0.9. So the final specific emissivity is better than 0.9, and the shape of variable-temperature extend black body is shown in Fig. 6.

Fig. 6. The model diagram of blackbody

The calibration method is technically mature, and the main difficulty lies in the uniformity and accuracy of the temperature control of the blackbody. The copper material used in the blackbody has good temperature uniformity (the temperature fluctuation is within ±0.1 K). In addition, two sets of temperature-controlled film heating sheets and two sets of temperature-measuring Pt100-type platinum resistors are uniformly attached to the back surface of the honeycomb bottom plate, as shown in Fig. 7.

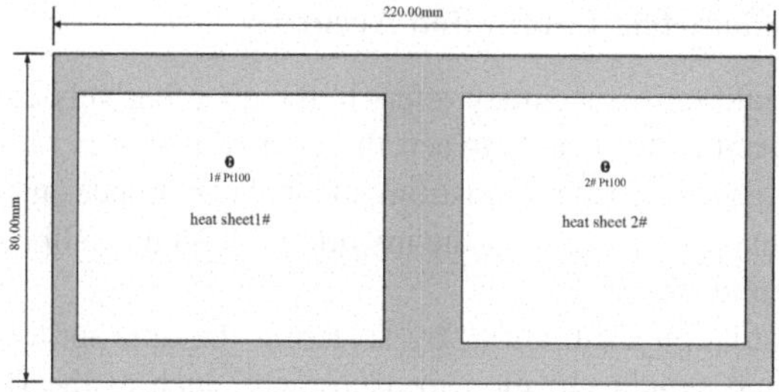

Fig. 7. The distribution of temperature measurement points and blackbody heating sheets

The honeycomb blackbody is heated by heat conduction to further improve the uniformity and accuracy of temperature control. The blackbody temperature control adopts the natural cooling method after active heating. During the cooling process, 3∼4 temperature points are selected for thermal insulation heat balance, and the midpoint of the highest temperature-equilibrium temperature is selected as temperature control point. Each temperature control point is held about 5 min. The temperature control mode adopts the switching mode, the frequency is about 2 Hz, and the temperature control precision is ±0.1 K. In the process of temperature balance, the system continuously collects relevant data.

3 Calibration Accuracy Analysis

Comprehensive user requirements and platform conditions, the main technical indicators of WIS calibration system are as follows: (1) VINR spectral accuracy: center wavelength positioning is better than 2.5 nm; (2) absolute radiation calibration accuracy: VNIR and SWIR is better than 10%, and TIR is better than 2 K (300 K blackbody).

3.1 Spectral Calibration Accuracy

The spectral calibration accuracy refers to the deviation between the center wavelength and bandwidth of each detection channel and the nominal value of the indicator. There are 14 programmable detection channels in WIS VNIR module, and the offset of the center wavelength of each channel is corrected by in-orbit spectral calibration. Spectral calibration accuracy is an important indicator for assessing the correction effect. In the in-orbit spectral calibration, the edge of the spectral absorption peak of the Pr-Nd glass is sampled, and the in-orbit spectral position is determined by comparing the envelope with the absorption peak position curve of the glass before the emission. The difference of the absorption peak of the Pr-Nd glass underground and space conditions is negligible, and the accuracy of the in-orbit spectral calibration is determined by the envelope comparison algorithm and the obtained signal-to-noise ratio of the spectral curve.

3.2 Absolute Radiometric Calibration Accuracy

Absolute radiation calibration accuracy refers to the quantitatively calculated difference of energy. More specifically, it is between the energy which is radiated by the target ground object and received by the instrument at the entrance after the atmospheric correction, and the energy inversed by the instrument with the calibration coefficient for the same target ground object.

In the orbital radiation calibration, by selecting the appropriate object target and combining the ground synchronization measurement, such as the Dunhuang radiation correction field, the instrument can run the VNIR and SWIR channel in-orbit radiation calibration, and by selecting Qinghai Lake radiation correction field, it can run the in-orbit radiation calibration of TIR channel.

In the case of good weather conditions, the area with relatively uniform radiometric correction field is determined. When WIS is transiting, ground synchronization detection is adopted to obtain the absolute calibration coefficient of the instrument, and the coefficient is then simultaneously used with WIS's own calibration coefficient for the same target in the inversion of a comprehensively revised calibration factor. The coefficient is used to evaluate the calibration coefficient obtained by the calibration source on the star for a period of time, and the accuracy of absolute radiation calibration is calculated. Absolute radiation calibration accuracy is mainly guaranteed by the output stability of the on-board radiation calibration source.

3.3 Error Control Requirements

The accuracy of spectral calibration and absolute radiation calibration mainly depends on the strict control of various factors, including pointing mirror, Pr-Nd glass, blackbody and light source attenuation monitoring detector.

On-board calibration is achieved by rotating the pointing mirror. Considering the non-uniformity of the target source (especially the light source), it is necessary to ensure the consistency of the position of the pointing mirror, that is, the repeatability. The calibration accuracy of the pointing mirror is less than or equal to 5%, the motion range is $0 \sim 135^0$, and the positioning time is less than 5 min.

The spectral absorption characteristics of Pr-Nd glass are an important characterization of spectral positioning accuracy, and its wavelength drift is the basis of the spectral positioning reference. The uncertainty of the wavelength centering of the Pr-Nd glass is less than or equal to 0.4 nm.

The variable temperature extend blackbody is used as the calibration source of the TIR band radiation, and its change must be measured in real time. The temperature measurement accuracy of the platinum resistance is required to be less than or equal to ±0.1 °C.

For the VNIR and SWIR bands, since the sun cannot be used as calibration source, and artificial light source has some changes more or less, so it is necessary to use high stability detectors for monitoring, and the response stability requirement is better than 1%.

4 Orbital Calibration Result

4.1 In-Orbit Spectral Calibration

WIS performed two in-orbit full-frame spectral calibrations on February 23 and March 8, 2017. The calibration data of VNIR sub-module1 was compared with the full-frame spectral calibration data of the laboratory in May 2015. As shown in Fig. 8. The red, blue, and green curves in the figure are the spectral curves obtained in May 2015, February 23, 2017 and March 8, 2017 respectively. X-axial is pixel number, and Y-axial is the output signal of pixels. The wavelength offset of the three calibrations is described in Fig. 9. X-axial is pixel number, and Y-axial is the wavelength offset. The blue part indicates the wavelength offset of the two in-orbit calibrations, which is not exceeding 0.02 nm. The red part indicates the wavelength offset of in-orbit and the laboratory, which is not exceeding 0.32 nm. The total wavelength offset is 0.34 nm.

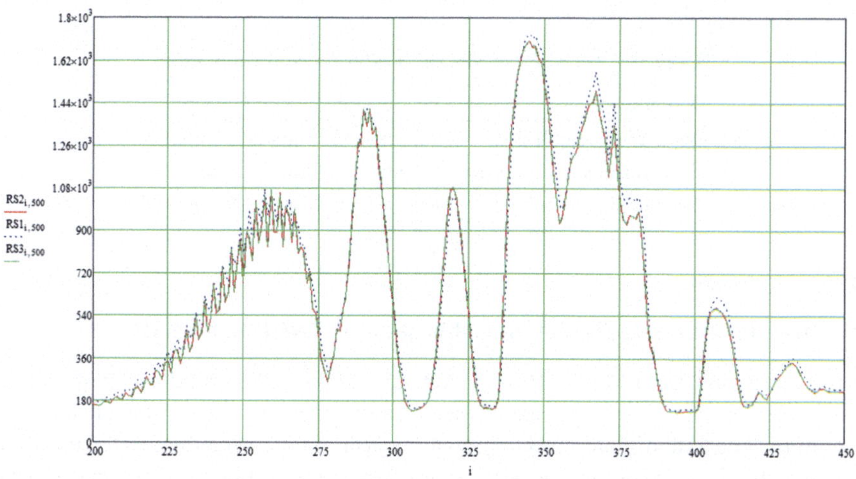

Fig. 8. Comparison of full-frame spectral curves before and after launch

In-orbit spectral calibration accuracy is obtained by calculating on-board spectral positioning error, which is the sum of the wavelength offset before and after launch and the laboratory spectral positioning error. The laboratory spectral positioning error is 1.7 nm.

So the conclusions can be drawn from Table 1 as follows: in-orbit spectral calibration accuracy of WIS does not exceed 2.5 nm, which meets the requirements of the index.

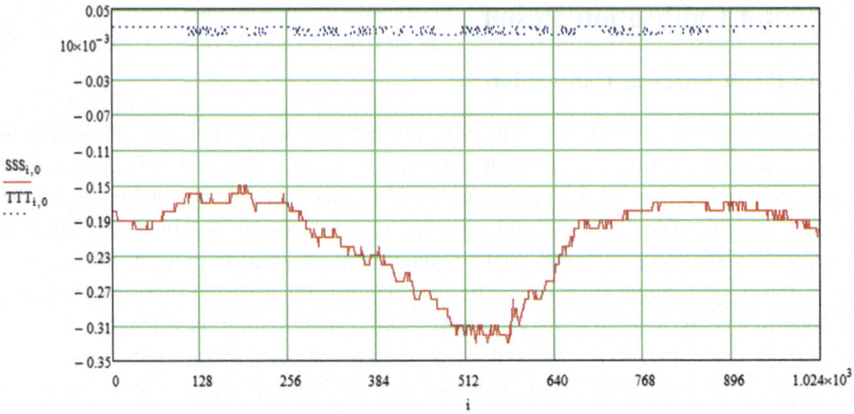

Fig. 9. Wavelength offset

Table 1. Accuracy results of in-orbit spectral calibration

Modules	1	2	3
Max Wavelength drift	0.34 nm	0.25 nm	0.21 nm
Spectral positioning error	1.7 nm	1.7 nm	1.7 nm
On-satellite error	2.04 nm	1.95 nm	1.91 nm

4.2 In-Orbit Absolute Radiation Calibration of VNIR and SWIR

During the period from March 10 to March 25, 2017, WIS carried out four simultaneous measurements. The experimental sea area includes sea in east of Wenchang, sea in south of Sanya, and the connecting area between the two regions. Referring to the requirements of SeaWiFS and MODIS marine vicarious calibration, the performance evaluation of WIS selected the experimental data on March 22, 2017. The radiance of each waveband of WIS is compared with the measured value of MODIS by using the forward algorithm of atmospheric radiative transfer. The deviation is shown in Table 2.

From the above table, it can be seen that the average absolute percentage deviation of 16 spectral bands of VNIR and SWIR (413–1640 nm) is about 5.742%; the average absolute percentage deviation of visible light (413–980 nm) is about 5.239%; and the calibration accuracy of VNIR and SWIR is 10% better than index requirements.

Table 2. Radiance and deviation based on Radiation transmission algorithm

Band (μm)	Lt_TG2 (w/m2/sr/μm)	(w/m2/sr/μm)	Relative percentage deviation (%)	Absolute percentage deviation (%)
0.413	56.888	59.075	−3.702	3.702
0.443	51.659	51.133	1.029	1.029
0.49	42.82	41.606	2.918	2.918
0.52	33.307	31.598	5.409	5.409
0.565	25.315	23.153	9.338	9.338
0.62	17.301	16.455	5.141	5.141
0.665	13.704	13.472	1.722	1.722
0.683	12.638	11.764	7.429	7.429
0.75	8.908	8.865	0.485	0.485
0.82	4.757	5.071	−6.192	6.192
0.865	5.261	5.723	−8.073	8.073
0.905	3.158	3.089	2.234	2.234
0.94	1.459	1.62	−9.938	9.938
0.98	3.090	2.816	9.730	9.730
1.24	1.730	1.919	−9.328	9.328
1.64	0.684	0.76	−9.211	9.211

4.3 In-Orbit Absolute Radiation Calibration of TIR

The main factors affecting in-orbit absolute radiation calibration of TIR are the stability of the in-orbit output signal, including the difference stability and absolute value stability. The difference stability characterizes the stability of the response rate, and the absolute stability characterizes the stability of the background. WIS performed two on-board TIR radiation calibration tests on February 23 and March 8, 2017. Figure 10 is the calculation result of the response rate stability of TIR submodule1, and Fig. 11 is the calculation result of the background stability of TIR submodule1.

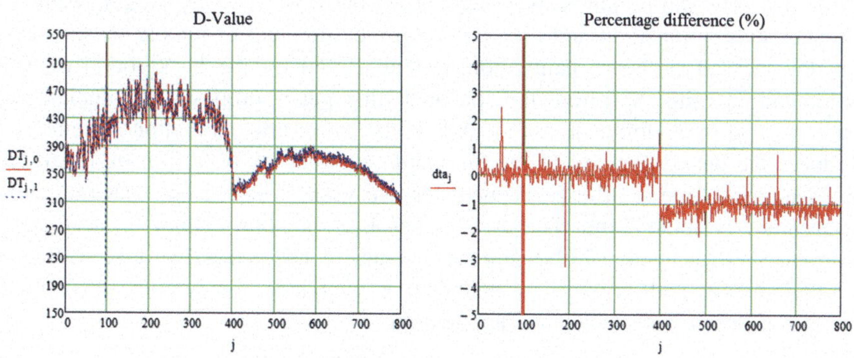

Fig. 10. Response rate stability curve

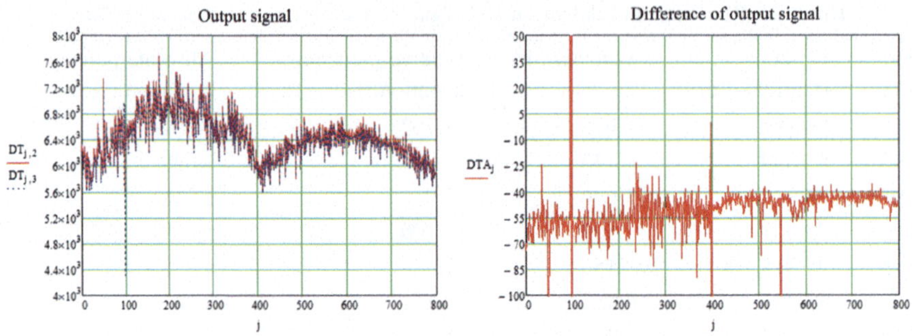

Fig. 11. Background stability curve

Table 3. In-orbit absolute radiation calibration accuracy results of TIR

TIR	submodule1		submodule2	
	band1	band2	band1	band2
Absolute radiation accuracy (K)	1.22	0.98	1.42	1.14

According to the data of the response rate change, the background radiation stability and the noise equivalent temperature difference obtained by twice in-orbit radiation calibration, the absolute radiation calibration accuracy of TIR can be calculated, as shown in Table 3. It can be seen from Table 3 that the absolute radiation calibration accuracy of WIS TIR is 1.42 K, which is better than the 2 K (300 K target) index requirement.

5 Conclusion

At present, Chinese ocean color satellite remote sensing has made great progress in satellite platform and remote sensor technology, but it is relatively lagging behind in the quantitative application and high-precision measurement of remote sensors, especially the lack of on-board calibration system. Based on the background of TG-2 "Wide-band Imaging Spectrometer" payload, this paper proposes a complete full-detection band radiation calibration and VNIR spectrum calibration scheme, and introduces the design method of the on-board calibrator of WIS. The on-board calibrator has undergone in-orbit assessment for two years. The average spectral calibration accuracy is about 2 nm. The average absolute radiation calibration accuracy of VNIR and SWIR is about 5.724%, and the accuracy of TIR is less than 1.42 K. It meets the requirements and provides a quantitative guarantee for remote sensing data inversion of ocean water color and temperature.

Acknowledgements. Thanks to China Manned Space Engineering for providing Wide-band Imaging Spectrometer data products of Tiangong-2.

References

1. Hongbo, W.: Research on In-orbit spectral calibration and polarization correction technology of visible/near-infrared push broom imaging spectrometer. Graduate School of Chinese Academy of Sciences (Shanghai Institute of Technical Physics) (2016)
2. Liang, Ma., Jun, Dian, Xiaoxian, Huang, et al.: Laboratory spectral calibration method of grating dispersive imaging spectrometer. Laser Optoelectron. Prog. **02**, 198–202 (2013)

Pre-launch Calibration of the Multi-Azimuth UV Imager on Tiangong-2

Xiaohu Yang[1(✉)], Shurong Wang[1], Yu Huang[1], Zihui Zhang[1], Guanyu Lin[1], Zhanfeng Li[1], Qingjun Ma[1], Daren Lv[2], Minzheng Duan[2], Wenxing Zhang[2], and Xuemei Zong[2]

[1] Changchun Institute of Optics, Fine Mechanics and Physics, Chinese Academy of Sciences, Changchun 130033, China
yangxiaohu861106@163.com
[2] Institute of Atmospheric Physics, Chinese Academy of Sciences, Beijing 100029, China

Abstract. The Multi-Azimuth UV Imager (MAVI) is a totally new passive sensor onboard. MAVI gives real time spectral scattering information from multi-direction of ±(70°–73°) limb and ±5° nadir field. Introduction of MAVI has been stated in detail. Based on the design of MAVI, the wavelength and radiance calibration have been respectively and specifically described, which include the approach, the measured setup and the calibration results. The method, first segmenting and then combining fields, has been proposed and adopted in the whole radiometric calibration. Analysis shows that the absolutely combined standard uncertainties of three bands are respectively 5.4%, 5.1%, 4.4%.

Keywords: Tiangong-2 · MAVI · Limb · Segment and combine Radiometric calibration · Uncertainty analysis

1 Introduction

The Multi-Azimuth UV Imager (MAVI), one of the two main payloads of the Limb UV exploration project on Tiangong-2(TG-2), is an upgrade sensor of the balloon six-band imager of the national "863" project. As a successor, MAVI has a simplification on both bands and azimuths, which can make MAVI more compact and reliable [1].

MAVI is developed and applied respectively by two institutes from CAS (Chinese Academy of Sciences), the CIOMP (Changchun Institute of Optics, Fine Mechanics and Physics) taking charge with the imager developing and the IAP (Institute of Atmospheric Physics) for the data retrieval. MAVI is designed to collect global limb spectral data and establish/validate a suit of atmospheric species retrieval algorithm. On the other hand, the global limb spectral data can make up of an UV background radiation field, which can be used in object recognition and tracking [1, 2].

Y. Gu et al. (Eds.): *Proceedings of the Tiangong-2 Remote Sensing Application Conference*, LNEE 541, pp. 40–48, 2019.
https://doi.org/10.1007/978-981-13-3501-3_4

MAVI, successfully launched on September 15th, 2016 with TG-2 spacecraft, is all along working properly from then on [3].

2 The MAVI Instrument

MAVI is a totally new remote sensor because of its unique multi-azimuth observation in limb and nadir field. It has three homologous bands, which are selected by 3 specific filters associated with the ozone absorption in atmosphere. Each band adopts the structure of 3-axis attitude sensor. The optical system and its onboard observation modes are shown in Fig. 1.

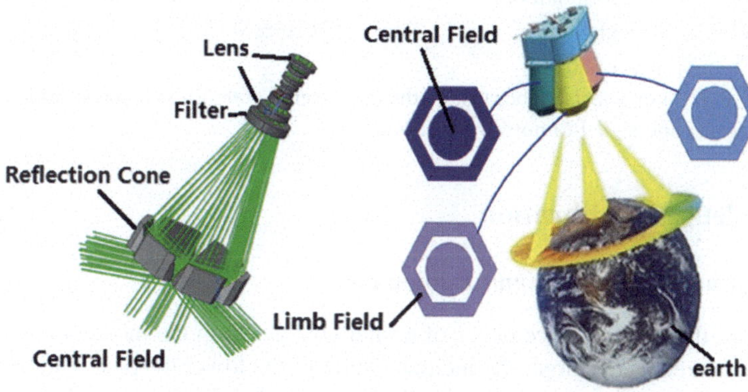

Fig. 1. Optical system and onboard observation modes of MAVI

From Fig. 1, we can see the optical system mainly consists of a 6-reflection-cone, a filter and 6 lenses. As we know, the filter and the 6 lenses form a concentric optical system, and the 6-reflection-cone compresses the 6-azimuth limb ±(70°–73°) field into the concentric optical system with ±(13°–16°), leaving a ±5° central field without vignetting and shading. According to the optical system imaging principle, the ±5° central field and the 6-azimuth limb ±(70°–73°) field will image into a circle and a 6-azimuth ring on the detector respectively. When MAVI is onboard, every circle represents the distributing of atmospheric radiance or ground brightness in ±5° nadir field, while every ring represents the earth radiance varying with altitude and azimuth in ±(70°–73°) limb field [4, 5].

The central wavelengths of the 3 specific filters are approximately 265 nm, 295 nm and 360 nm. The detectors are all e2v technologies' high-performance, advanced inverted mode operational, and back-illuminated CCD 47-20. Because of the small signal of band-265 and band-360, the CCDs are working on −13 °C, while CCD of band-295 is working without refrigeration.

Table 1 gives a mass of key parameters of MAVI.

Table 1. Key parameters of MAVI

Number	Item	Value
1	Central wavelength/nm	265/295/360
2	Bandwidth/nm	(15 ± 1)/(12 ± 1)/(6 ± 1)
3	Focal length/mm	22.37
4	F-number	3.3
5	Field of view/°	±5 (center)/± (70–73)(limb)
6	Limb spatial resolution/km	≤4[a]
7	Limb attitude range/km/400 km	10–80/10–80/10– 60[b]
8	Stray light	≤1%
9	Dynamic range	10^3
10	Dimension/mm	275 × 256 × 230
11	Weight/kg	9.10 ± 0.18

[a](3 × 3) pixels binning;
[b](10–60) km detectable because of the large signal, (60–80) km undetectable because the signal-to-noise ratio (SNR) < 1.

3 Wavelength Calibration

3.1 Calibration Approach and Equipment

As we know, the transmittance curve of a filter changes with the incident angle, which can be generalized as the larger the incident angle is, the lower the peak will be, and the further the peak's shortwave shifting will be, but the bandwidth will stay unchanged. On the other hand, MAVI is a passive filtered imager, whose multi-field corresponds to different incident angles. In fact, every pixel corresponds to its own field, which means a different responsibility from any other different pixels [6, 7].

Considering the millions of pixels, measurement of one pixel by one pixel is definitely unattainable. However, there are some relations among the pixels' responsibility according to the upper view. For each wavelength with same intensity, every pixel has a responsibility varying gradually with incident angle. In case of this, we can just measure some representative pixels wavelength by wavelength and then fill the others with spectral surface fittings by MATLAB.

Figure 2 shows the detailed procedures of wavelength calibration for both central and limb fields. The Xenon lamp with high stability and intensity focuses on the entrance slit of a McPherson monochromator by a reflecting system. The monochromatic light paralleled by another reflecting system, enters into the different fields of MAVI. The McPherson monochromator can supply a series monochromatic light correcting by a standard Si detector in the sensor's responded bandwidth. The difference between limb and central wavelength calibration is just the absence of the 6-demension rotator. By the way, Fig. 2 also gives a schematic diagram of the measured points of wavelength calibration.

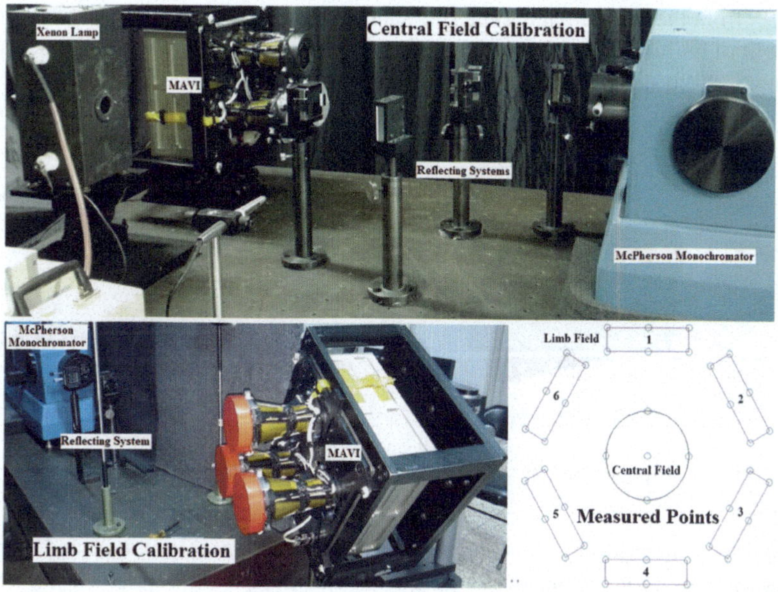

Fig. 2. Detailed procedures of wavelength calibration and the measured points

3.2 Wavelength Calibration Results

Figure 3 shows some normalized spectral transmittance curves of MAVI's 3 bands. We can see a similar shortwave shifting from center to limb transmittance curves. Besides, the transmittance curves of 3 bands are unsymmetrical in shape, which indicates that the radiance received by all fields should be an integrating of all wavelengths' response.

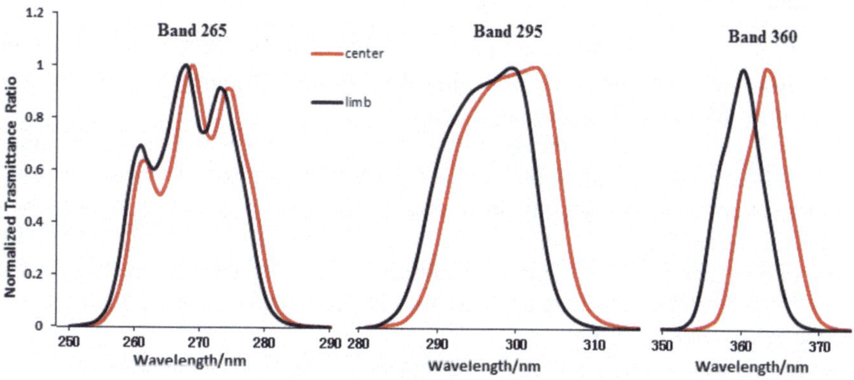

Fig. 3. Some normalized spectral transmittance curves of MAVI's 3 bands

Figure 4 shows a typical spectral surface fitting for center and limb field of band 295 using the measured points. From center field of Fig. 4, we can see the peak of the relative spectrum deviates off the central pixel ([512,512]), which means a slant of the filter-295 from the axis of band 295. This slanting adjustment is intentional to avoid the stray light of multiple reflections of the filter. On the other hand, we can also see arcs of relative spectrum distribution of limb field, which illustrates the decreasing of relative spectrum according to incident angles.

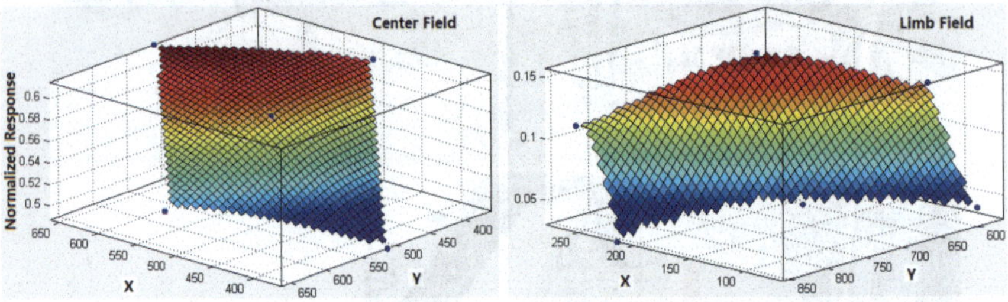

Fig. 4. Typical spectral surface fitting for center and limb field of band 295

In these surface fitting procedures, the independent variables are X and Y, which illustrate the location of the pixels uniquely. The dependent variable is the relative spectrum of the measured points in Fig. 2. In consideration of discontinuities of all fields, the 3 image planes are divided into 6 limb fields and one center field, which can improve the surface fittings' accuracies. The whole fields' relative spectrum distribution can be achieved by combination of these divided surface fittings, one spectrum by one spectrum.

4 Radiance Calibration

4.1 Calibration Approach and Equipment

As a radiance sensor, MAVI is usually calibrated by an integrating sphere in Laboratory. However, there are some limitations to a usual integrating sphere, which are unsuitable for MAVI's radiance calibration [8].

Figure 5 shows a schematic geometric dimension of calibration by integrating sphere. In Fig. 5, the 6-azimuth limb fields of MAVI are ±(70°–73°), which make the limb projection areas distributing in a ring of 750 mm diameter at a distance of 100 mm. Though there is nothing difficult in building an integrating sphere with 1650 mm diameters, MAVI's limb fields almost observe the edge of aperture, and the uniformity and lambert characteristic are both far from satisfaction for high precision radiance calibration [1, 9, 10].

As for the wavelength calibration, the radiance calibration can be also divided into 7 parts, including the 6-azimuth limb parts and the central part. Each part calibrates individually but all uses the center of the integrating sphere's aperture with small off-axis angles. Figure 6 shows the detailed segmental calibration of MAVI's radiance calibration.

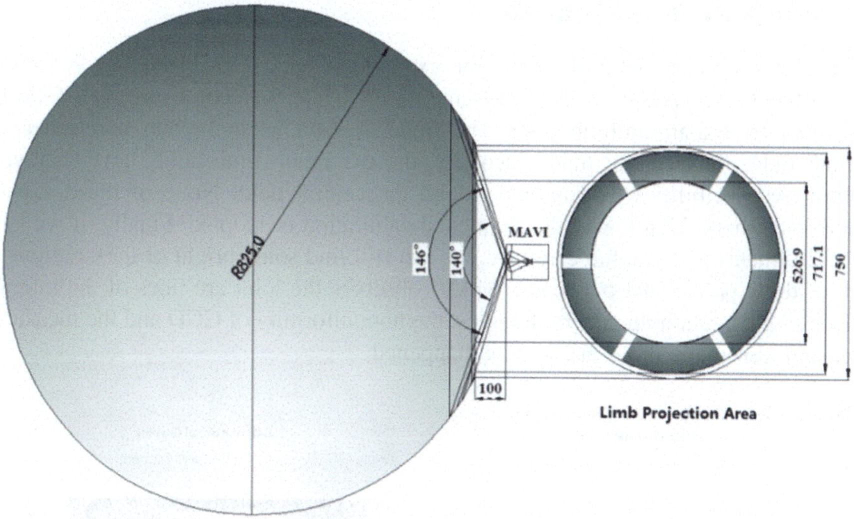

Fig. 5. Schematic geometric dimension of calibration by integrating sphere

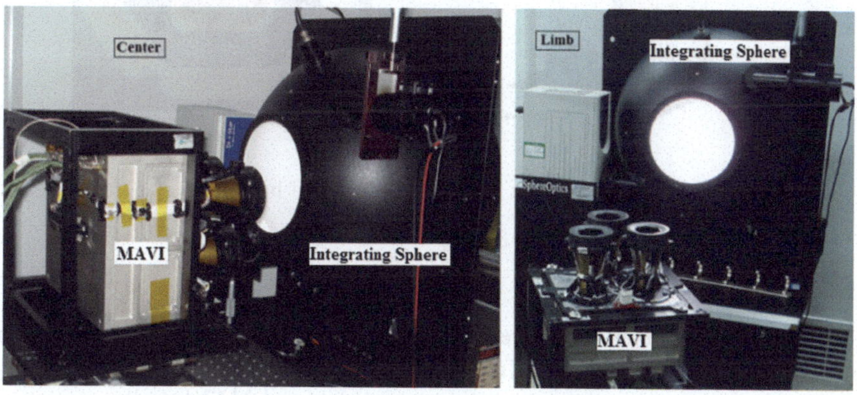

Fig. 6. Detailed calibration of MAVI's 7 segmental fields

In Fig. 6, calibration of the central part is so easy with alignment of the axis of MAVI and sphere. Calibrations of the 6-azimuth limb parts are more complicated than central part because they should adjust the axis of limb fields to sphere with the help of a tilt table and a turn table. The central field's projection is a circle, while the 6-azimuth limb fields' are corresponding to 6 arc-shaped trapeziums. This segmental calibration reduces not only the source uncertainty of surface uniformity to a great extent but also the source uncertainty of angular uniformity in limb calibration. Furthermore, the integrating sphere is no need to be too large, which will save a lot of money.

4.2 Radiance Calibration Results

Figure 7 gives a schematic of calibration example of band 295. From Fig. 7, we can see the band responsibilities are in a same order, while the central part is a little bit larger than the 6-azimuth limb fields. The biggest reason lies in the one more reflection of limb fields than central field. Secondly, the 6-azimuth limb fields have relatively uniform responsibilities distributing around the central field. None of these 7 parts disturb the others. That's why the segmental calibration is adopted. Finally, if we look at the 2-dimensional graphic carefully, we can still find some bright or dark regions in each of the 7 parts. That is because MAVI observes the joint arc-lines of the internal surface of the integrating sphere. Besides, the nonuniformity of CCD and the measured noise can also make these phenomena happened.

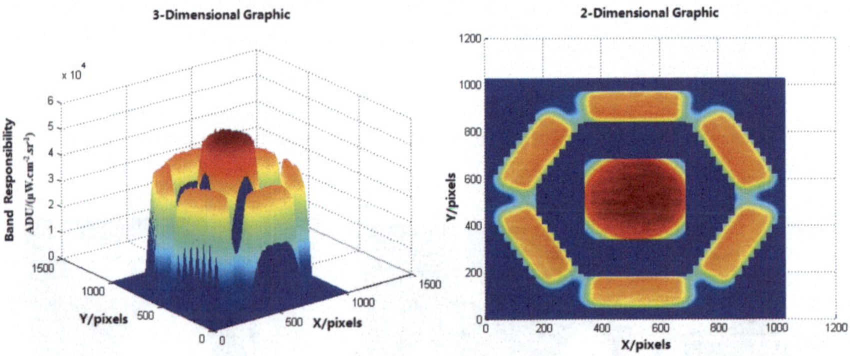

Fig. 7. Graphics of calibration result of band 295

5 Uncertainty Analysis

Table 2 shows the detailed sources and values of absolutely combined standard uncertainty. As can be seen from Table 2, the sources mostly originate from 3 parts, including the integrating sphere, the environment state and MAVI itself.

- The radiance of the integrating sphere transfers from the NIST quartz-tungsten-halogen standard lamp and a spectralon panel diffuser with FY-3 SBUS (Solar Backscattered Ultraviolet Spectroradiometer) sensor. The uncertainties are respectively 5%, 4.7%, 4% for 3 bands;
- The surface uniformity of the integrating sphere is measured by a Si detector with small aperture on a 2-dimensional moving table. The biggest bias is less than 1% [11];
- The angular uniformity of the integrating sphere is measured by a Si detector with small aperture on a 2-dimensional rotating table. The biggest bias is also less than 1% [11];
- The environment state mostly refers to the temperature, and the uncertainty is estimated with 0.5% as empirical;

Table 2 Absolute Combined Standard Uncertainty of MAVI's Calibration

Uncertainty Sources		Value		
		265	295	360
Integrating Sphere	Radiance	5%	4.7%	4%
	Surface Uniformity		1%	
	Angular Uniformity		1%	
Environment State			0.5%	
MAVI	Spectral Shift	1%	0.6%	0.3%
	Repeatability		0.5%	
	Stray Light	0.5%	0.6%	1%
	Readout Noise of CCD	0.1%	1%	0.1%
Absolutely Combined Standard Uncertainty		5.4%	5.1%	4.4%

- The spectral shift of MAVI derives from the wavelength calibration. The estimated values are about 1%, 0.6%, and 0.3%;
- The repeatability of MAVI is 0.5%;
- The stray light of MAVI comes from the outside and the field interference, and the values are measured as 0.5%, 0.6%, and 1%;
- The readout noise of CCD relates to the CCD itself and the electrical design. CCDs of band 265 and band 360 are working on minus 13 °C, while CCD of band 295 is working without refrigeration. These kinds of uncertainties are respectively 0.1%, 1%, and 0.1%.

According to uncertainties analysis above and the comprehensive uncertainty formula (1), TN1297 of NIST [12]

$$\mu_c^2(y) = \sum_{i=1}^{N} \left(\frac{\partial f}{\partial x_i}\right)^2 \bullet \mu^2(x_i) + 2\sum_{i=1}^{N-1} \sum_{j=i+1}^{N} \frac{\partial f}{\partial x_i} \bullet \frac{\partial f}{\partial x_j} \bullet \mu(x, y) \quad (1)$$

Based on the uncorrelated uncertainties in pairs, formula (1) is transformed into formula (2) as below

$$\mu_c^2(y) = \sum_{i=1}^{N} \left(\frac{\partial f}{\partial x_i}\right)^2 \bullet \mu^2(x_i) \quad (2)$$

The absolutely combined standard uncertainties of three bands are respectively calculated to 5.4%, 5.1%, and 4.4% by formula (2).

6 Conclusion

MAVI is totally a new experimental sensor in atmospheric probing. Based on the structure characteristic, there are so many difficulties in the whole designing period. The filtered design not only makes the wavelength calibration fussy, but also affects the precision of atmospheric species' retrieval. The super large limb fields bring out inconvenience not only in radiance calibration, but also in other measurements and calibrations. Even so, MAVI is a splendid idea which will guide the subsequent instruments for too many years.

Acknowledgments. MAVI is part of a joint project of CIOMP and IAP from CAS, funded by the major national special projects. Thanks to Chinese Manned Space Engineering for providing space science and application data products of Tiangong-2.

References

1. Xiaohu, Y.: Study on Performance Evaluation and Radiometric Calibration of the Earth Limb Annular Imager, pp. 20–102. Changchun Institute of Optics, Fine Mechanics and Physics, Chinese Academy of Sciences, Changchun (2012) (in Chinese)
2. Zhang, J., Wang, S.-R., Huang, Y., Xue, Q.-S., Li, B.: Status and development of limb imaging spectrometers. Chin. Opt. **6**(5), 692–700 (2013) (in Chinese)
3. LNCS Homepage. http://www.xinhuanet.com/science/2017-09/15/c_136600362.htm. Last accessed 23 Aug 2018 (in Chinese)
4. Xiangyang, Y., Shurong, W., Yu, H., Guanyu, L.: Development of ultraviolet annular imager for earth limb and nadir sounding simultaneously. Acta Opt. Sin. **32**(12), 1211004 (2012) (in Chinese)
5. Zhang, J., Wang, S.-R., Li, B., Xue, Q.-S., Huang, Y.: Design of UV annular imaging system for atmospheric limb detection. Spectrosc. Spectr. Anal. **34**(3), 843–846 (2014) (in Chinese)
6. Yongqiang, L., Ruimin, F., Yongxiang, G., Xiaoying, Z.: Measurement and revision for ultra-narrow bandpass filter spectral characteristics. Spacecr. Recovery Remote Sens. **34**(2), 25–29 (2013) (in Chinese)
7. Xuguo, Z., Dakai, Z., Li, W., Zhaorong, L.: Influences on image quality of filter installation errors. Spacecr. Recovery Remote Sens. **33**(6), 74–79 (2012) (in Chinese)
8. Heath, D.F.: Large aperture spectral radiance calibration source for ultraviolet remote sensing instruments. In: Proceedings of SPIE, vol. 4891, pp. 335–342 (2003)
9. Zhang, J.: The study on UV panoramic imaging system for middle atmosphere sounding in limb view, pp. 10–94. University of Chinese Academy of Sciences, Changchun (2014) (in Chinese)
10. Xiaohu, Y., Shurong, W., Yu, H.: Research on radiation characteristic of the multi-azimuth/multi-waveband earth limb ultraviolet imager. Acta Opt. Sin. **34**(8), 0801001 (2014) (in Chinese)
11. Xiaohu, Y., Shurong, W., Yi, Q., et.al.: Research of UV spectral radiance characteristics of the integrating sphere based on calibration by the absolute radiance radiometer. Chin. J. Lasers **38**(10), 1008007-1-6 (2011) (in Chinese)
12. Jiapeng, W., Shurong, W., Lingdi, X., et.al.: Wavelength calibration for ultraviolet spectroradiometer and analysis of its uncertainty. OptoElectron. Eng. **35**(6), 42–47 (2008) (in Chinese)

Pre-launch Calibration of the Tiangong-2 Front-Azimuth Broadband Hyperspectrometer

Zhanfeng Li, Shurong Wang(✉), Yu Huang, Qingjun Ma, Qingsheng Xue, and Zhigang Li

State Key Laboratory of Applied Optics, Changchun Institute of Optics, Fine Mechanics and Physics, Chinese Academy of Sciences, Changchun 130033, China
srwang@ciomp.ac.cn

Abstract. The Front-Azimuth Broadband Hyperspectrometer was launched on board the Tiangong-2 Space Laboratory on 15 Sep 2016 and had successfully measured spectral radiance of Earth limb atmosphere, to provide profiles of the volumetric concentration of ozone, nitrogen dioxide and other minor trace gases. In order to ensure the high accuracy and stability of data, The Front-Azimuth Broadband Hyperspectrometer was calibrated in laboratory and in orbit accurately. This paper mainly introduced the facility, method and results of the pre-launch calibration. Finally, the uncertainty of the calibration results was analyzed, and the results met the requirements of the data application.

Keywords: Tiangong-2 · ABHI · On-ground calibration · Smear Stray light correction

1 The FABH Instrument

The Front-Azimuth Broadband Hyperspectrometer (FABH) is a compact and miniaturized imaging spectrometer for observing limb-scattered solar radiation. Figure 1 shows physical picture of FABH. Similar to Limb Profiler in Ozone Mapping & Profiler Suite (OMPS), the FABH is developed mainly to provide the amounts and vertical distributions of ozone by limb observation mode. The optical layout of FABH is shown in Fig. 2. It consists of a telescope and a spectrometer which utilizes a prism as the dispersive element to achieve wavelength range from 290 nm to 1000 nm. To achieve the spatial resolution of 3 km and the spectral resolutions from 1 nm in the UV to 25 nm in the NIR, three off-axis aspheric mirrors are used to compensate aberration in the FABH. An off-axis parabolic mirror is used for the telescope, and an ellipsoidal mirror and a hyperboloid mirror are used respectively for collimating mirror and imaging mirror in the prism spectrometer. An enhanced UV coated frame transfer CCD is used to increase the signal-to-noise in the UV. Table 1 gives the important performance parameters of FABH.

Y. Gu et al. (Eds.): *Proceedings of the Tiangong-2 Remote Sensing Application Conference*, LNEE 541, pp. 49–60, 2019.
https://doi.org/10.1007/978-981-13-3501-3_5

Table 1. Parameters of Front-Azimuth Broadband Hyperspectometer

Parameter	Value
Probe limb height	10–60 km
Spectral range	290–1000 nm
Spectral resolution	1.2 nm@290 nm–23 nm @1000 nm
Field of view	2.4° × 0.02°
F number	6.92
Entrance pupil diameter	10 mm
Dynamic Range	>10^6
Stray light coefficient	3%@600 nm
Spatial resolution	3 km
Weight	11.8 kg
Size	393 mm × 328 mm × 174 mm

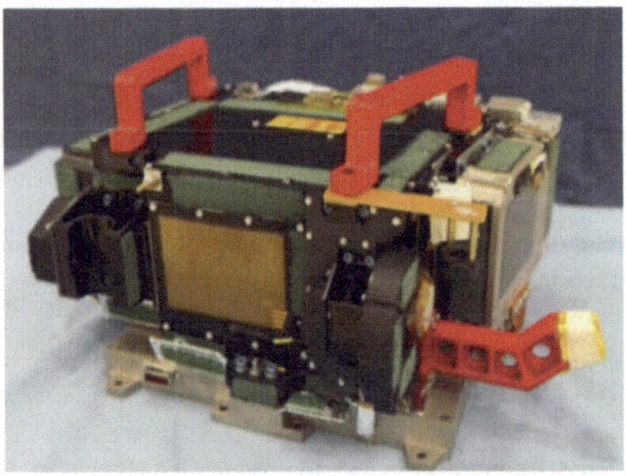

Fig. 1. Physical picture of the FABH

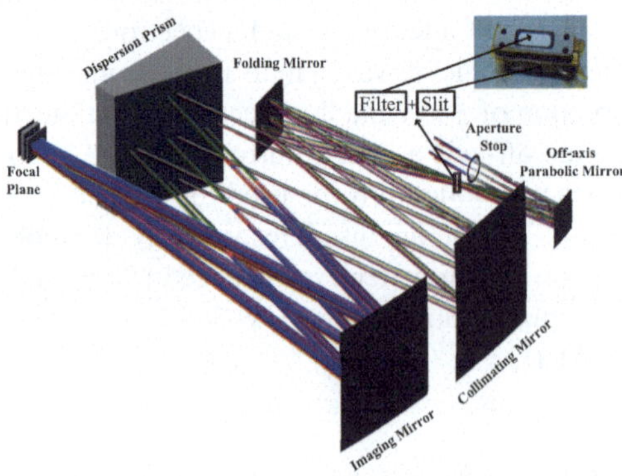

Fig. 2. Optical layout of the FABH

2 Pre-launch Calibration

The aim of the spectrometer calibration is to establish a quantitative relationship between the raw data (electronic signal) and its corresponding spectral radiation of targets [1]. Therefore the calibration is a crucial prerequisite for any successful retrieval of atmospheric geophysical parameters [2]. The space-borne spectrometer calibration includes pre-launch calibration and in-flight calibration. The in-flight calibration of FABH uses a common method, namely solar calibration, which is beyond the scope of this paper. This paper mainly introduces the pre-launch calibration of FABH.

The pre-launch calibration formula for the FABH is

$$I(\lambda) = \frac{S_{\text{det}}[pixel_i(\lambda), pixel_j(w)] - DC - S_{elec} - S_{stray}(\lambda)}{R(\lambda)} \quad (1)$$

where $I(\lambda)$ is the spectral radiation of targets, S_{det} is the raw signal measured on the detector, $pixel_i(\lambda)$ is the relational function between pixel column number $pixel_i$ and wavelength λ, $pixel_i(w)$ is the relational function between pixel row number $pixel_j$ and field of view w, DC is the total dark signal, S_{elec} is detector effects such as smear, $S_{stray}(\lambda)$ is the spectral stray light and $R(\lambda)$ is the spectral response function.

According to Eq. (1), the pre-launch calibration of FABH is achieved by applying a complex sequence of individual calibration steps to raw data S_{det}, which mainly includes detector corrections S_{elec}, spectral calibration $pixel_i(\lambda)$, stray light correction $S_{stray}(\lambda)$ and radiometric calibration $R(\lambda)$.

2.1 Detector Corrections

The detector used in FABH is a frame transfer Charge-Coupled Device (CCD47-20 BI AIMO) manufactured by e2v technologies. Its general data and typical performance are show in Table 2.

Table 2. Parameters of e2v CCD47-20 BI AIMO

Parameter	Value
Pixel size	13 × 13 μm
Image area pixels	1024 × 1024
Storage area pixels	1024 × 1033
Maximum readout frequency	5 MHz
Output responsivity	4.5 μV/e$^-$
Peak signal	100 k e$^-$/pixel
Dynamic range (at 20 kHz)	~50 000:1
Spectral range	200–1100 nm
Readout noise (at 20 kHz)	2.0 e$^-$ rms

In all effects of detector in the FABH, smear has the greatest impact on measurement accuracy, especially when the integration time is short. Smear is a common feature in frame transfer CCD, which occurs because the detection area is exposed to light during the transfer of charges from the detection area to the readout buffer [3].The amount of smear influence is inversely proportional to the integration time and the read frequency. To eliminate the smear effect, a simple algorithm is proposed and demonstrated.

According to the operational principle of roller blind frame transfer CCD, the smear signals of all pixels in a column in the frame transfer direction are the same. The smear signal ΔS(x,y) can be expressed as follows.

$$\Delta S(x,y) = \sum_{x=1}^{m} k\eta T_{ti} I(x,y) \tag{2}$$

Where $\Delta S(x,y)$ is the smear signal, x is pixel row number, y is pixel column number, k is the instrument transmittance, η is the photoelectric conversion efficiency of the CCD, T_{ti} is the frame transfer clock cycle, $I(x,y)$ is the incoming intensity of pixel (x,y). And the unsmeared signal $S(x,y)$ of pixel (x,y) is

$$S(x,y) = k\eta T_{\exp} I(x,y) \tag{3}$$

where T_{exp} is integration time for target detection. The following equation can be obtained from Eqs. (2) and (3),

$$\Delta S(x,y) = \frac{T_{ti}}{T_{\exp}} \sum_{x=1}^{m} S(x,y) \tag{4}$$

After preprocess including dark signal subtraction and signal nonlinear correction, the CCD image signal with smear is

$$S'(x,y) = S(x,y) + \Delta S(x,y) \tag{5}$$

where $S'(x,y)$ is the CCD image signal after preprocessing, which is calculated from the raw output signal of CCD. And the smear signal $\Delta S(x,y)$ can be obtained from Eqs. (4) and (5).

$$\Delta S(x,y) = \frac{K}{(1+m\times K)} \sum_{x=1}^{m} S'(x,y) \tag{6}$$

Where $K = T_{ti}/T_{exp}$, which is defined as smear coefficient. Finally, the unsmeared signal $S(x,y)$ is

$$S(x,y) = S'(x,y) - \frac{K}{(1+m\times K)} \sum_{x=1}^{m} S'(x,y) \tag{7}$$

In order to prove the effectiveness of the smear correction algorithm, an experiment was carried out using the FABH, an integrating sphere and a deep cutoff high pass filter. In the wavelength band less than 500 nm, the transmittance of the filter was less than 0.01%, which prevented the optical signal with wavelength of less than 500 nm from entering the CCD image plane. Therefore, the residual signal of the CCD at the corresponding wavelength position was the smear signal. Figure 3(a) showed the CCD image obtained in the experiment. Obviously, there was a smear signal effect. The image after smear removal by this algorithm was shown in Fig. 3(b). After calculation, the residual smear effect was less than 0.5% of the true signal. The main source of residual effect is detector noise, and we are currently studying how to eliminate it [4].

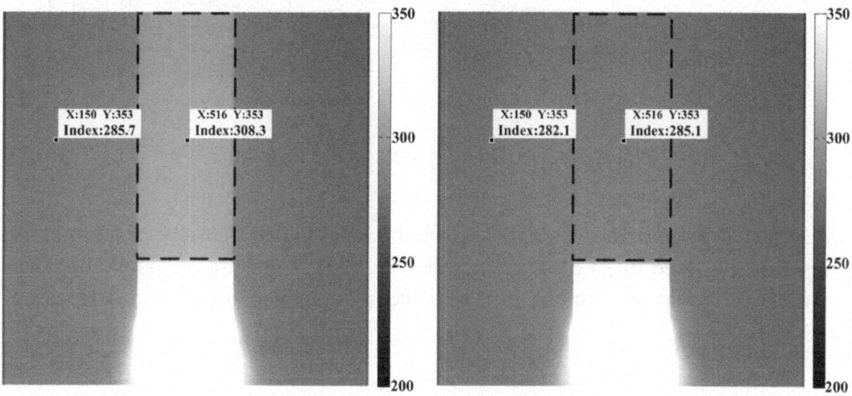

(a) CCD image with smear (b) CCD image after smear removal

Fig. 3. CCD image before and after smear removal

2.2 Spectral Calibration

For the spectral calibration of FABH, a calibration device was established in the laboratory. Figure 4 shows the schematic of the device. It comprises a wide spectral band and high stability light source, a convergence system, a high precision monochromator, an integrating sphere system and an absolute radiometer. The monochromator is McPherson Model 209 produced in the United States. Its wavelength accuracy is 0.05 nm and repeatability is ±0.005 nm.

Due to the nonlinear dispersion of the prism in FABH, the spectral response functions are calibrated sequentially for every pixel on the image plane [5]. And the weighted fitting was used as processing method of spectral calibration data. The following is a brief introduction to the processing method.

The center wavelength λ_c of a fixed pixel (column number i) is

$$\lambda_c = \frac{\int_{\lambda_1}^{\lambda_2} \lambda R_i(\lambda) d\lambda}{\int_{\lambda_1}^{\lambda_2} R_i(\lambda) d\lambda} \tag{8}$$

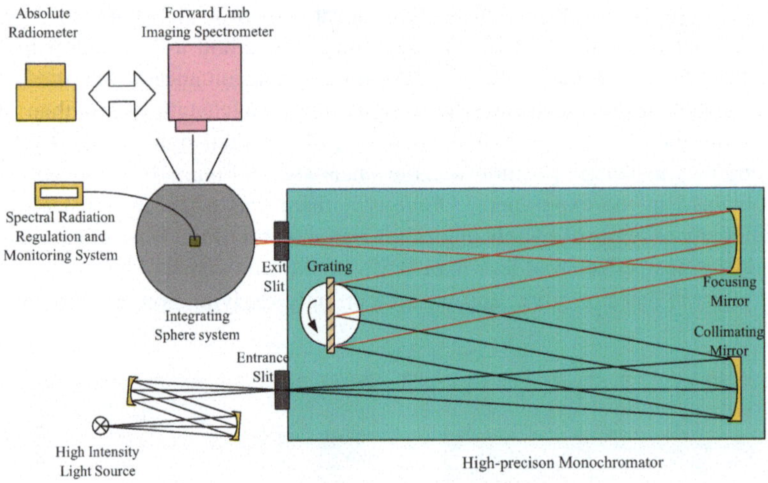

Fig. 4. Device diagram for spectral calibration

where λ_1 and λ_2 is cutoff wavelength, in which signal is less than 5% of the maximum response for the fixed pixel (column number *i*), $R_i(\lambda)$ is spectral response function in the fixed pixel (column number *i*). The spectral response function is expressed as follows.

$$R_i(\lambda) = \frac{S_i(\lambda)R_s(\lambda)}{S_s(\lambda)} \tag{9}$$

Where $R_i(\lambda)$ is the output signal in the fixed pixel (column number *i*), S_s (λ) is output signal of the absolute radiometer, R_s (λ) is the absolute radiometer's response function.

The spectral calibration results of FABH are shown in the Figs. 5 and 6. Figure 5 shows the spectral response functions at some typical wavelengths. Figure 6 shows relational curve between wavelength, spectral resolution and pixel column number at 0° field of view, and the relational function is a sixth degree polynomial.

In the spectral calibration of FABH, smile is a problem that could not be ignored [6]. The smile property is a center wavelength shift with field of view. Table 3 shows spectral calibration result of FABH for different field of view. In addition, the effect of temperature on spectral calibration is also verified in the experiment. In the temperature range of 18 ± 5 °C, the maximum wavelength shift could reach 0.4 pixels. According to the on-orbit working environment, the spectral calibration results would be corrected accordingly.

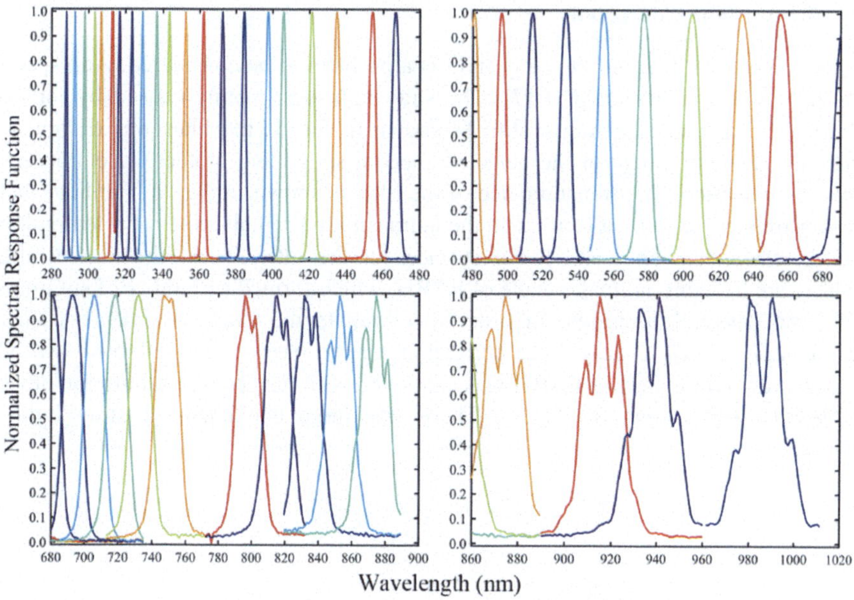

Fig. 5. Spectral response function of FABH

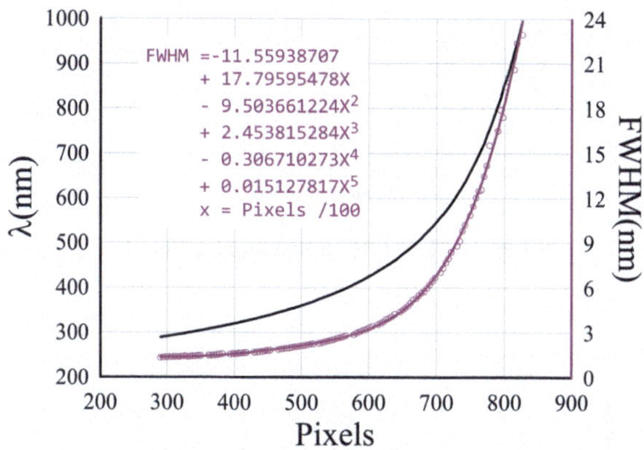

Fig. 6. Relational curve between wavelength, spectral resolution and pixel column number

Table 3. Spectral calibration result of FABH for different field of view

Field of view	Center wavelength/nm (pixel column number *i*)				
	i = 308	*i* = 576	*i* = 736	*i* = 800	*i* = 836
−1.2°	290.55	399.01	599.67	806.51	996.65
0°	290.45	398.77	598.85	805.40	995.18
+1.2°	290.55	398.98	599.70	806.52	996.62
Maximum deviation/nm	0.10	0.24	0.85	1.02	1.47

2.3 Stray Light Correction

In spectrometer calibration, the stray light is a problem to be considered. And spectral stray light was found in FABH [7]. The stray light was mainly caused by multiple reflections of a filter at entrance slit. As shown in Fig. 1, the filter was designed to reduce the dynamic range of the detected spectral signal and protect the entrance slit from contamination. The transmittance of the filter is shown in Fig. 7. The filter had a low transmittance at 500–800 nm, and the untransmitted signals were reflected multiple times between the upper and lower surfaces of the filter. Then, the untransmitted signals were incident on image plane of FABH, which formed a plurality of slit images at different spectral positions. In order to eliminate the extra slit images (namely spectral stray light), a stray light correction matrix method was adopted [8]. The spectral stray light matrix of FABH was calibrated with the spectral calibration device. The results were shown in Fig. 8. There was large coefficient of stray light in 500–800 nm.

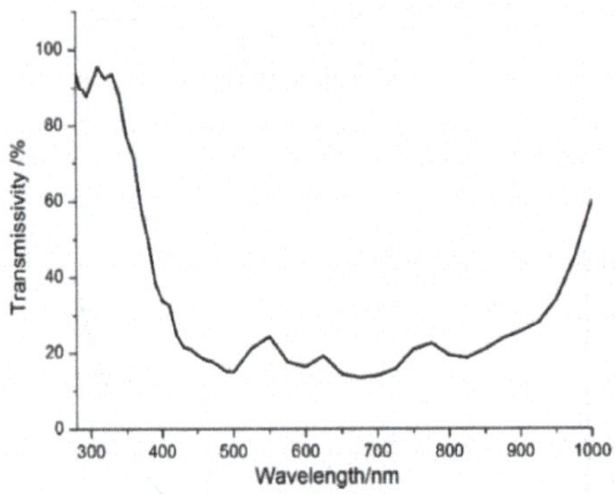

Fig. 7. The filter transmittance curve

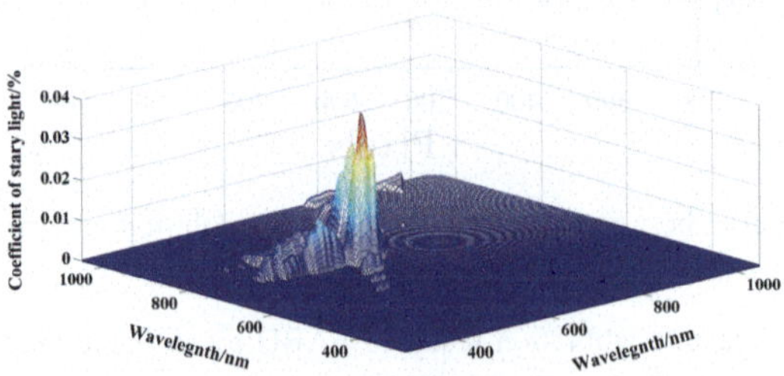

Fig. 8. Spectral stray light matrix of FABH

The effectiveness of the stray light correction matrix was verified by an external field experiment of atmosphere detection. In the experiment, a high-precision CAS 140 CT spectrometer manufactured by German INSTRUMENT SYSTEMS GmbH was used as a reference. The stray light of the CAS spectrometer was less than 10^{-4}. To eliminate the difference in absolute radiometric calibration, the CAS spectrometer and FABH were calibrated using the same NIST standard source (standard tungsten halogen lamp F661). Then the same area of clear sky was simultaneously observed by the two spectrometers. The measured spectral radiances of atmosphere were shown in Fig. 9. After stray light correction, the measured radiances difference between the CAS spectrometer and FABH was reduced from 20% to 3%. The difference was caused by the calibration error of the stray light matrix, which could be further reduced by improving the calibration device and algorithm.

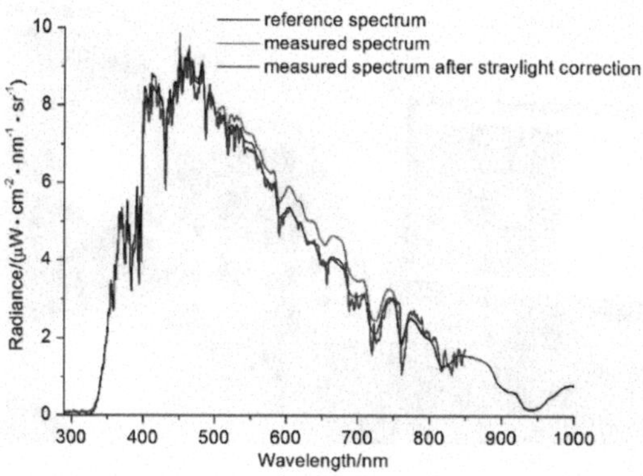

Fig. 9. Measured atmospheric radiance before and after stray light correction

2.4 Radiometric Calibration

There are two traditional methods for spectral radiance calibration of a spectrometer. The first method is that uses a NIST-calibrated lamp illuminating a calibrated Spectralon diffuser to generate a light source with known brightness. The largest source of error for this method is the bidirectional reflectance distribution function (BRDF) calibration of the diffuser. And this calibration method is not suitable for spectrometers with large apertures or large fields of view. In order to overcome these shortcomings, the Walker sphere calibration method was developed [9]. The Walker sphere calibration method uses large-area integrating sphere source as the standard radiance source viewed directly by the instrument. The Walker sphere calibration method not only eliminates the influence of the diffuser, but also can increase the light level of 250–400 nm by using a combination of xenon and tungsten lamps.

The radiometric calibration source of FABH is a UV integrating sphere TX-20Z-UV manufactured by Sphereoptics. As shown in Fig. 10, the diameter of the integrating sphere is 50 cm and the opening diameter is 20 mm. The output uniformity of the sphere is better than 98% in ±15° viewing field. The integrating sphere is equipped with a xenon lamp and a tungsten halogen lamp. For calibration in 290–400 nm, the xenon lamp is used and its stability is better than 0.5%/h. And the tungsten halogen lamp is used for calibration in 400–1000 nm. The power supply of the tungsten halogen lamp is the Agilent E3634A with current stability 0.1%/h. The standard spectral radiance of the integrating sphere is obtained from NIST standard lamp F661 by the Walker sphere calibration method. And its uncertainty is better than 3%.

In radiometric calibration, the FABH was placed in the center of the integrating sphere opening. And according to the output signal of FABH, a reasonable integration time was set to ensure that the SNR was greater than 100. Finally the spectral radiance response of FABH was obtained, as shown in Fig. 11.

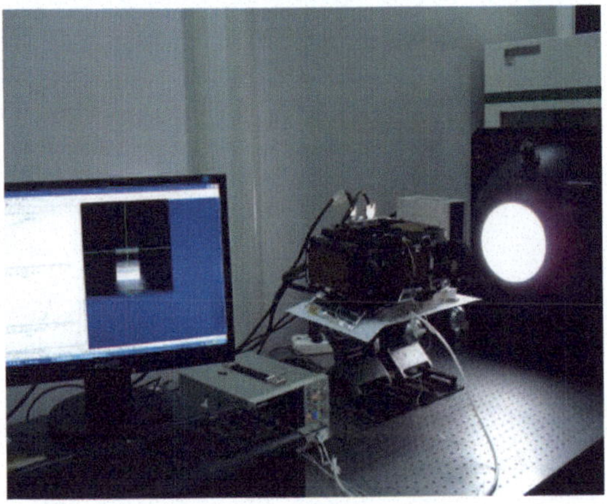

Fig. 10. Spectral radiance calibration of FABH

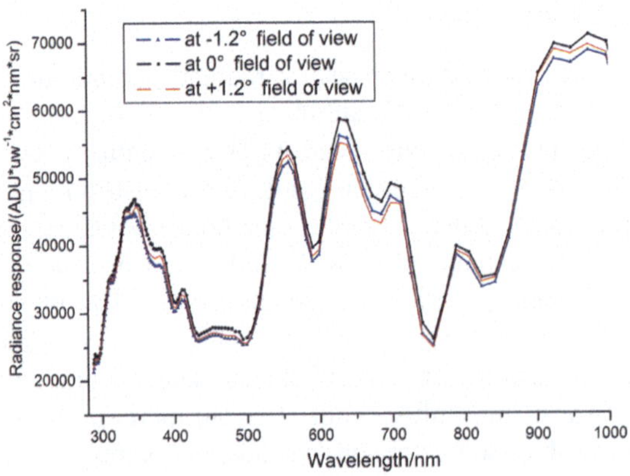

Fig. 11. Spectral irradiance calibration of FABH

The uncertainty for spectral radiance response of FABH is listed in Table 4. The main uncertainty is derived from the transfer system for standard radiation. In the transfer system, the uncertainty of the NIST lamp's standard irradiance is 1.4% (k = 2) at 290–1000 nm. The uncertainty caused by distance error between the lamp and the diffuser is about 0.3%. In addition, the transfer radiometer contributes to an uncertainty of 2%. The uncertainty of the spectral mismatch caused by spectral calibration error is 1.2%. Residual stray light is also an important source of uncertainty, which is about 1.8%. Since the instrument's field of view angle is 2.4° and the entrance pupil diameter is 20 mm, the estimated output uniformity of the integrating sphere error is 1.5%. And the combined uncertainty is 4.2%.

Table 4. Uncertainty for spectral radiance response of FABH

Component	Uncertainty (k = 2)
NIST lamp standard irradiance	1.4%
The transfer radiometer	2.0%
Distance measurement	0.3%
Spectral mismatch	1.2%
Stray light correction	1.8%
Smear correction	0.5%
Uniformity of the integrating sphere	1.5%
Stability of the integrating sphere	0.5%
Repeatability of FABH	2.0%
Combined uncertainty	4.2%

3 Conclusions

In the pre-launch calibration of FABH, we mainly carried out smear correction, spectral calibration, stray correction and radiometric calibration. A simple smear correction algorithm was used to reduce the residual smear effect to 0.5%. The pixel-by-pixel spectral calibration ensured that the spectral accuracy of the FABH reached 0.5 nm. The stray light correction algorithm reduced the influence of stray light from 20% to 3%. By using an integrating sphere traced to the NIST standard, radiometric calibration of FABH was carried out. Finally, the uncertainty of the pre-launch calibration was analyzed. And the combined uncertainty (4.2%) met the accuracy requirement (5%) of FABH's calibration.

Acknowledgements. Thanks to China Manned Space Engineering for providing space science and application data products of Tiangong-2.

References

1. Werij, H., Kruizinga, B., Olij, C., et al.: Calibration aspects of remote sensing spaceborne spectrometer. In: SPIE, vol. 2820, pp. 126–137 (1996)
2. Frerick, J., Bovensmann, H., Noel, S., et al.: SCIAMACHY on-ground/in-flight calibration, performance verification and monitoring concepts. In: SPIE, vol. 3117, pp. 176–187 (1997)
3. Ruyten, W.: Smear correction for frame transfer charge-coupled-device cameras. Opt. Lett. **24**(13), 878–880 (1999)
4. Barducci, A., Mecocci, A.: Theoretical and experimental assessment of noise effects on least-squares spectral unmixing of hyperspectral images. Opt. Eng. **44**(8), 087008-1–087008-17 (2005)
5. Zhang, R., Bayanheshig, Yin, L., et al.: Wavelength calibration model for prism-type echelle spectrometer by reversely solving prism's refractive index in real time. Appl. Opt. **55**(15), 4153–4158 (2016)
6. Yokoya, N., Miyamura, N., Iwasaki, A.: Preprocessing of hyperspectral imagery with consideration of smile and keystone properties. In: SPIE, vol. 7857, pp. 78570B-1–78570B-9 (2010)
7. Zhanfeng, L.: Spectral stray light correction of forward limb imaging spectrometer. http://kns.cnki.net/kcms/detail/31.1252.O4.20180627.1743.054.html (2018)
8. Tarrant, A.W.S.: Optical techniques for studying stray light in spectrophotometers. Opt. Acta **25**(12), 1167–1174 (1978)
9. Heath D.F., Wei Z., Fowler W.K.: Comparability of spectral radiance calibrations of large aperture earth observing instruments based upon diffuse reflective panels and internally illuminated spherical integrator techniques. In: SPIE, vol. 2209, pp. 148–159 (1994)

Tiangong-2 Ground Data Processing and Service System Towards Multiple Categories of Payloads Product Processing

Shengyang Li, Wanfeng Zhang(✉), Zhiwen Liu, Hong Tan, Haijun Yu, Bo Wang, Lingli Mu, and Xuan Li

Key Laboratory of Space Utilization, Technology and Engineering Center for Space Utilization, Chinese Academy of Sciences, Beijing, China
{shyli, wfzhang, zwliu, tanhong, hjyu, wangbo}@csu.ac.cn

Abstract. Tiangong-2 ground data processing and service system is a specialized software which has been used for processing the products of three earth observation payloads and other several space science experiments. In the course of geometric rectification on the Wide-band Imaging Spectrometer data, a high-precision sensor model of the wide-width virtual camera is established to improve the geometric positioning precision. Towards the Interferometric imaging radar altimeter (InIRA) product processing, a series of steps should be performed to generate backscatter coefficient images and surface topography data, such as imaging processing, interferometric processing, radiometric correction and geometric location. After the production of these categories of products, several indicators included signal to noise ratio, clarity, contrast, comentropy and the radiometric accuracy were adopted to assess and control the products quality. As a part of complete process, the space application data promoting service platform for China Manned Space Engineering provide all the products of earth observation payloads and other space science experiments.

Keywords: Tiangong-2 · Ground data processing · Data storage
Data sharing service

1 Introduction

To execute multiple categories of space science experimental projects, Tiangong-2 space laboratory was designed and launched as a prototype of space station. The main purposes of this space mission are to obtain the breakthrough and mastery of spaceflight vehicles' space rendezvous & docking and assembly control technologies. There were three categories of earth observation payloads, including ***the Wide-band Imaging Spectrometer (WIS)***, ***the Interferometric imaging radar altimeter (InIRA)***, and ***the Multi-Azimuth UV Imager (MAVI)*** on Tiangong-2 space laboratory. The functions of Tiangong-2 ground data processing and service system (TG2-GDPSS) covered various products processing and archiving automatically, products integrating immediately and visualization effectively, and products distribution and sharing on-line. The parallelized architecture had been accepted and applied generally in the field of space science data and earth observation data processing extensively [1–5].

Y. Gu et al. (Eds.): *Proceedings of the Tiangong-2 Remote Sensing Application Conference*, LNEE 541, pp. 61–70, 2019.
https://doi.org/10.1007/978-981-13-3501-3_6

The TG2-GDPSS had been studied and developed based on parallelized architecture for many types of payloads space science products processing by Technology and Engineering Center for Space Utilization (CSU). Main function of the TG2-GDPSS including data retrieval, ***the WIS*** product processing, ***the InIRA*** product processing, ***the MAVI*** product processing, and computational task controlling. These three types of payloads were brand new and the processing algorithms of products were developed newly. The TG2-GDPSS had integrated into several workflows for various levels of products processing. Several indicators were imported to assess and control the quality of WIS and InIRA products. Besides, a website named ***Space application data promoting service platform for China Manned Space Engineering*** (http://www.msadc.cn/) was developed to provide products sharing and service.

2 Architecture of TG2-GDPSS

The framework of space science data processing and distribution system had evolved from standalone pattern into distributed parallelized processing schema on account of isometric clusters [6–9]. Therefore, the TG2-GDPSS was designed and developed upon cluster and parallel file system (Fig. 1).

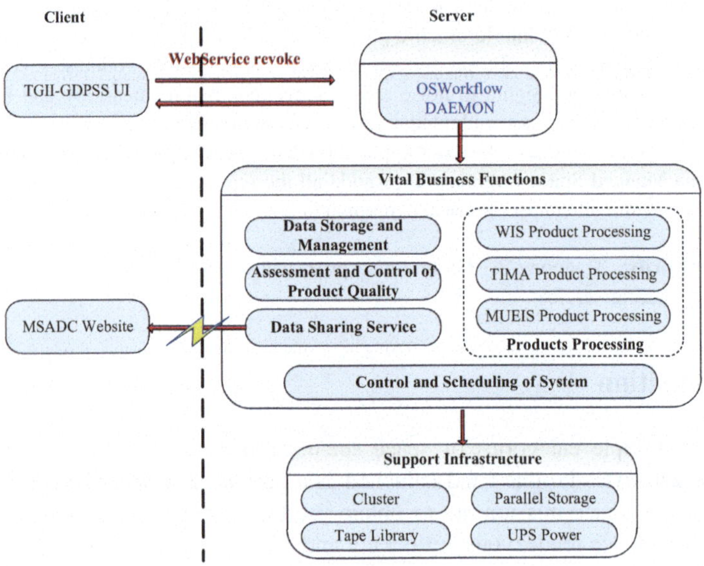

Fig. 1. The Architecture of Tiangong-2 ground data processing and service system

The cluster, parallel storage, tape library, and UPS power consist of the support infrastructure. All the nuclear software modules were involved in the Vital Business Functions to provide products processing, data processing and management, assessment and control of products quality, and data sharing service. These functions modules would be released through the OSWorkflow daemon which is an open source workflow tool used to manage a large number of products processing tasks. The working mode of TG2-GDPSS covers full-automatic and manual operation. Under normal circumstances, all three types of products processing are operated full-automatically.

In order to fulfill the functions of TG2-GDPSS at the side of internal and external users, several sets of software were developed to improve the processing efficiency. There would be more details about these software later in this article.

3 Description of Tiangong-2 Payloads

3.1 The Wide-Band Imaging Spectrometer

The WIS is a new generation of wide-band, wide-field and image-spectrum merging optical payload. The WIS adopts linear push broom imaging in visible near-infrared, short-wave infrared and thermal infrared. There are 7 well-designed detectors inside the wideband imaging spectrometer, 3 of which are visible near-infrared detectors, 2 are short-wave infrared detectors, and 2 are thermal infrared detectors. The field-of-view is spliced and extend to 42° by combining these similar spectrum detectors. The visible near-infrared detectors, short-wave infrared detectors, and thermal infrared detectors have 14, 2, and 2 spectral segments, separately. The registration accuracy between the detection channels is high, which is convenient for data fusion to extract more abundant information.

3.2 The Interferometric Imaging Radar Altimeter (InIRA)

The InIRA on the Tiangong-2 space laboratory is a new type of microwave remote sensor for oceanography and hydrology. It uses a short baseline to measure sea surface height as well as land topography at near-nadir angles with wide swath for the first time in the world. Compared with traditional radar altimeters, InIRA has advantages of higher spatial resolution, wide swath, and land-sea compatibility.

The InIRA operates at Ku-band with a baseline of 2.3 m and with looking angles ranging from 2.5 to 7.5°. It provides topography data of the oceans on a ten square kilometers grid with a relative height precision of 20 cm and with a swath of ~35 km. The imaging geometry in the cross-track plane for the right-looking InIRA is shown in Fig. 2.

Areas in potential applications include mainly: monitoring of submesoscale (<100 km) eddies, currents, and inner waves, etc., retrieval of significant wave heights, wave directions, and sea surface wind speeds, etc. [10, 11], retrieval of sea surface slopes, and velocities and vorticities of the geostrophic current vectors, etc., over oceans; monitoring the change of the distribution, the elevation, the water flow, and the water storage of inland lakes, rivers, and wetlands, etc., monitoring the change of the height of the continental ice sheets, etc., over lands.

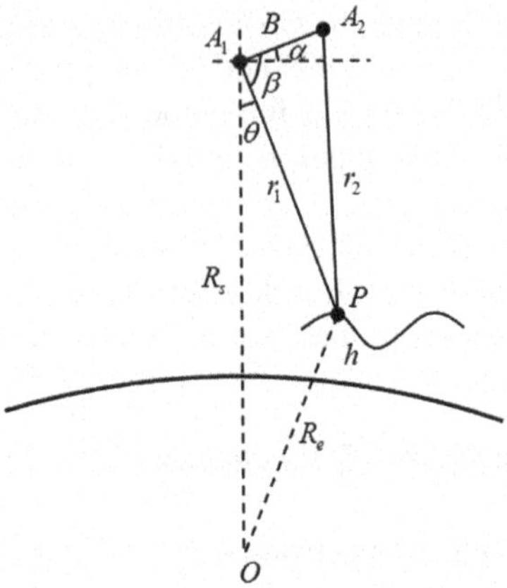

Fig. 2. A figure caption is always placed below the illustration. Short captions are centered, while long ones are justified. The macro button chooses the correct format automatically.

3.3 The Multi-Azimuth UV Imager

The Multi-Azimuth UV Imager include an ultraviolet ring imager and an ultraviolet forward spectrometer. The ultraviolet ring imager uses a large field of view (360° ring) to detect ultraviolet radiation characteristics of the global middle atmosphere. And the ultraviolet forward spectrometer comprehensively detects each spectral segments of the forward limb atmosphere by a fine spectral resolution.

4 Tiangong-2 Ground Data Processing and Service System

4.1 Multiple Categories of Payloads Product Processing

The Processing Algorithm of WIS Data Processing
The WIS adopts an integrating sphere whose spectral radiance was accurately calibrated as the light source to determine the radiation response coefficient. The comprehensive calibration error is 4.25%, and the non-uniformity correction accuracy is between 0.5% and 2%.

A high-precision sensor model of the wide-width virtual camera is established to perform geometric correction processing. The positioning accuracy of WIS images is within 8 pixels without GCPs, and 1 pixel with GCPs [12]. The geometric correction result is shown in Fig. 3.

The Interferometric Imaging Radar Altimeter (InIRA) Data Processing
The raw data acquired by the InIRA have to be processed with a series of steps to generate backscatter coefficient images and surface topography data, such as imaging processing, interferometric processing, radiometric correction and geometric location [13, 14].

Fig. 3. The WIS data processing.

The imaging processing with the raw data generates two channels of complex images as well as backscatter coefficient images over a same observed area after range and azimuth compression. Then the two complex images are put into the interferometric processing, where surface topography data over both ocean and continent can be produced after images coregistration, interferometric phase retrieving, filtering, unwrapping, absolute phase determination, and phase to elevation conversion. Figure 4 shows the main chains of the interferometric processing taking measurements over a mountain near Chamical in Argentina as an example.

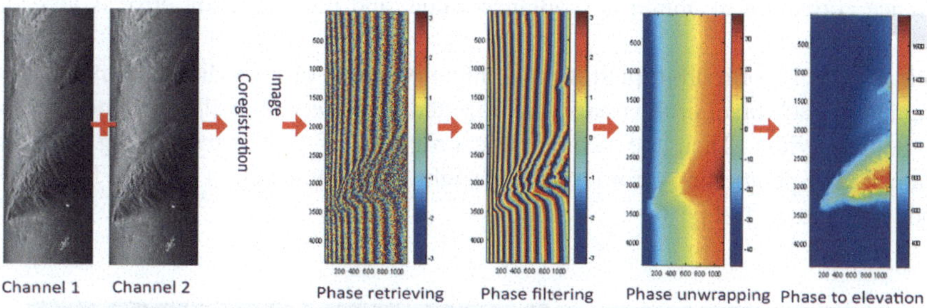

Fig. 4. Interferometric processing chains.

The radiometric correction removes spatial distortions in the backscatter coefficient images caused by the antenna pattern, the range attenuation, and the variation of the incidence angles, as well as the Auto Gain Control (AGC) and a constant calibration factor. The antenna pattern and the constant calibration factor have to be measured after launching through external calibration by collecting InIRA measurements over the Amazon rainforest or a calibration site where man-made corner reflectors are deployed. Figure 5 displays the effect of the radiometric correction.

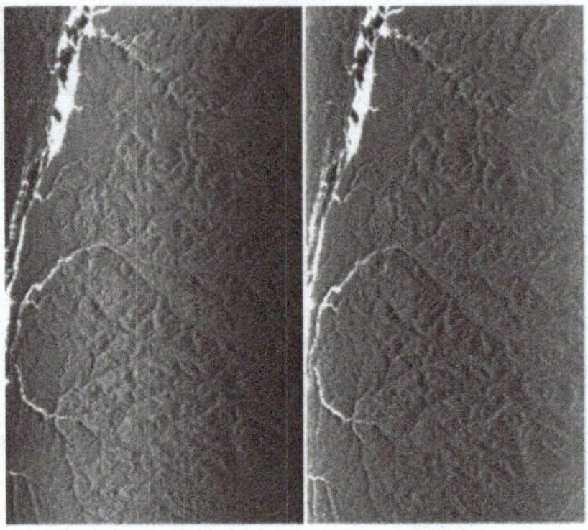

(a) (b)

Fig. 5. Backscatter coefficient image over the Amazon rainforest before (a) and after (b) radiometric correction (down is the azimuth direction and right is the range direction).

A high-precision geometric correction model of Interferometric imaging radar altimeter (InIRA) is proposed, which reduces the geometric positioning error caused by a small near-nadir incidence, and makes the integrated geometric positioning error within 3 pixels [15].

The Multi-Azimuth UV Imager Data Processing

The MAVI data is processed through dark signal correction, smear correction and radiance correction to obtain the radiance value, and the correction error is less than 3%.

Based on platform orbit, platform attitude, pixel field of view distribution and other parameters, each pixel of the images is projected onto the virtual ellipsoid model along the field of view. Then the geometric coordinates of the corresponding pixel are obtained. The positioning error along the height direction is within 1 pixel. The MAVI product is shown in Fig. 6.

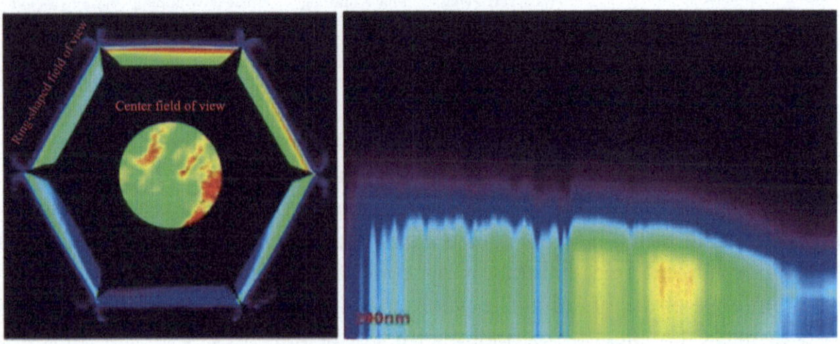

Fig. 6. MAVI Product.

4.2 Assessment and Control of Product Quality

To quantitatively assess the quality of the images acquired by the Wide-band Imaging Spectrometer, four indicators are measured: Signal to Noise Ratio (SNR), clarity, contrast, and comentropy.

Backscatter coefficient images and surface topography data over both ocean and continent are produced through the measurements of the InIRA. To quantitatively assess the products quality, several indicators are measured:

- The radiometric accuracy: including the relative accuracy, the absolute accuracy, and the radiometric resolution. The relative accuracy is measured by data over the Amazon rainforest. The absolute accuracy is measured by comparison with corner reflectors with precisely known Radar Cross Sections (RCS). The radiometric resolution is measured by data over uniform distributed targets.
- The planar accuracy: including the spatial resolution, the geometric location accuracy, both of which are measured by comparison with corner reflectors or ground control points with precisely known positions.
- The altimetry accuracy: measured by comparison with corner reflectors or ground control points with precisely known elevations.

Specially, the altimetry accuracy of the InIRA over ocean [16] and over inland lake [17] have been assessed through simulations and real acquisitions. All of the aforementioned indicators are measured to evaluate the product quality and in return to help select proper methods and parameters in data processing.

4.3 Data Storage and Management

Space application data involves many subject areas such as space experiment, earth science and space environment. The data obtained directly and through post-processing are increasing. Data resolution, data level and other related data description information are also different. It is necessary to develop the optimal data storage strategy, establish a reasonable data management mechanism to achieve efficient organization and management of space application data [18]. We designed a data storage and management system for massive, multi-source and heterogeneous space application data.

Data Storage: As the basic data support module, the main function of the data storage is to store the distributed file system and the data information database after processing of all kinds of data and products such as the original payload data, the standard product data, the data quality evaluation result and so on. The entire data storage process is performed automatically based on the message triggering mechanism, and the process and content are visually displayed through the visual interface [19].

Data Visualization Management: Practice shows that intuitive visual information is more helpful for people to improve work efficiency [20]. Combined with the spatial characteristics of space application data itself, we designed a space application data management mechanism based on data visualization technology. The mechanism realized the integrated visualization management of multi-type space application data, providing multi-source, multi-dimensional, multi-scale and multi-temporal space application data flexible query, positioning, browsing, download and other services, as well as aircraft simulation, vector plotting, space measurement.

The data storage and management system has been successfully applied to the comprehensive storage and management of the space application data products of Tiangong-1 aircraft and Tiangong-2 aircraft, which satisfies the multi-level requirement of data management. The data visualization management is shown in Fig. 7.

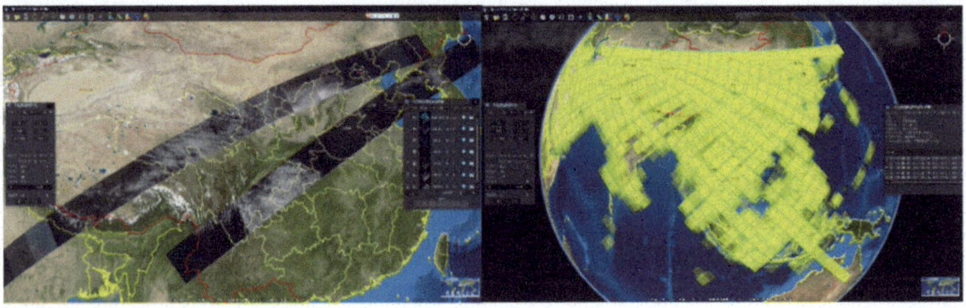

Fig. 7. The data visualization management.

4.4 Data Sharing Services

For the massive manned space application data, based on the actual demand of data sharing, this paper designs and implements the space application data promoting service platform for China Manned Space Engineering (http://www.msadc.cn). The platform is based on B/S structure, with SSH as the framework, with Oracle as the background database, using Nginx for load balancing, and provided effective security protection measures [21, 22].

According to the platform construction goals and requirements, the platform is composed of two parts, including the foreground portal and the backstage management. The foreground portal mainly realizes the space application data on-line inquiry, the retrieval and downloading, as well as the application result display and the popular science propaganda and so on. The backstage management mainly realizes the space application data dissemination and the management, as well as the user registers, the data order management and the analysis and so on.

The platform has been successfully applied to online sharing of space application data products of Tiangong-1 aircraft, Tiangong-2 aircraft and Tianzhou-1 aircraft. Since launching, the platform has operated steadily and normally, effectively met numerous and different types of data online release and user query and retrieval requests, and realized the convenient online sharing of the space application data products.

5 Conclusions

As what mentioned above, the TG2-GDPSS was an integrated technical system for China Manned Space ground data processing. Various categories of products volume exceeding 100 terabytes had been distributed to a lot of research institutes, universities, governmental agencies, and individual users based on the TG2-GDPSS so far. Several applications such as land utilization, forestry monitoring, and urban development had been performed to play valuable roles in lots of fields. With the continuously production of these products, more and more worthy application results would be brought forth in many fields.

Acknowledgements. Thanks to China Manned Space Engineering for providing space science and application data products of Tiangong-2.

References

1. Buyya, R.: High Performance Cluster Computing: Architectures and Systems, vol. 1. Prentice Hall, USA (1999)
2. Ma, G., Lu, P.: PBSWeb: a web-based interface to the portable batch system. In: Proceedings of the 12th IASTED International Conference on Parallel and Distributed Computing and Systems (PDCS), Las Vegas, Nevada, USA (2000)
3. Bode, B., Halstead, D., Kendall, R., Lei, Z., Jackson, D.: The portable batch scheduler and the maui scheduler on Linux clusters. In: Proceedings of the 4th Linux Showcase and Conference, Atlanta, USA (2000)
4. Ryu, K.D., Pachapurkar, M., Fong, L.L.: Adaptive memory paging for efficient gang scheduling of parallel applications. In: Proceedings of the 18th IEEE International Parallel and Distributed Processing Symposium (2004)
5. Thain, D., Tannenbaum, T., Livny, M.: Distributed computing in practice: the Condor experience. Concur. Comput. Pract. Exp. **17**(2–4), 323–356 (2005)
6. Zhou, S., Zheng, X., Wang, J., Delisle, P.: Utopia: a load sharing facility for large, heterogeneous distributed computer systems. Softw. Pract. Exp. **23**(12), 1305–1336 (1993)
7. Stonebraker, M., Devine, R., Kornacker, M., Litwin, W., Pfeffer, A., Sah, A.: An economic paradigm for query processing and data migration in Mariposa. In: Proceedings of 3rd International Conference on Parallel and Distributed Information Systems. IEEE CS Press, Austin, TX, USA (1994)
8. Waldspurger, C.A., Weihl, W.E.: Lottery scheduling: flexible proportional-share resource management. In: Proceedings of OSDI 94, pp. 1–12 (1994)
9. Stoica, I., Zhang, H., Ng, T.: A hierarchical fair service curve algorithm for link-sharing, real-time and priority service. In: Proceedings of ACM SIGCOMM 1997, pp. 162–173. Cannes, Frances (1997)
10. Jingsong, Y., Lin, R., Gang, Z.: The first quantitative ocean remote sensing by using Chinese interferometric imaging radar altimeter onboard TG-2. Acta Oceanol. Sin. **36**, 122–123 (2017)
11. Bao, Q., Lin, M., Zhang, Y., Jia, Y., Lang, S.: Wind speed inversion for imaging microwave altimeter. J. Remote Sens. **21**, 835–841 (2017)
12. Cheng, Y.F., Jin, S.Y., Wang, M.: An high accuracy image mosaicking approach of optical remote sensing satellite for multi-camera system. Acta Opt. Sin. **37**(8), 336–345 (2017)

13. Dong, X., Zhang, Y., Zhai, W.: Design and algorithms of the Tiangong-2 interferometric imaging radar altimeter processor. In: 2017 Progress In Electromagnetics Research Symposium - Spring, PIERS 2017, May 22, 2017–May 25, 2017, St. Petersburg, Russia, pp. 3802–3803 (2017)
14. Liu, Z., Tan, H., Li, S.: Direct geometric positioning for the three dimensional imaging microware altimeter. In: 2017 2nd International Conference on Computer Engineering, Information Science and Internet Technology, Chang Sha, China (2017)
15. Liu, Z.W., Li, S.Y.: High precision limb height calculation of ultraviolet forward spectrometer of TianGong-2. Opt. Precis. Eng. **26**(6), 1517–1523 (2018)
16. Kong, W., Chong, J., Tan, H.: Performance analysis of ocean surface topography altimetry by Ku-Band Near-Nadir interferometric SAR. Remote Sens. **9**, 933 (2017)
17. Tan, H., Li, S.Y., Liu, Z.W.: An assessment of spaceborne near-nadir interferometric SAR performance over inland waters with real data. In: 2018 ISPRS TC III Mid-Term Symposium on Developments, Technologies and Applications in Remote Sensing, May 7, 2018–May 10, 2018, Beijing, China (2018)
18. Lv, X., Cheng, C., Gong, J., Guan, L.: Review of data storage and management technologies for massive remote sensing data. Sci. China Tech. Sci. **41**(12), 1561–1573 (2011)
19. Yang, J., Li, G.: The China GEO data center: bringing order to open earth-observation data. IEEE Geosci. Remote Sens. Mag. **5**(2), 77–85 (2017)
20. Sample, J.T., Ioup, E.: Tile-Based Geospatial Information Systems: Principles and Practices. Springer (2010)
21. Zhu, Y., Liu, R., Feng, M., et al.: Research on distributed earth system scientific data sharing platform. Comput. Eng. Appl. **45**(1), 245–248 (2009)
22. Ramapriyan, H.K., Pfister, R., Weinstein, B.: An overview of the EOS data distribution systems. Remote Sens. Digit. Image Process. **11**(3), 183–202 (2011)

Automatic Multi-spectral Image Registration for Tiangong-2 Wide-Band Imaging Spectrometer

Liangji Li[1,2], Xiaohua Liu[2(✉)], Xiaoxian Huang[2], and Yuyu Tang[2]

[1] University of Chinese Academy of Sciences, Beijing 100049, China
[2] Key Laboratory of Infrared System Detection and Imaging Technology, Shanghai Institute of Technical Physics of the Chinese Academy of Sciences, Shanghai 200083, China
xhliuz@mail.sitp.ac.cn

Abstract. Automatic multi-spectral image registration can improve the application efficiency and ability of remote sensing images while this technology is yet to be improved. Based on the remote sensing data of Tiangong-2 wide-band imaging spectrometer, the differences and similarities between multi-spectral images were analyzed in this paper. An edge extraction method with edge symmetry and non-directionality was proposed according to the neighborhood structure of image pixels. Using the edge information images, multi-spectral image registration was achieved through the measurement of mutual information. A contrast experiment was also set up to compare the effects of the proposed edge extraction method with those of several traditional edge extraction methods in image registration. The results showed that the registration precision of this method is higher, indicating that the extracted edge features were more stable among multi-spectral images.

Keywords: Tiangong-2 · Wide-band imaging spectrometer
Image registration · Edge information · Mutual information

1 Introduction

In remote sensing image processing, image registration is a basic and critical preprocessing step in applications such as image fusion, change detection and target recognition. Multi-spectral image registration is the process of matching different images captured by different sensors for the same scene and establishing a coordinate conversion relationship between images. In recent decades, many researchers have devoted themselves to the research of image registration and have contributed to the development of this field. However, due to the difference in spectral characteristics at different wavelengths for the same object, the automatic registration of multi-spectral images is still a difficult problem in the field of remote sensing image processing.

To cope with the image registration with different characteristics, many image registration methods have been proposed and can be divided into three categories: grayscale-based image registration, feature-based image registration and transform-domain-based image registration.

Y. Gu et al. (Eds.): *Proceedings of the Tiangong-2 Remote Sensing Application Conference*, LNEE 541, pp. 71–81, 2019.
https://doi.org/10.1007/978-981-13-3501-3_7

Grayscale-based image registration. This is a classic method for solving multimodal medical image registration problems. The method directly uses the grayscale statistical information of the image to establish a similarity measurement between the reference image and the image to be registered, determining the image registration result by the best similarity measure. This class of methods includes methods such as cross-correlation, mutual information, and invariant moments. Wu et al. use the mutual information of the gradient to achieve the registration of single aerial image and LiDAR (Light Detection and Ranging) point cloud [1]. There are also registrations for multi-spectral images using normalized cross-correlation after Fourier-Merlin transformation [2].

Feature-based image registration. Registration using local features of the image is currently the most common method in the field of image registration in the same band. This kind of method usually adopts different feature extraction operators to extract representative feature points in the image, before using them to obtain image registration results. Common feature extraction operators are: Harris, SIFT, SURF and ORB. In order to solve the problem that the direction and amplitude of the gradient change nonlinearly in multispectral image registration, many researchers have improved the SIFT operator, such as GOM-SIFT [3], LC-SIFT [4] and NR-SIFT [5]. Meanwhile, there are also studies on multi-spectral image registration using SURF [6].

Transform-domain-based image registration. The main representative in this category is the method based on Fourier transform and wavelet transform. When applying wavelet transform, the most important part is the selection of wavelet coefficients. The method of applying the maximization of HL and LH wavelet coefficients in wavelet transform to register remote sensing data images is proposed by Jacqueline et al. [7]. Although the image registration technique based on wavelet transform has the advantage of multi-resolution analysis, the adaptive selection of wavelet base and wavelet decomposition scale needs further research.

These three kinds of methods have contributed to the development of the remote sensing image registration. However, the research on wide-band spectral image registration of visible and near-infrared, short-wave infrared and thermal infrared is yet to be extended.

This paper analyzes the commonalities and differences of remote sensing images of different spectrums in the Tiangong-2 wide-band imaging spectrometer. We also propose an image registration method based on image edge information for automatic registration of multi-spectral images without the help of geographic coordinate information.

2 Data and Methods

2.1 Data Sources

Tiangong-2 is the second space laboratory independently researched and developed by China. It possesses a wide-band imaging spectrometer containing 18 channels of data from three spectrum of visible and near-infrared, short-wave infrared and thermal infrared. Its observation data plays an important role in fishery guidance, weather forecasting, land and resources classification and sudden disaster monitoring [8]. The data in this experiment are obtained from the original remote sensing image of the WIS on August 8th, 2017 and December 30th, 2017.

2.2 Edge Information Extraction

In multi-spectral images, the grayscale information between images from different bands changes nonlinearly with the wavelength and the direction of the gradient in these images also changes in multiple directions [9]. This makes many traditional methods (such as SIFT, SURF, etc.) that use image local gradient statistics for feature extraction no longer work well. On the other hand, the edge features of objects are relatively stable in images from different bands. This paper takes advantage of this and proposes an image edge extraction method based on pixel neighborhood information.

In the image, when the grayscale is discontinuous or locally abrupt between two adjacent regions, an edge is formed. The gradient information of the edge can be calculated from the value of grayscale difference between a pixel and its neighborhood. Inspired by this principle, a non-directional feature information is designed to describe the edges of images from different bands.

Neighborhood of pixels. In digital images, pixels are discrete. The spatial relationship between pixels can be described by the discrete distances. With the discrete distances, various neighborhoods of pixels can be defined [10]. The 4-neighborhood $N_4(p)$ of a pixel $p(x_p, y_p)$ can be defined as:

$$N_4(p) = \{r|d_4(p,r) = 1\} \tag{1}$$

where the pixel coordinates represented by r are (x_r, y_r), and $d_4(p,r)$ is the urban distance, which is defined as:

$$d_4(p,r) = |x_p - x_r| + |y_p - y_r| \tag{2}$$

Similarly, the r pixel contained in the diagonal neighborhood $N_D(p)$ of a pixel $p(x_p, y_p)$ can be expressed as:

$$N_D(p) = \{r||x_p - x_r| = |y_p - y_r| = 1\} \tag{3}$$

The 8-neighborhood of a pixel $p(x_p, y_p)$ is the union of its 4-neighborhood and diagonal neighborhood, expressed as:

$$N_8(p) = N_4(p) \cup N_D(p) \tag{4}$$

The specific pixel neighborhood structure is shown in Fig. 1. Based on the above-mentioned neighborhood structure, the proposed method extracts the edge information by using the grayscale information of each pixel and its 8-neighborhood pixels.

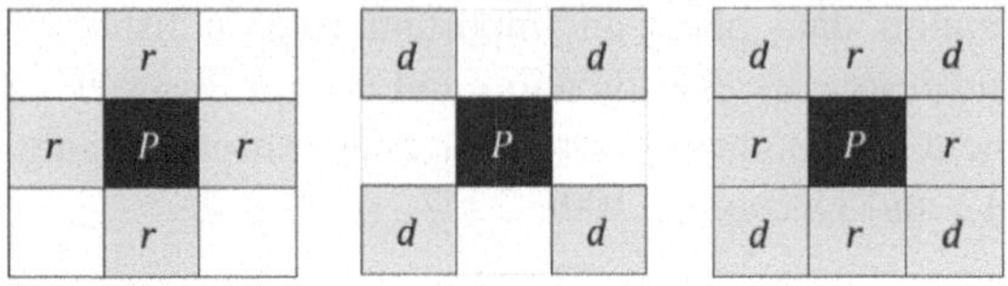

Fig. 1. Schematic diagram of pixel neighborhood structure. $N_4(p)$ of pixel P is the position of r pixel, $N_D(p)$ of pixel P is the position of d pixel, and $N_8(p)$ is the union of $N_4(p)$ and $N_D(p)$

Edge feature extraction. For each image from different bands, based on the pixel 8-neighborhood structure described above, the parameter $Edge(x_p, y_p)$ which evaluates the edge attribute of the pixel $p(x_p, y_p)$ is calculated by the grayscale information $I(x, y)$ in the image. The calculation method is shown as:

$$Edge(x_p, y_p) = \max(Grad(p, r)), r \in N_8(p) \tag{5}$$

where the gradient attribute parameter $Grad(p, r)$ is solved as:

$$Grad(p, r) = \begin{cases} |I(x_r, y_r) - I(x_p, y_p)|, r \in N_4(p) \\ k|I(x_r, y_r) - I(x_p, y_p)|, r \in N_D(p) \end{cases} \tag{6}$$

where k is the diagonal gradient coefficient adopted to balance the influence of the diagonal neighborhood and 4-neighborhood discrete distance on *Grad*, which is set to 0.7 in this experiment (the ratio of the 4-neighborhood distance to the diagonal neighborhood distance of the pixel is about 0.7:1).

Based on the above description, it can be inferred that the edge features extracted by the method have two characteristics: non-directionality and edge symmetry. This means that the edge parameter $Edge(x_p, y_p)$ of each pixel represents the gradient magnitude of the pixel and has no directionality, which prevents the extracted edge feature from being affected by the change of the gradient direction of the multi-spectral image. At the same time, the two pixels of the image maintain the property of being 4-neighbor or diagonal neighbor mutually. These two pixels also have a symmetric relationship in the $Grad(p, r)$ function. Therefore, the obtained feature images have the characteristic of edge symmetry, which makes the edge feature more stable in multi-spectral images with non-linear changes in grayscale.

Taking the local image of the short-wave infrared spectrum S2 channel (band range: 1.63–1.65 μm) taken by the Tiangong-2 wide-band imaging spectrometer at a certain time on August 8, 2017 as an example, the edge information image based on the edge features extracted by the above method is shown in Fig. 2. As shown, the boundaries between the waters and the land are clearly extracted. Different types of land cover are also detected accordingly.

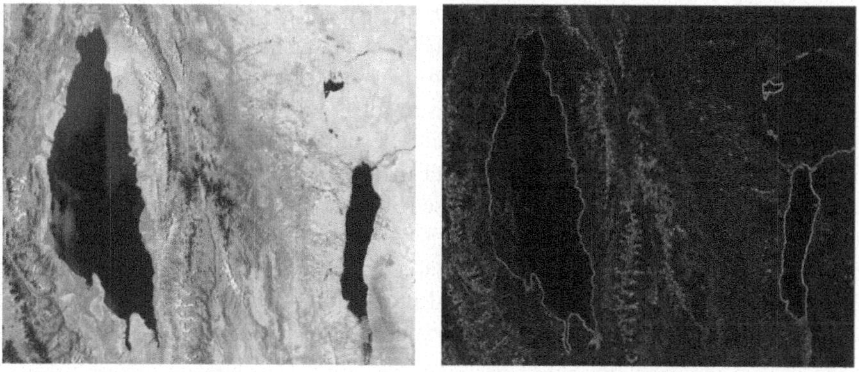

(a) (b)

Fig. 2. Edge feature extraction result of image. (a) original wide-band imager short-wave infrared remote sensing image; (b) Edge information images of (a) extracted by the method

2.3 Image Registration Based on Mutual Information of Feature Images

The mutual information of the two images can be calculated by the information entropy for both images and their joint entropy.

Image information entropy. Information entropy is an important concept in information theory and is a measure used to describe system uncertainty. A digital image X can be regarded as a set of discrete data whose information entropy $H(X)$ is defined as:

$$H(X) = -\sum_{i=1}^{N-1} p_i log p_i,\ p_i = h_i / \sum_{i=1}^{N-1} h_i \tag{7}$$

where h_i represents the total number of pixels whose grayscale value is i in the image X, N represents the gray level of the image X, and p_i represents the probability that the gradation i appears in the image.

Joint entropy of image. Joint entropy can reflect the correlation of two random variables. For the joint information entropy $H(X, Y)$ of the two images X and Y, the overlapping part of the two images is used, and the calculation method is:

$$H(X, Y) = -\sum_{x,y} p_{XY}(x, y) log p_{XY}(x, y) \tag{8}$$

$$p_{XY}(i,j) = \frac{h_{XY}(i,j)}{\sum_{i=0}^{N-1} \sum_{j=0}^{M-1} h_{XY}(i,j)} \tag{9}$$

where $p_{XY}(x, y)$ is the gray joint probability distribution of images X and Y, which can be calculated from their joint histogram $h_{XY}(i,j)$. N and M are respectively the number of gray levels of images X and Y.

In practical registration, the image to be registered is often used as a template to calculate the mutual information between the template image and the sub-images of the reference image. Then, the sub-image with the largest mutual information of the reference image is regarded as the registration result. As a parameter of statistic between two images, the mutual information mainly evaluates the amount of information contained between the two images, without taking the spatial position information of the image into account. Therefore, it is sensitive to the geometric distortion of the image and is mainly suitable for image registration with translation, micro rotation and size change. After scaling the images from the three different spectrums of Tiangong-2 wide-band imaging spectrometer in a fixed proportion, they exactly meet the requirements of mutual information image registration.

3 Experimental Results and Analysis

To perform image registration experiments between the spectrums using the proposed method, one representative channel data was selected from each spectrum segment of the Tiangong-2 wide-band imaging spectrometer. The parameter information of the wide-band imaging spectrometer is shown in Table 1.

Table 1. Wide-band imaging spectrometer parameter information index

Imager spectrum	Visible and near infrared	Short wave infrared	Thermal infrared
Spectral range/μm	0.4–1.0	1–1.7	8–10
Number of channels	14	2	2
Spatial resolution/m	100	200	400

The image data used in the experiment is shown in Table 2. The V8 channel is selected in the visible and near-infrared spectrum with the band range as 0.655–0.675 μm, which represents the red channel information of visible light. In the short-wave infrared spectrum, the S2 channel is selected with the band range as 1.63–1.65 μm. The T2 channel is selected to represent the thermal infrared spectrum with the band range as 8.925–9.275 μm. The original images of each band have undergone pre-processing operations such as non-uniformity correction. In this experiment, in order to enhance the visibility of the image, the image is histogram equalized.

Table 2. Experimental image data index

Image spectrum	Visible and near infrared	Short wave infrared	Thermal infrared
Channel type	V8	S2	T2
Center wavelength/μm	0.665	1.64	9.1
Band width/μm	0.02	0.02	0.35

3.1 Multi-spectral Registration Experiment

For the short-wave infrared and thermal infrared bands, the image registration scheme is designed as shown in Fig. 3.

First, based on the spatial resolution of the image, the two-band image is fixed to the same spatial resolution; Then, taking the short-wave infrared image *S*2 as a reference image as shown in Fig. 4(a), the thermal infrared spectrum image *T*2 is taken as a partial sub-image *T*2*i* as a to-be-registered image as shown in Fig. 4(c). *Edge_S*2 and *Edge_T*2*i* are feature images obtained by the edge feature extraction method introduced in Sect. 2.2, which are represented by Fig. 4(b) and (d) respectively. Finally, *Edge_T*2*i* is used as a template to calculate the mutual information of each sub-image with *Edge_S*2. The maximum point of mutual information is regarded as the matching point.

The cross-information distribution images of *Edge_S*2 and *Edge_T*2*i* are shown in Fig. 4(e). When the pixel coordinates $(x = 81, y = 268)$ of *Edge_S*2 are located, the mutual information is taken to a maximum value of 4.045. Figure 4(f) shows the registration result of these two edge feature images. The final registration area of the feature sub-image *Edge_T*2*i* is marked by the red square.

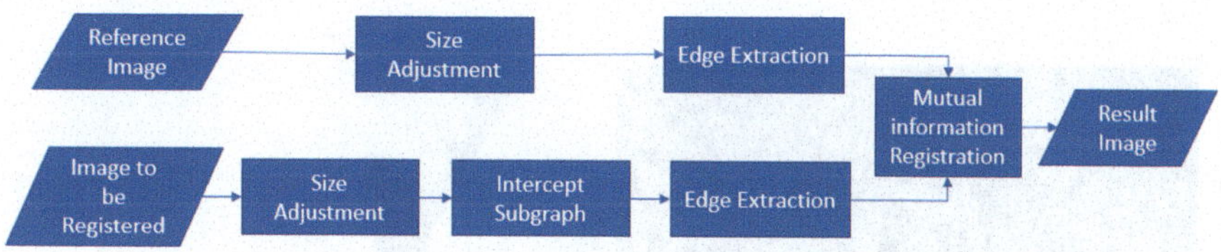

Fig. 3. Multi-spectral image registration flow chart

The short-wave infrared and thermal infrared image matching results are shown in Fig. 4(g). The red square marked area is the merged pseudo color images, where T2 represents R channel and S2 represents G, B channel. As can be seen from the figure, since the grayscale of the cloud is darker in the thermal infrared image and the grayscale of the land is brighter, the cloud is shown as blue-green and the land is shown as light red in the pseudo color images. The pseudo color images are very clear on the boundary between the cloud and the land, indicating that the two-band image registration result is ideal.

Registration experiments for short-wave infrared and visible and near-infrared spectrum was also performed by the same way. Still taking the image S2 of Fig. 4(a) as a reference, a partial sub-image V8i of channel V8 in the visible and near-infrared spectrum is shown in Fig. 5(a) while Fig. 5(b) shows the registration result of V8i and S2. In the area marked by the red square, there is a fused pseudo color image, in which V8i represents R channel, and S2 represents G, B channels. An ideal result is obtained. Because the imaging results of the V8 band and the S2 band are similar with minor differences in the overall gray level and the imaging characteristics of certain areas, this fusion area shows an overall blue-green color with certain areas appearing red.

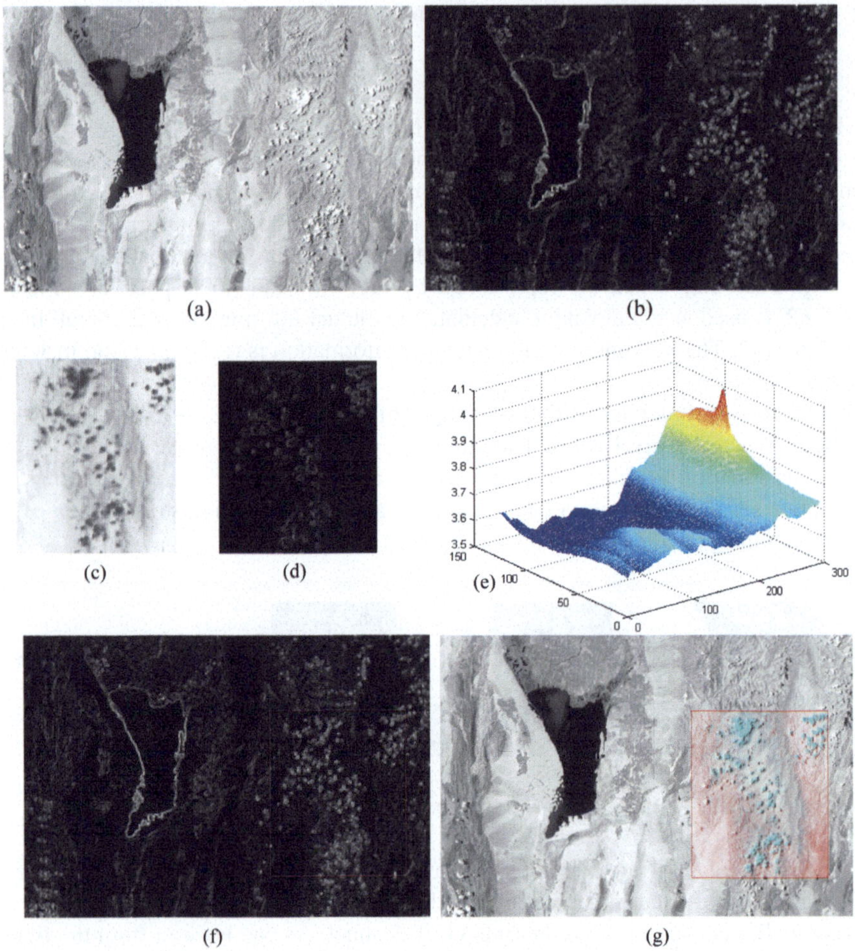

Fig. 4. Experimental procedure and results of S2 and T2 remote sensing image registration. (a) short-wave infrared remote sensing image S2; (b) edge information images of S2; (c) thermal infrared remote sensing image to be registered T2i; (d) edge information images of T2i; (e) mutual information distribution in the registration process; (f) registration result of edge information images; (g) image registration and preliminary fusion results of S2 and T2i.

3.2 Contrast Experiment

In order to further verify the advantages of the edge feature extraction method proposed in this paper compared with the traditional method in multi-spectral image registration, the contrast experiment was performed using the image data of the Tiangong-2 wide-band imaging spectrometer.

In the registration experiment, the S2 channel image in short-wave infrared band is used as the georeferenced image, and the V8 channel image in visible band and T2

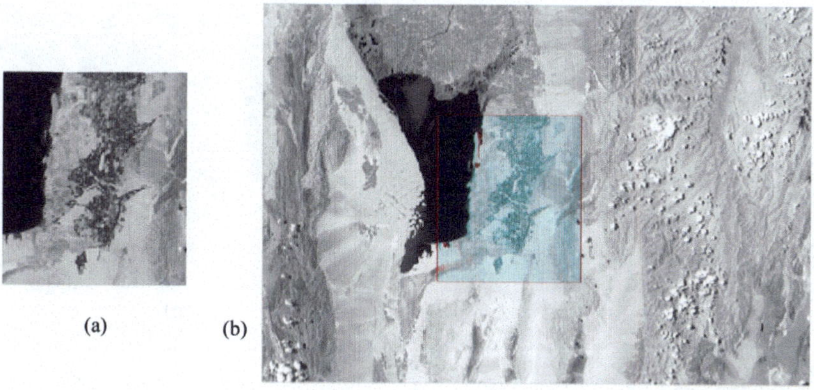

Fig. 5. Registration results of short-wave infrared and visible and near-infrared images. (a) Visible and near infrared image V8i; (b) image registration and preliminary fusion results of V8i and S2;

chan-nel image in thermal infrared band is warped with 50 control points respectively as shown in Fig. 6.

Based on the above experimental data, each group of images are processed by the edge extraction method proposed in this paper and the classic edge extraction methods such as LOG, Sobel and Roberts. Then the mutual information method is used to register based on the obtained edge information image.

The Mean (mean distance of control points) and RMSE (Root Mean Square Error) of the 50 selected control points are used to evaluate the registration errors. The results are shown in Table 3.

In Table 3, it can be seen that the Mean and RMSE of the four methods are similar in the registration experiment between the short-wave infrared and visible band. This is because the images of short-wave infrared and visible band are similar. Therefore, different methods can get similar registration results.

In the registration experiments between the short-wave infrared and thermal infrared band, the proposed method has obvious advantages in the evaluation of Mean and RMSE values compared with the other three methods. This is because the radiation characteristics of ground objects in the short-wave infrared and thermal infrared bands are quite different. Gradient inversion and gradient amplitude change often occur between the feature points of different images. In this case, the edge information extraction method with non-directionality and edge symmetry proposed in this paper can make the edge information more stable between different bands resulting in better registration results. (It needs to be mentioned that since the 50 control points in the experiment are manually selected instead of using the geographical coordinate information, there is a systematic certain error in the experiments. Therefore, the value of Mean and RMSE in the experiment are both on the high side. However, the effect of this error has the same impact on each registration method.)

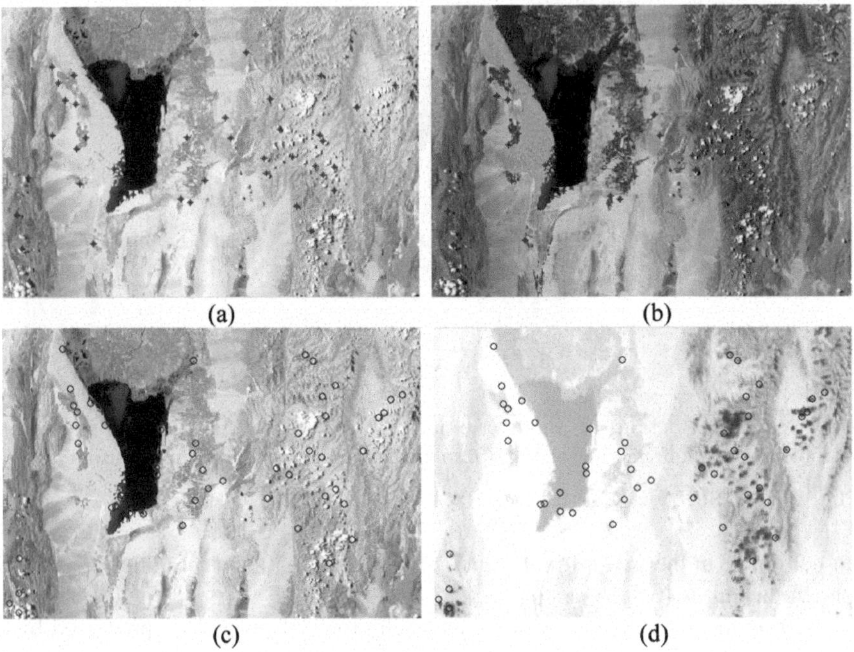

Fig. 6. Experimental image and annotation of the control points. (a) short-wave infrared image and the control points marked with '*'; (b) visible light image and the control points marked with '*'; (c) short-wave infrared image and the control points marked with 'o'; (d) thermal infrared image and the control points marked with 'o'.

Table 3. Experimental registration error

Experimental category	Registration method	Mean	RMSE
S2-V8 registration experiment	LOG + MI	0.613	0.825
	Sobel + MI	0.714	0.911
	Roberts + MI	0.707	0.902
	This methods	0.635	0.815
S2-T2 registration experiment	LOG + MI	1.328	1.493
	Sobel + MI	1.802	2.064
	Roberts + MI	1.293	1.114
	This methods	0.736	1.029

4 Conclusion

The method proposed in this paper extracts the edge features of the image through the pixel neighborhood information before utilizing the mutual information between the feature images to achieve image registration. Based on the data of Tiangong-2 wide-band imaging spectrometer, the image registration between visible and near infrared, short-wave infrared and thermal infrared was achieved. The registration results were tested by preliminary fusion. The proposed method is also compared with other edge extraction and image registration methods. The results show that this method has higher registration precision in the multi-spectral image registration of Tiangong-2.

Acknowledgements. Thanks to China Manned Space Engineering for providing space science and application data products of Tiangong-2.

References

1. Wu, J., Cao, Y.S., Yang, H.S.: Automatic registration of high-resolution multispectral imageries released by low-altitude unmanned airship. J. Remote Sens. **16**(3), 625–643 (2012)
2. Wu, J., Rao, Y., Hu, Y.J.: Automatic registration of single aerial image with LiDAR data based on "pin-hole" imaging simulation and iterative computation. J. Remote Sens. **20**(1), 80–93 (2016)
3. Yi, Z., Zhiguo, C., Yang, X.: Multi-spectral remote image registration based on SIFT. Electron. Lett. **20**(1), 80–93 (2008)
4. Sajid, S., Robert, S.: A modified SIFT descriptor for image matching under spectral variations. In: International Conference on Image Analysis and Processing 2013. LNCS, vol. 8165, pp. 652–661. Naples, Italy (2013)
5. Sajid, S., Robert, S.: A robust SIFT descriptor for multispectral images. IEEE Signal Process. Lett. **21**(4), 400–403 (2015)
6. Lu, Y.F., Zhao, H.M., Li, B.: Multi-spectral remote sensing image registration based on SURF. In: Deng, W. (ed.) AASRI Conference on Artificial Intelligence and Industry Application 2011, Proceedings Paper, pp. 236–239 (2010)
7. Moigne, J.L., Campbell, W.J., Cromp, R.F.: An automated parallel image registration technique based on the correlation of wavelet feature. IEEE Trans. Geosci. Remote Sens. **40** (8), 1849–1864 (2002)
8. Luo, Q.: Black Technology in "Tiangong-2". China Policy Review **2016**(10), 111–114 (2016)
9. Yu, H.: Fast and Accuracy Registration of Multispectral Image, pp. 2–4. Beijing University of Posts and Telecommunications, Beijing (2014)
10. Zhang, Y.J.: Image Engineering (II)-Image Analysis, 3rd edn. Tsinghua University Press, Beijing (2012)

A Cloud Detection Algorithm for Tiangong-2 Remote Sensing Data Over the Tibetan Plateau

Qing Chu, Yanan Liu(✉), and Guangjian Yan

State Key Laboratory of Remote Sensing Science, Jointly Sponsored by Beijing Normal University and Institute of Remote Sensing and Digital Earth of Chinese Academy of Sciences, Beijing 100875, China
yananliu@mail.bnu.edu.cn

Abstract. Cloud plays an essential role in energy budget and climate change. As the "Third Pole" of the earth, accurate cloud monitoring over the Tibetan Plateau is indispensable. The Tiangong-2 Space Laboratory provides multi-spectral remote sensing data with relatively high spatial resolution and is useful to get an accurate spatial distribution of cloud. We propose the single band and multiband threshold methods for cloud detection using Tiangong-2 data. To generate appropriate thresholds, the spectral reflectance of the thick cloud, the thin cloud, and land objects were analyzed based on the data from visible to shortwave infrared bands. It was found that the cloud can be accurately distinguished from snow and other land objects including water, ice, bare soil, plant and so on. Compared with Tiangong-2, the geostationary satellite Himawari-8 significantly overestimated the cloud cover over the Tibetan Plateau. The False Alarm Rate (FAR) and the Cloud Hit Rate (CHR) were 29.12% and 94.77%, respectively. Within the incorrectly classified pixels of Himawari-8 data, the cloud types were mainly altocumulus and cumulus. The main reason is probably the difference in classification method and spatial resolution.

Keywords: Cloud cover · Cloud detection · Tiangong-2 · Himawari-8

1 Introduction

Cloud is one of the most vital forces of Earth's energy balance and hydrothermal cycle. However, it is difficult to be described accurately. The accuracy of cloud cover detection widely affects the research of radiation, precipitation, climate change, and so on [1–4].

The Tibetan Plateau (across 25°–40°N, 74°–104°E), which is also called "Third Pole", is the highest and largest plateau in the world. Cloud monitoring over the Tibetan Plateau is quite crucial since this unique area has significant influence on the regional monsoon and global climate [5–10]. Afterwards, the quick development of satellites enabled the generation of many cloud products at a different resolution. For example, the Moderate Resolution Imaging Spectroradiometer (MODIS) provides cloud products with

Y. Gu et al. (Eds.): *Proceedings of the Tiangong-2 Remote Sensing Application Conference*, LNEE 541, pp. 82–93, 2019.
https://doi.org/10.1007/978-981-13-3501-3_8

1-km-pixel resolution [11], Himawari-8 geostationary satellite provides the cloud products with a spatial resolution of 2 km/5 km and a temporal resolution of 10 min [12, 13], and the International Satellite Cloud Climatology Project (ISCCP) generates cloud cover data with 30 km/280 km spatial resolution [14]. However, the complex shape of clouds and the confusion caused by cloud-like targets (snow, ice, desert, et al.) make it difficult to achieve accurate cloud data [15]. Thus, the research of cloud detection with high-spatial-resolution satellite imagery is necessary [16, 17].

Tiangong-2 Space Laboratory, which carries Wide-band Imaging Spectrometer (WIS) and many other payloads, was launched on 15 September 2016. It is designed to perform scientific and spacecraft-related experiments and test systems for a future space station. The WIS captures the spectral information of the earth's surface including visible and near-infrared spectrum range (VNI), short wavelength infrared spectrum range (SWI), and thermal infrared spectrum range (INF) with 100 m, 200 m, 400 m spatial resolution respectively. Compared with Tiangong-2, Himawari-8 has a very high temporal resolution which is good for cloud research [18]. In order to make full use of these data, we used the high spatial resolution data from Tiangong-2 to evaluate the cloud cover data of Himawari-8.

It is necessary to develop a convenient and practical cloud detection method because of the absence of the Tiangong-2 cloud product. Researchers have proposed numerous methods of cloud detection like the threshold, radiative transfer and statistical methods, which have been successfully used to detect clouds from satellite data [19, 20]. The threshold-based cloud detection method is widely used because of its high efficiency, simplicity, physical significance, and high precision [21, 22].

This study aims to develop a cloud detection method for Tiangong-2 WIS over the Tibetan Plateau and make a comparison with the cloud product of Himawari-8. The article is organised as follows. Section 2 introduces the satellite data used in this work. Section 3 analyses the spectral reflectance of land objects and clouds in 16 bands of WIS, and proposes suitable thresholds for cloud detection. In Sect. 4, we evaluate the classification results and make a detailed comparison with Himawari-8. Section 5 summarises the main contribution of this study.

2 Data

In this research, the data from Tiangong-2 WIS and Himawari-8 cloud product were used to analyze the cloud cover over the Tibetan plateau region from January to June 2018. The data of 16 bands from visible to shortwave infrared of WIS of Tiangong-2 were used to retrieve high-resolution cloud coverage. The band information of WIS is shown in Table 1. The spectral radiance was converted into the apparent reflectance in cloud detection.

Table 1. Bands information of the Tiangong-2 WIS

Bands	Wavelength range (Central wavelength) (nm)	Resolution (m)
VNI bands		
1	970-990 (980)	100
2	930-950 (940)	
3	895-915 (905)	
4	845-885 (865)	
5	810-830 (820)	
6	740-760 (750)	
7	677.5-687.5 (682.5)	
8	655-675 (665)	
9	610-630 (620)	
10	555-575 (565)	
11	510-530 (520)	
12	480-500 (490)	
13	433-453 (443)	
14	403-423 (413)	
SWI bands		
1	1230-1250 (1240)	200
2	1628-1652 (1640)	
INF bands		
1	8125-8825 (8475)	400
2	8925-9275 (9100)	

Himawari-8 is the 8th geostationary satellite, launched on 7 October 2014. The Advanced Himawari Imager (AHI) carried on the Himawari-8 facilitates the generation of the cloud property product (CLP). The CLP consists of the data of cloud types (ISCCP Definition) and four cloud parameters which including cloud optical thickness, cloud effective radius, cloud top temperature, and cloud top height, with a 10-min temporal resolution for day-time. The cloud types data (CLTYPE) is generated as a grid scale of 0.05°, and each grid is classified into nine cloud types (Table 2) according to its cloud microphysical properties. The CLP data could be downloaded from the official website (https://www.eorc.jaxa.jp/ptree/) and was used as the cloud mask in this study.

Because of the high temporal resolution of AHI, it is easy to match the passing time of WIS data. However, there are not always available matched data of two sensors every day. After removing the clear sky images, there were 18 images over the Tibetan Plateau from January to June in 2018 used in this study.

Table 2. The cloud types in Himawari-8 product

Cloud types		
0	Clear	Clear
1	Ci	Cirrus
2	Cs	Cirrostratus
3	DC	Deep convection
4	Ac	Altocumulus
5	As	Altostratus
6	Ns	Nimbostratus
7	Cu	Cumulus
8	Sc	Stratocumulus
9	St	Stratus

3 Cloud Detection Method

3.1 Spectral Dataset Analysis

To figure out the difference between land cover types in the spectral data of WIS, the apparent reflectance of typical land cover types in study area including cloud, snow, lake, bare soil, and plant have been graphically described in Fig. 1. The reflectance of 16 wavelengths was derived from WIS in the visible, near infrared, and shortwave infrared.

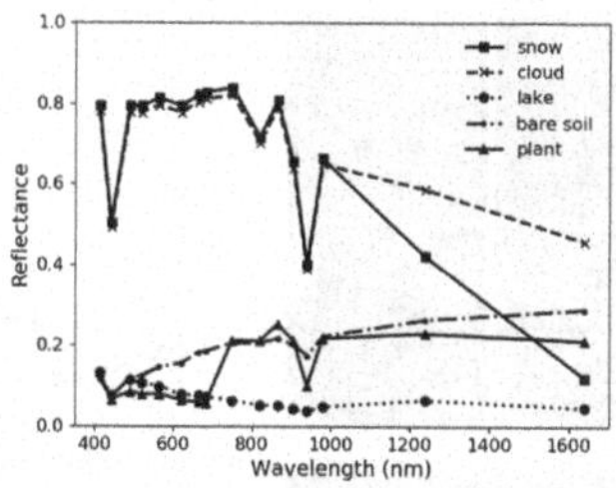

Fig. 1. Spectral curves of typical pixels in 14 visible and near-infrared bands and 2 short-wave infrared bands of Tiangong-2 WIS. The figure shows the reflectance of four types of land cover (snow, lake, bare soil, plant) and cloud

As the figure shows, the reflectance of the cloud is much higher than lake, bare soil and plant in the visible and near-infrared wavelengths. The reflectance of snow is very similar with cloud but is much lower in two SWI bands. These features can be used to differentiate cloud from the land objects. However, the reflection of the ground is more complicated in most cases. There are not always the pure pixels of land objects. To achieve appropriate thresholds, more reflectance of land cover and clouds were extracted from 16 WIS bands to find patterns of different objects (Table 3). The false-color images in Table 3 were composited from near-infrared (band1), red (band 7),

Table 3. Spectral reflectance of different land objects and clouds

Objects	False-color Image	Spectral Reflectance
Thick cloud	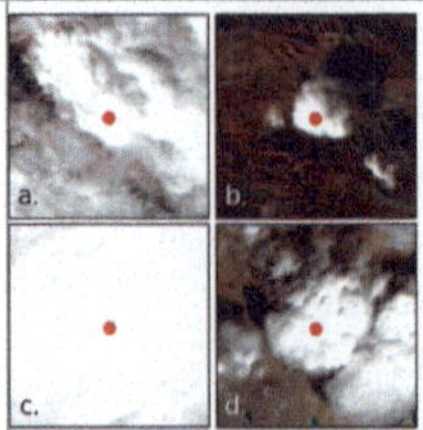	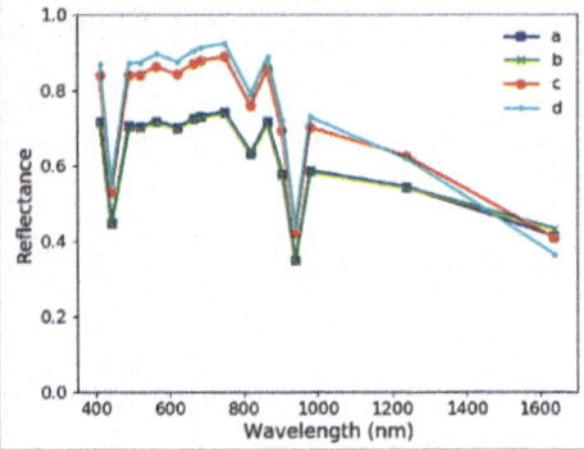
Thin cloud	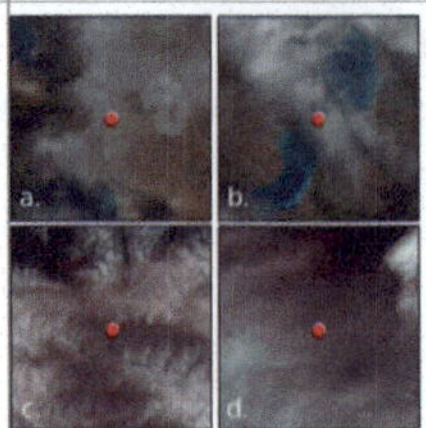	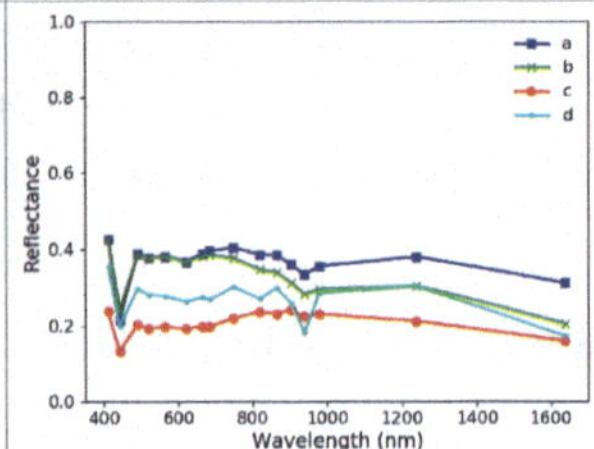
Bare soil	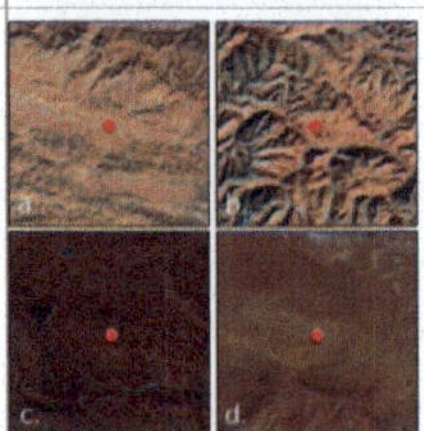	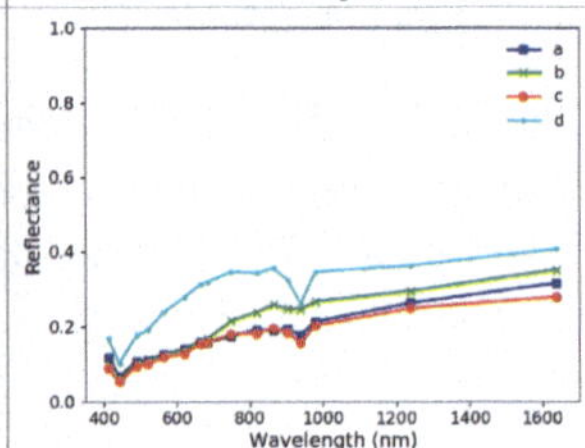
Plant	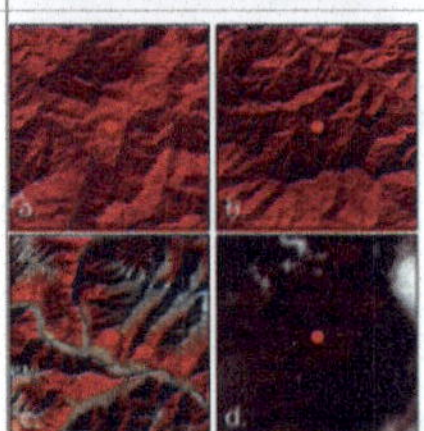	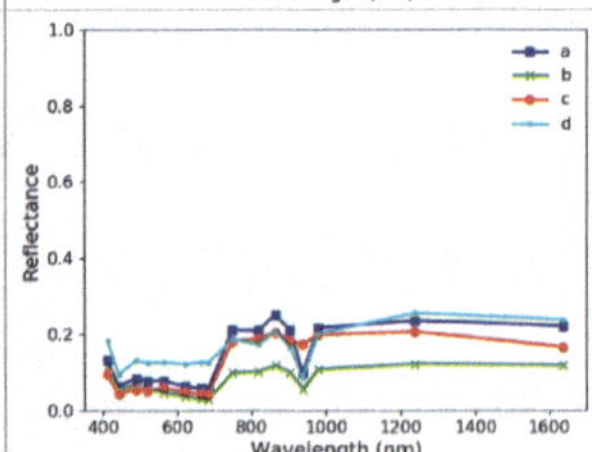
Lake	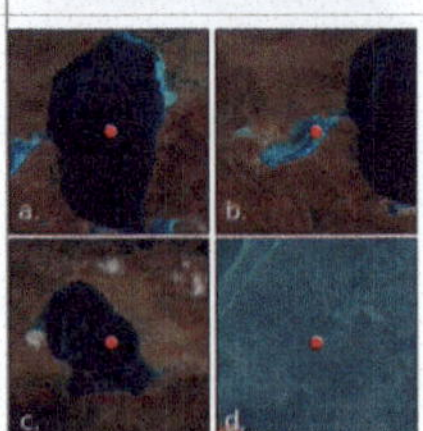	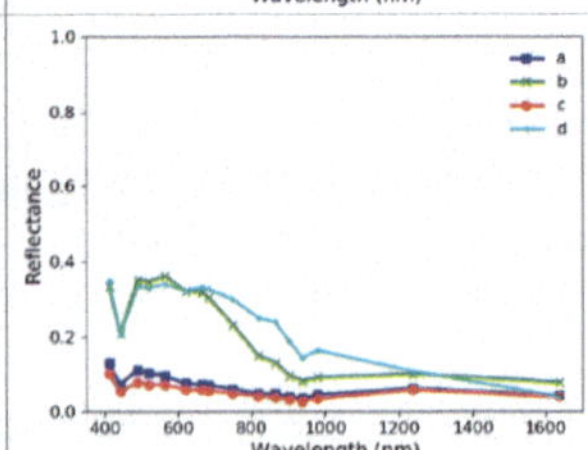
Snow	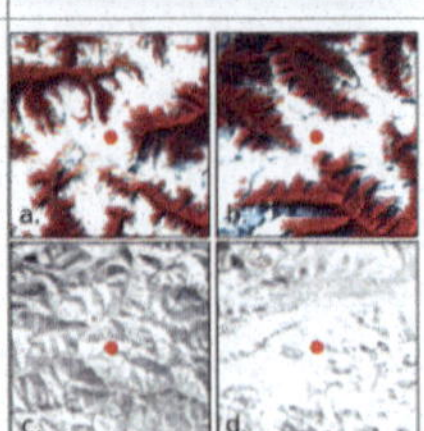	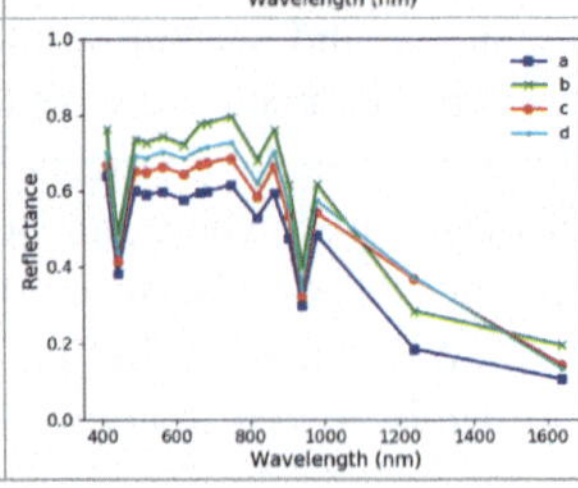

green (band 10) as RGB. The spectral reflectance was derived at the four locations (red point) in the images.

As the reflectance datasets are shown in Table 3, the lake shows low reflectance at all channels, but it may cause confusion when ice exists and causes high reflection in the visible wavelength (Table 3 Water: b, d). The reflectance of thick cloud and snow are both high and similar in visible light and near-infrared wavelength, while they have a difference in shortwave infrared wavelength. The difference in shortwave infrared wavelength can be used to distinguish snow and cloud. We will discuss this in Sect. 3.2.

The four samples of thin clouds in Table 3 are the thin cloud above the bare soil (a), lake (b) and plant (c, d). Comparing to the reflectance of bare soil, lake and plant, the thin cloud samples show the rule that the value of VNI band 4 (865 nm) is slightly higher than the value of VNI band 9 (620 nm) and 12 (490 nm). After analyzing a large number of pixels, we found that the VNI-4/VNI-9 ratio of cloud pixels is generally in the range from 1 to 1.1, and the difference between VNI-4 and VNI-12 of cloud pixels is usually between 0 and 0.017.

3.2 Snow Detection

Cloud usually has higher reflectivity than snow in the short-wave infrared band. Consequently, the Normalized Difference Snow Index (NDSI) is widely used to identify snow and cloud [23, 24]. NDSI is a standardized ratio of the difference between the reflectance at visible and shortwave infrared bands.

$$\mathrm{NDSI} = (\mathrm{R}_{vis} - \mathrm{R}_{swir})/(\mathrm{R}_{vis} + \mathrm{R}_{swir}) \tag{1}$$

In this study, the reflectance of VNI-10 band (565 nm) and SWI-2 band (1640 nm) were used as R_{vis} and R_{swir}. According to the numerous samplings and statistics in study area, 0.4 is chosen to be the threshold of NDSI. The NDSI of snow is generally bigger than 0.4 while the NDSI of the cloud is mostly lower than 0.4.

The snow and cloud can be distinguished clearly with the NDSI threshold in most of area. However, there still exists some misclassification for some mixed pixels of snow and thin cloud. From Fig. 2b we can see that the cloud is underestimated with the NDSI threshold. Many cloud pixels are misclassified as snow. To get more accurate

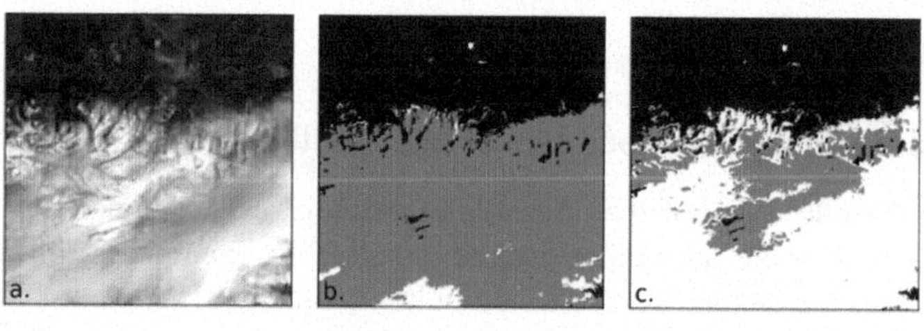

Fig. 2. The results of snow-cloud detection. The three images present: a. the false color image; b. the detected snow (gray) and cloud (white) with the NDSI threshold; c. the detected snow (gray) and cloud (white) with the NDSI threshold and the threshold of VNI-6/SWI-2 ratio.

distinguishing result of snow and cloud, a new index was added. The index is the ratio of the reflectance of band6 in visible and band2 in the shortwave infrared spectrum. As Fig. 3 shows, the ratio of VNI-6/SWI-2 of snow is bigger than the cloud. As a result, the threshold of the ratio is set as 3.9 over the Tibetan plateau. The cloud detection result is consistent with the visual interpretation from the false-color image for most of the area after combining the NDSI threshold and the VNI-6/SWI-2 threshold (Fig. 2c). However, little misclassification still exists around the edge of the snow.

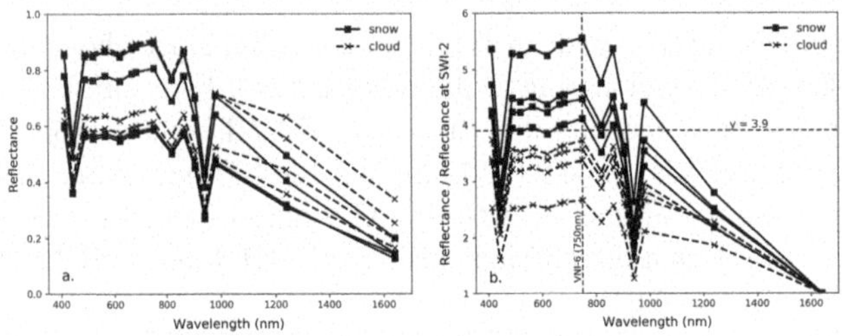

Fig. 3. The reflectance samples of snow and cloud pixels extracted from misclassified pixels (a) and the ratio of band reflectance/ reflectance at SWI-2 (b).

According to the above analysis, the single band and multiband thresholds were proposed to differentiate cloud from bare soil, plant, water, ice, snow (Fig. 4).

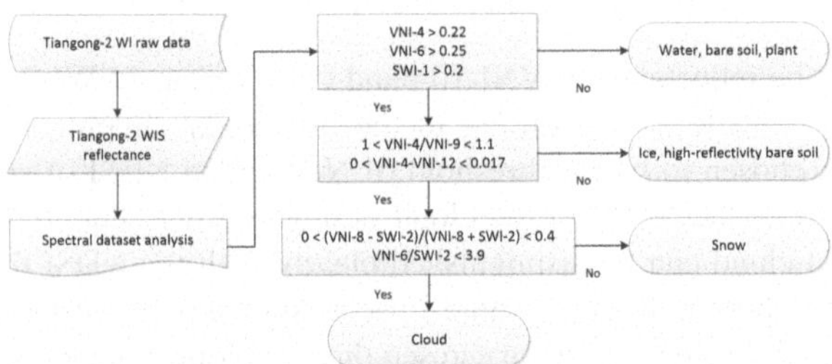

Fig. 4. The flowchart of the cloud detection algorithm

4 Results

4.1 Visual Evaluation of Cloud Detection Algorithm

To evaluate the effectivity of the threshold algorithm, several cloud detection results were compared with the false-color images visually. An example is shown in Fig. 5. From the figure, the clouds were detected quite accurately. The result shows that the threshold algorithm works well in our study area. Then, the Tiangong-2 images were processed to be cloud mask at 100 m resolution with the proposed thresholds and were compared with the AHI cloud product.

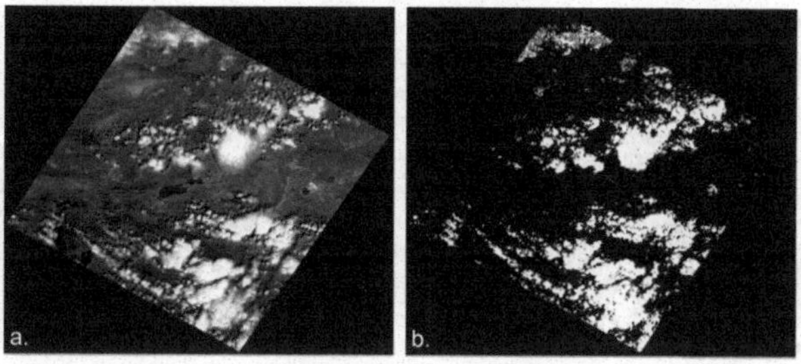

Fig. 5. The false-color image (a) and cloud detection result (b). The time of this image is around 05:00 UTC on March 13, 2018. The resolution of the data is 100 m.

4.2 Comparison with the Himawari-8 Cloud Product

The CLTYPE data classifies cloud into nine types. The pixels of all cloud types are considered to be cloud pixels. Figure 6 shows the cloud mask derived from CLTYPE data. The cloud cover from WIS of Tiangong-2 was sampled to the same resolution with AHI CLTYPE to calculate cloud fraction. We marked the pixel as cloud pixel when the cloud fraction is bigger than 0.9 and clear pixel when the cloud fraction is smaller than 0.1. After comparing the two cloud cover data, we could see that the difference mostly occurred around the edge of clouds. The cloud recognition results of AHI are highly overestimated, which may be caused by the coarse spatial resolution and cloud classification methods. The criterion of AHI 5 km cloud types is according to the cloud's microphysical properties including cloud top pressure and optical thickness, which may bring the inversion errors of these parameters into the cloud classification results and lead to significant deviation.

4.3 Analysis of Cloud Cover Difference in Himawari-8

To quantitatively evaluate the accuracy of the cloud cover derived from the AHI CLTYPE, two statistical indicators were used based on the cloud detection results of Tiangong-2. The False Alarm Rate (FAR) represents the fraction of clear pixels incorrectly classified as cloud [25]. The Cloud Hit Rate (CHR) represents the fraction of cloud pixels which are correctly classified [26]. The calculation formulas are as follows:

$$FAR = \frac{Himawari - 8_{cloud} \,\&\, Tiangong - 2_{clear}}{Himawari - 8_{cloud}} \tag{2}$$

$$CHR = \frac{Himawari - 8_{cloud} \,\&\, Tiangong - 2_{cloud}}{Tiangong - 2_{cloud}} \tag{3}$$

Where $Himawari - 8_{cloud}$ is the number of cloudy pixels in the Himawari-8 AHI cloud mask, and $Tiangong - 2_{cloud}$ and $Tiangong - 2_{clear}$ is the number of cloudy

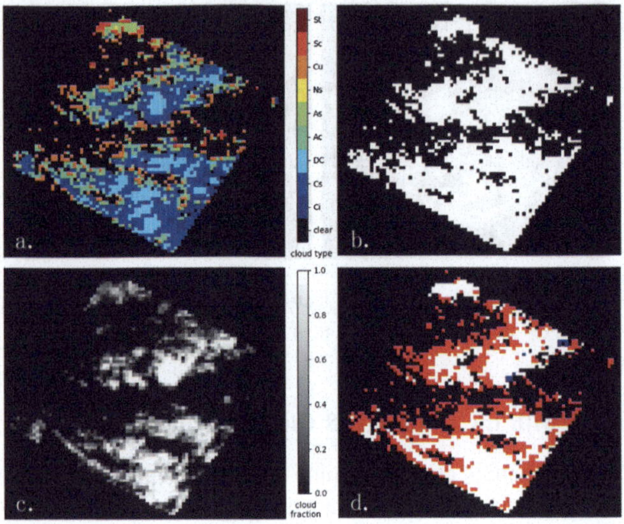

Fig. 6. The AHI cloud cover and the difference with Tiangong-2 WIS. (a) the cloud types from the CLTYPE data; (b) the cloud mask derived from CLTYPE; (c) the Tiangong-2 cloud fraction data in the same resolution with AHI CLTYPE; (d) the difference between the two cloud cover data, the red pixels are the clear pixels (cloud fraction < 0.1) but classified as cloud in AHI, the blue pixels are the cloud pixels (cloud fraction > 0.9) but classified as clear pixels in AHI.

pixels (cloud fraction > 0.9) and clear pixels (cloud fraction < 0.1) in the cloud detection result from WIS of Tiangong-2.

The FAR and CHR of all available data in this study are 29.12% and 94.77% respectively. The CHR is closed to 100%, which means almost all cloudy pixels are correctly classified as cloud in CLTYPE data. The FAR is a little large which hints that there are a lot of clear pixels incorrectly classified as cloud.

The AHI cloud pixels are not always the pure cloudy pixels. Figure 7 shows the cloud fraction distribution of all AHI cloudy pixels. There are almost 30% AHI cloud pixels which are clear and quite a large part of AHI cloud pixels is partly cloudy.

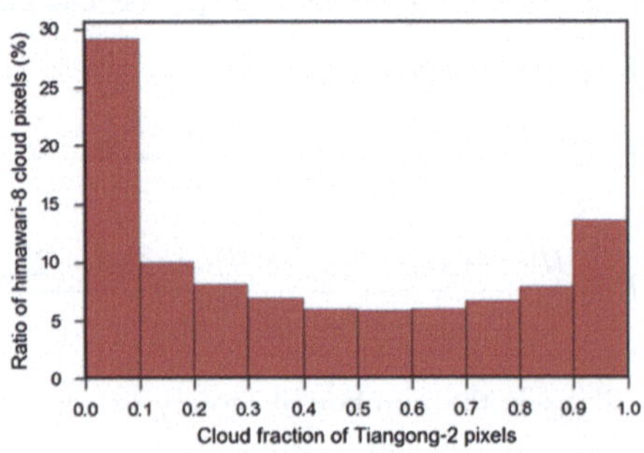

Fig. 7. The cloud fraction of all Himawari-8 AHI cloud pixels.

To figure out the cloud detection accuracy for all of the cloud types by AHI, we count the ratio of incorrectly-detected pixels and correctly-detected pixels in each cloud type (Fig. 8). It shows that the Altocumulus (Ac) and Cumulus (Cu) are mainly detected incorrectly in AHI product. From Fig. 6a, these two types of cloud are mainly located at the edge of cloud. The ratio of Ac in correct pixels is also high, but compared with the ratio of wrong detection, the Ac detection by AHI is not so reliable. The figure shows that the pixels of Cirrostratus (Cs), deep convection (DC), Nimbostratus (Ns) have the higher credibility to be clouds.

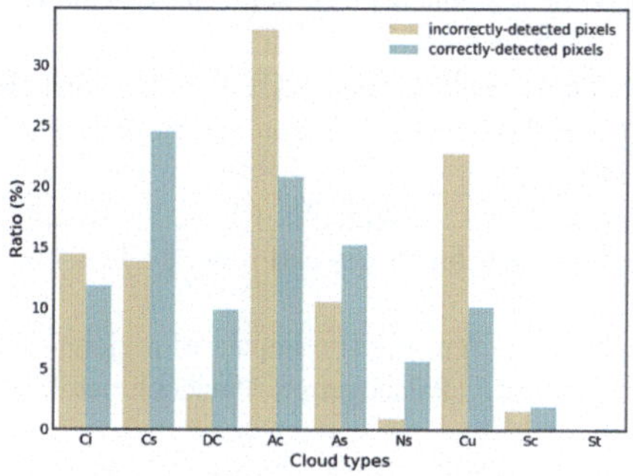

Fig. 8. The ratio incorrectly-detected and correctly-detected pixels for each cloud type

5 Conclusion

In this study, the cloud cover over the Tibetan Plateau was analyzed by the data from Tiangong-2 and Himawari-8 AHI product. The apparent reflectance of Tiangong-2 WIS was used to develop the threshold algorithm to detect cloud. Spectral reflectance of different land objects was extracted from the data in the study area. We proposed the thresholds of single band and multiband to detect cloud. With these thresholds, the clouds could be detected quite well in the study area.

The AHI CLP product contains the cloud type data, which can be used to get cloud mask. We compared the cloud cover from these two datasets and found that the cloud cover from the AHI data is overestimated. The cloud pixels of WIS of Tiangong-2 were mostly within the scope of the AHI cloud mask.

The two statistical indicators (FAR, and CHR) are 29.12% and 94.77% respectively for AHI products, which means that in AHI data almost all cloud pixels were correctly classified as cloud, and some clear pixels were also classified as cloud. The incorrect pixels are mostly around the edge of clouds, and are mainly classified as the altocumulus and cumulus.

Acknowledgments. This work was supported by the key program of NSFC (Grant No. 413311771). Thanks to the China Manned Space Engineering for providing space science and application data products of Tiangong-2. The CLP data used in this paper was supplied by the P-Tree System, Japan Aerospace Exploration Agency (JAXA).

References

1. Baker, M.: Cloud microphysics and climate. Science **276**, 1072–1078 (1997)
2. Cheng, G., Wu, T.: Responses of permafrost to climate change and their environmental significance, Qinghai–Tibet Plateau. J. Geophys. Res. Earth Surf. **112** (2007)
3. Ni, J.: A simulation of biomes on the Tibetan plateau and their responses to global climate change. Mt. Res. Dev. **20**, 80–89 (2000)
4. Wang, Z., Li, Z., Xu, M., Yu, G.: River Morphodynamics and Stream Ecology of the Qinghai-Tibet Plateau. CRC Press (2016)
5. Dim, J., Murakami, H., Nakajima, T., Nordell, B., Heidinger, A., Takamura, T.: The recent state of the climate: driving components of cloud-type variability. J. Geophys. Res. Atmos. **116** (2011)
6. Favre, A., Päckert, M., Pauls, S.U., Jähnig, S.C., Uhl, D., Michalak, I., Muellner-Riehl, A. N.: The role of the uplift of the Qinghai–Tibetan plateau for the evolution of Tibetan biotas. Biol. Rev. **90**, 236–253 (2015)
7. Yang, K., Wu, H., Qin, J., Lin, C., Tang, W., Chen, Y.: Recent climate changes over the Tibetan Plateau and their impacts on energy and water cycle: a review. Global Planet Change **112**, 79–91 (2014)
8. Zhang, R., Jiang, D., Zhang, Z., Yu, E.: The impact of regional uplift of the Tibetan Plateau on the Asian monsoon climate. Palaeogeogr. Palaeoclimatol. Palaeoecol. **417**, 137–150 (2015)
9. Kang, S., Xu, Y., You, Q., Flügel, W.-A., Pepin, N., Yao, T.: Review of climate and cryospheric change in the Tibetan Plateau. Environ. Res. Lett. **5**, 015101 (2010)
10. Kutzbach, J., Prell, W., Ruddiman, W.F.: Sensitivity of Eurasian climate to surface uplift of the Tibetan plateau. J. Geol. **101**, 177–190 (1993)
11. Frey, R.A., Ackerman, S.A., Liu, Y., Strabala, K.I., Zhang, H., Key, J.R., Wang, X.: Cloud detection with MODIS. Part I: Improvements in the MODIS cloud mask for collection 5. J. Atmos. Ocean. Technol. **25**, 1057–1072 (2008)
12. Bessho, K., Date, K., Hayashi, M., Ikeda, A., Imai, T., Inoue, H., Kumagai, Y., Miyakawa, T., Murata, H., Ohno, T.: An introduction to Himawari-8/9—Japan's new-generation geostationary meteorological satellites. J. Meteorol. Soc. Japan. Ser. II, **94**, 151–183 (2016)
13. Ishida, H., & Nakajima, T.Y.: Development of an unbiased cloud detection algorithm for a spaceborne multispectral imager. J. Geophys. Res. Atmos. **114** (2009)
14. Rossow, W.B., Schiffer, R.A.: ISCCP cloud data products. Bull. Am. Meteor. Soc. **72**, 2–20 (1991)
15. Wang, Y., Zhao, C.: Can MODIS cloud fraction fully represent the diurnal and seasonal variations at DOE ARM SGP and manus sites? J. Geophys. Res. Atmos. **122**, 329–343 (2017)
16. Foga, S., Scaramuzza, P.L., Guo, S., Zhu, Z., Dilley, R.D., Beckmann, T., Schmidt, G.L., Dwyer, J.L., Joseph Hughes, M., Laue, B.: Cloud detection algorithm comparison and validation for operational landsat data products. Remote Sens. Environ. **194**, 379–390 (2017)
17. Lin, C.-H., Lin, B.-Y., Lee, K.-Y., Chen, Y.-C.: Radiometric normalization and cloud detection of optical satellite images using invariant pixels. ISPRS J. Photogramm. Remote Sens. **106**, 107–117 (2015)
18. Shang, H., Letu, H., Nakajima, T.Y., Wang, Z., Ma, R., Wang, T., Lei, Y., Ji, D., Li, S., Shi, J.: Diurnal cycle and seasonal variation of cloud cover over the Tibetan Plateau as determined from Himawari-8 new-generation geostationary satellite data. Sci. Rep. **8**, 1105 (2018)

19. Chauvin, R., Nou, J., Thil, S., Traoré, A., Grieu, S.: Cloud detection methodology based on a sky-imaging system. Energy Procedia **69**, 1970–1980 (2015)
20. Dürr, B., Philipona, R.: Automatic cloud amount detection by surface longwave downward radiation measurements. J. Geophys. Res. Atmos. **109** (2004)
21. Sun, L., Mi, X., Wei, J., Wang, J., Tian, X., Yu, H., Gan, P.: A cloud detection algorithm-generating method for remote sensing data at visible to short-wave infrared wavelengths. ISPRS J. Photogramm. Remote Sens. **124**, 70–88 (2017)
22. Sun, L., Wei, J., Wang, J., Mi, X., Guo, Y., Lv, Y., Yang, Y., Gan, P., Zhou, X., Jia, C., Tian, X.: A universal dynamic threshold cloud detection algorithm (UDTCDA) supported by a prior surface reflectance database. J. Geophys. Res. Atmos. **121**, 7172–7196 (2016)
23. Hall, D.K., Riggs, G.A., Salomonson, V.V.: Development of methods for mapping global snow cover using moderate resolution imaging spectroradiometer data. Remote Sens. Environ. **54**, 127–140 (1995)
24. Roy, D.P., Kovalskyy, V., Zhang, H., Vermote, E.F., Yan, L., Kumar, S., Egorov, A.: Characterization of Landsat-7 to Landsat-8 reflective wavelength and normalized difference vegetation index continuity. Remote Sens. Environ. **185**, 57–70 (2016)
25. Hutchison, K.D., Iisager, B.D., Hauss, B.: The use of global synthetic data for pre-launch tuning of the VIIRS cloud mask algorithm. Int. J. Remote Sens. **33**, 1400–1423 (2012)
26. Mace, G.G., Jakob, C., Moran, K.P.: Validation of hydrometeor occurrence predicted by the ECMWF model using millimeter wave radar data. Geophys. Res. Lett. **25**, 1645–1648 (1998)

Evaluation of Several Filtering and Unwrapping Methods for the Interferometric Imaging Radar Altimeter

Hong Tan[1,2(✉)], Shengyang Li[1,2], Zhiwen Liu[1,2], Wanfeng Zhang[1,2], and Leijuan Li[1,2]

[1] Key Laboratory of Space Utilization, Chinese Academy of Sciences, Beijing 100094, China
tanhong@csu.ac.cn
[2] Technology and Engineering Center for Space Utilization, Chinese Academy of Sciences, Beijing 100094, China

Abstract. Radar altimeters using interferometric synthetic aperture radar (InSAR) techniques at near-nadir angles have been proposed to measure elevations of the ocean and the inland water with wide swath and high spatial resolution. The Ku-band Interferometric Imaging Radar Altimeter (InIRA) on the Tiangong-2 space laboratory, which was launched on 15 September 2016, is the first spaceborne instrument of such kind. Another similar instrument is the forthcoming Ka-band Radar Interferometer (KaRIn) of the Surface Water and Ocean Topography (SWOT) mission. In the interferometric processings, which have to be done to derive the elevations with the interferometric altimeter, the phase filtering and the phase unwrapping are two key steps. There are many filtering and unwrapping methods for conventional InSAR published in literature, but very few for interferometric altimeters. In this paper, several common filtering and unwrapping methods were evaluated with real InIRA acquisitions over three different scenes, including an open ocean, a coastal area, and a land with a mountain. The phase filtering was evaluated by the pseudo-correlation coefficient, the residue number, and the Root Mean Square (RMS) of the difference between the filtered phase and the original phase. The phase unwrapping was evaluated by the statistics of the difference between the rewrapped phase and the input wrapped phase, the ε value, and the discontinuity number. The evaluation results were shown and some recommendations were made on the method selection and the parameter setting for interferometric altimeters.

Keywords: Tiangong-2 · Altimetry · Interferometry · Phase filtering
Phase unwrapping · Synthetic aperture radar

1 Introduction

Radar altimeters using the interferometric synthetic aperture radar (InSAR) techniques have been studied recently [1–4], which can be referred as interferometric altimeters. They show potentials in providing topography information of the oceans and the inland

Y. Gu et al. (Eds.): *Proceedings of the Tiangong-2 Remote Sensing Application Conference*, LNEE 541, pp. 94–103, 2019.
https://doi.org/10.1007/978-981-13-3501-3_9

waters in high spatial resolution and with wide swath. One example is the Ku-band Interferometric Imaging Radar Altimeter (InIRA) on the Tiangong-2 space laboratory of China [5], which was launched on 15 September 2016 and is the first interferometric radar altimeter operating in space. Another example is the Ka-band Radar Interferometer (KaRIn) on the National Aeronautics and Space Administration (NASA) and the Centre National d'Etudes Spatiales (CNES) wide swath altimetry mission Surface Water and Ocean Topography (SWOT) [6], which is planned to be launched in 2021.

The InIRA works at Ku-band with a baseline of 2.3 m and with look angle ranging from 2.5 to 7.5°. It provides topography data of the oceans on a 10-km^2 grid with a relative height precision of 20 cm and with a swath of 35 km. It also provides elevations of the continent waters on a 200-m^2 grid and the precision is currently not defined.

For oceans, InIRA data can provide information on submesoscale (<100 km) eddies, currents, and inner waves, etc., which can be used in retrieving significant wave heights, wave directions, sea surface wind speeds, [7, 8], sea surface slopes, and velocities and vorticities of the geostrophic current vectors [9], etc. For continents, InIRA data can provide information on heights, width, surface slope, and discharge of rivers and elevations and storage of large lakes [10, 11], etc. Benefiting from the high spatial resolution, InIRA can give some insights on the transport processes in the coastal ocean [9].

However, as a new generation of altimetry instrument, meeting the requirements of the potential scientific applications is not easy for the InIRA for several reasons. Firstly, the near-nadir observing geometry makes the InIRA quite different from the conventional InSAR which is only used in terrain mapping with large incidence angles. Secondly, the severe lay-over effect and the low signal-to-noise ratio on land make the data processing on land much more difficult than that on ocean. Thirdly, the spatial resolution requirement of the surface height product is about 200 m on land which is much finer than that on ocean as 10 km.

To retrieve the topography data, a series of interferometric processing steps have to be done, where the interferometric phase filtering and unwrapping are two key steps that would influence significantly the derived data quality. The phase filtering reduces the instrumental noises. The phase unwrapping generates continuous phases from the wrapped phases. Different methods have different performances on different targets, such as the deep ocean, the coastal area, and the continent.

In this paper, we check several filtering and unwrapping methods that are commonly used in conventional InSAR and conduct a large number of tests and quantitative evaluations with real acquisitions to find the proper methods and the optimal corresponding parameters for the InIRA.

2 Interferometric Processing

2.1 Main Steps

The interferometric processing starts with inputting two channels of complex images and ends with producing surface height data, which is followed by images coregistration, interferometric phase retrieving, filtering, unwrapping, absolute phase determination, and phase to elevation conversion.

2.2 Phase Filtering Methods

There are three types of methods such as the spatial filtering, the frequency domain filtering, and time-frequency analysis filtering.

In spatial filtering, the output can be the phase of a mean or a median of a group of neighboring pixels where complex values are considered as in multilooked filtering or only phases are considered as in vectorial filtering. The mean is more efficient in computation than the median. The vectorial approach is not affected by the image amplitudes, thus it is more preferable when the phases are actually the main concern. The vectorial mean filter is

$$\hat{\phi}(x, y) = \text{Angle}\left(\frac{1}{MN}\sum_{c=0}^{M-1}\sum_{k=0}^{N-1}\exp\{j[\phi(x+c, y+k)+n(x+c, y+k)]\}\right) \tag{1}$$

where $\phi(x, y)$ is the ideal interferogram, $n(x, y)$ is the phase noise.

In frequency domain filtering, the noise is reduced by processing the power spectrum. There are several methods, such as the adaptive alpha filter [12], the main frequencies extraction, the weighting of frequency spectrum by the sinus function, the power function, or the exponential function. Unlike the adaptive alpha filter and the spectrum weighting, the main frequencies extraction does not change the relative spectrum power. As a result, the main frequencies extraction performs well in phase preserving and filtering. The main frequencies extraction can be written as

$$f(Z(u, v)) = \begin{cases} Z(u, v), & \mathit{if}\ |Z(u, v)|/\text{mean}(|Z(u, v)|) > T \\ 0, & \text{else} \end{cases} \tag{2}$$

where $Z(u, v)$ is the two-dimensional FFT of the exponential of the interferogram as

$$Z(u, v) = \text{FFT}\{\exp\{j[\phi(x, y) + n(x, y)]\}\} \tag{3}$$

In time-frequency analysis filtering, methods with complicated algorithms are designed to handle situations of dense fringes and rapid spatial variations in local regions of the interferogram, such as the wavelet filtering, which is still under development [13].

2.3 Phase Unwrapping Methods

There are many unwrapping methods published in literature, which can be classified into two types as the local approaches and the global approaches.

Local approaches, such as the classical branch cut method [14] and the quality map guide method, etc., look for optimized integrate path to prevent errors caused by residues from spreading to other pixels and sometimes incorporate artificial intelligence knowledge. However, when the interferogram quality is low or the residues are difficult to recognize, local methods would introduce globally-spreading integrate errors.

Global approaches, such as the minimum norm method, the least square method, and the minimum cost maximum flow method, etc., find the optimal solution based on the phase gradients, but may leave a local error in each pixel.

A new approach is proposed [13] that combines the quality map region growing and the moving surface phase fitting. This approach ensures the precise unwrapping in high quality regions with quality map region growing and unwraps the phases in low quality regions with the moving surface phase fitting. Thus, this approach performs unwrapping in the whole interferogram and still preserves the high quality in local regions.

3 Data and Evaluation Methods

3.1 Data Sets

To cover the variety of the observing targets of the InIRA, a total of three scenes were selected including an open ocean, a coastal area, and a land with a mountain as shown in Fig. 1. One significant difference between the InIRA and traditional SARs is that the intensity from oceans or inland waters is much larger than the intensity from the land whereas it is completely opposite for traditional SARs, as shown in Fig. 1 (b).

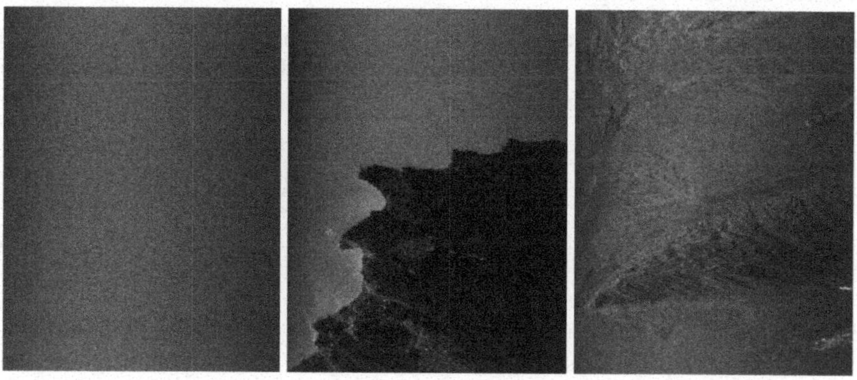

(a) (b) (c)

Fig. 1. Slant range images of the relative intensity of the primary channel of (a) an ocean, (b) a coastal area, and (c) a land with a mountain acquired by the InIRA in September 2016.

Each image has 1500 pixels in range and 4384 pixels in azimuth. In the range direction, the slant range is equally sampled with pixels spacing of 2.5 m. In the azimuth direction, 30 lines have been averaged resulting in a pixel spacing of 13.3 m. Each scene is an area of about 35 km (in range) × 58 km (in azimuth).

After images coregistration and interferometric phases retrieving, the computed interferograms are shown in Fig. 2, which are wrapped between -π and π and are contaminated by noises.

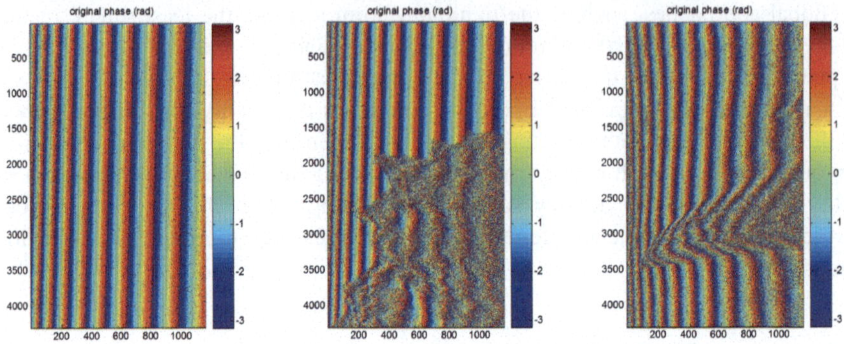

Fig. 2. Corresponding original interferograms of the same scenes as (a) an ocean, (b) a coastal area, and (c) a land with a mountain in Fig. 1.

3.2 Quality Indicators of Phase Filtering

Three indicators were selected to evaluate the phase filtering, such as the pseudo-correlation coefficient, the residue number, and the Root Mean Square (RMS) of the difference between the original phase and the filtered phase.

Pseudo-Correlation Coefficient. The pseudo-correlation coefficient is defined as (4). The larger the pseudo-correlation, the better the quality of the filtered interferogram is. In calculation, the window size k is set as 7.

$$\rho'(x,y) = \frac{\sqrt{\left(\sum_{i=-k}^{k}\sum_{j=-k}^{k}\cos\phi(x+i,y+j)\right)^2 + \left(\sum_{i=-k}^{k}\sum_{j=-k}^{k}\sin\phi(x+i,y+j)\right)^2}}{(2k+1)^2} \quad (4)$$

Residue Number. A residue is a pixel where the sum of the four phase changes is not zero along a square path that starts from the current pixel then meets the right one, the down-right one, the down one, and finally ends at the current pixel. The fewer the residue number, the better the quality of the filtered interferogram is.

RMS. The RMS of the difference between the original phase and the filtered phase is defined as (5). The smaller the RMS, the better the phase preserving performance is.

$$RMS = \sqrt{\frac{\sum_{i=0}^{M-1}\sum_{j=0}^{N-1}(\Delta\phi(i,j))^2}{MN}} \quad (5)$$

3.3 Quality Indicators of Phase Unwrapping

Three indicators were selected to evaluate the phase unwrapping, such as the statistics of the difference between the rewrapped phase and the input wrapped phase, the ε value [15], and the discontinuity number.

Statistics of the Phase Difference. When the unwrapped phases are rewrapped and compared with the original wrapped phases, the phase differences can be computed and the statistics such as the maximum, the minimum, the mean, and the median can be evaluated. The smaller the difference, the higher the unwrapping accuracy is.

ε Value. ε is defined as (6) to indicate the consistency of the unwrapped result with the original wrapped phase in the sense of phase gradients. The smaller the ε value, the higher the unwrapping accuracy is.

$$\begin{aligned}\varepsilon = & \frac{1}{MN}\sum_{i=0}^{M-2}\sum_{j=0}^{N-1} w^x(i,j)|\phi(i+1,j) - \phi(i,j) - \Delta^x(i,j)|^p \\ & + \frac{1}{MN}\sum_{i=0}^{M-1}\sum_{j=0}^{N-2} w^y(i,j)|\phi(i,j+1) - \phi(i,j) - \Delta^y(i,j)|^p\end{aligned} \tag{6}$$

Discontinuity Number. A discontinuity is a pixel whose phase has an absolute difference that is larger than π to its neighbor pixels. The fewer the discontinuity number, the higher the unwrapping accuracy is.

4 Test and Evaluation Results

4.1 Phase Filtering

We compared the performances of the vectorial mean filter with different averaging window size M and N in (1) and the main frequencies extraction with different thresholds T in (2). Here we set M equals N as using a square averaging window in vectorial mean filtering.

Figure 3 shows the residue number in filtered interferogram with different filtering methods and different parameters. It is obvious that under the same filtering condition, the ocean gives the fewest residues while the continent gives the moderate number of residues and the coastal area gives the largest number of residues. This is due to the slowly varied surface heights on ocean and the fast changed elevations at coastal areas as well as the very high SNR on ocean and the very low SNR on land.

It is also can be concluded that an averaging window size of 25 is the best for the vectorial mean filter and a threshold of 3.8 is the best for the main frequencies extraction.

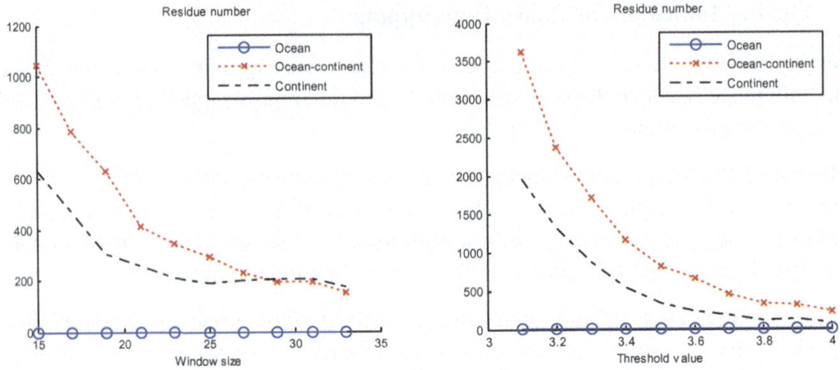

(a) (b)

Fig. 3. Residue numbers in the interferograms filtered by (a) the vectorial mean filter and (b) the main frequencies extraction over the three typical scenes in Fig. 1, where the ocean is indicated by the blue circle, the coastal area is indicated by the red cross, and the land is indicated by the black dash-dot.

The interferogram fringes over the inland mountain can be seen in Figs. 2 (c) and 4 (a), although they are not clear due to the low SNR. To compare the vectorial mean filter with a window size of 25 and the main frequencies extraction with a threshold of 3.8, evaluations over the inland mountain were performed as shown in Fig. 4 and Table 1. It can be concluded that the main frequencies extraction is preferred especially on land because of clearer detailed fringes as shown in Fig. 4 (c) and significantly reduced residues as well as comparable pseudo-correlation coefficients and RMS errors.

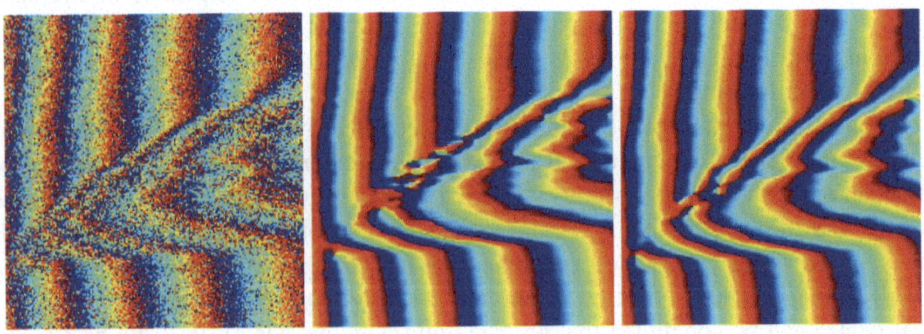

(a) (b) (c)

Fig. 4. Interferograms of a selected region in the scene of the inland mountain: (a) before filtering, (b) filtered by the vectorial mean filter with a window size of 25, and (c) filtered by the main frequencies extraction with a threshold of 3.8.

Table 1. Phase filtering performances over the inland mountain

Method	Parameter	Residue number	Average pseudo-correlation coefficient	RMS (deg)
Vectorial mean filter	25	188	0.998026	59.391
Main frequencies extraction	3.8	118	0.998450	59.387

4.2 Phase Unwrapping

We compared three phase unwrapping methods, such as the classical brand cut method, the quality map guide method, and a combined method using the quality map region growing and moving surface phase fitting. The evaluation results are shown in Table 2.

Table 2. Phase unwrapping performances over the three scenes, where B stands for the brand cut method, Q for the quality map guide method, and C for the combined method

Scene	Method	Maximum difference (rad)	Minimum difference (rad)	Average difference (rad)	Median difference (rad)	ε value	Discontinuity number
Ocean	B	6.283001	0	0.001591	0.000917	0.000010	0
	Q	6.283167	0	0.000105	0.000045	0.000005	0
	C	6.282928	0	0.001130	0.000629	0.000010	0
Coastal area	B	6.283097	0	0.003212	0.000288	0.022069	7628
	Q	6.283169	0	0.000195	0.000093	0.011964	7991
	C	6.283014	0	0.004404	0.001170	0.005280	2070
Land	B	6.283154	0	0.000674	0.000180	0.008784	1565
	Q	6.283166	0	0.000070	0.000035	0.002212	1462
	C	6.283009	0	0.001411	0.000527	0.001092	488

On oceans with high SNR and high correlation, the combined method shows negligible improvement in the statistics of the phase difference and shows no change in the ε value and the discontinuity number when compared with the brand cut method.

On coastal area and land with low SNR and low correlation, the combined method reduces the ε value and the discontinuity number significantly at a cost of acceptable increase in the phase differences.

As a result, the combined method is preferred in different targets including oceans and lands.

5 Conclusions

The Ku-band Interferometric Imaging Radar Altimeter (InIRA) on the Tiangong-2 space laboratory of China is a new generation of radar altimeter. To successfully derive the surface elevation of both the oceans and the lands, filtering and unwrapping of the interferogram with high quality are very important. Considering the special observing geometry and the potential scientific applications in both the oceanography and the hydrology which are all quite different from the conventional InSAR, we compared quantitatively several common filtering and unwrapping methods with real InIRA acquisitions.

To cover the variety of the observing targets of the InIRA, a total of three scenes were selected in an open ocean, a coastal area, and a land with a mountain. The phase filtering has been evaluated by the pseudo-correlation coefficient, the residue number, and the RMS of the difference between the original phase and the filtered phase. The phase unwrapping has been evaluated by the statistics of the difference between the rewrapped phase and the input wrapped phase, the ε value, and the discontinuity number.

The evaluation results have demonstrated that the main frequencies extraction filter is preferred than the vectorial mean filter and the optimized threshold is 3.8 for the InIRA. The results have also confirmed that the combined method of the quality map region growing and moving surface phase fitting is preferred in the large variety of targets including both oceans and lands.

Acknowledgements. Thanks to China Manned Space Engineering for providing space science and application data products. Thanks to Xiao Dong and Yunhua Zhang in the Key Laboratory of Microwave Remote Sensing, Chinese Academy of Sciences, for their discussions on the instrument design and for their help in generating the complex images.

References

1. Enjolras, V., Vincent, P., Souyris, J.-C., Rodriguez, E., Phalippou, L., Cazenave, A.: Performances study of interferometric radar altimeters: from the instrument to the global mission definition. Sensors. **6**(3), 164–192 (2006)
2. Esteban-Fernandez, D., Fu, L.L., Rodriguez, E., Brown, S., Hodges, R.: Ka-band SAR interferometry studies for the SWOT mission. In: Proceedings of the IEEE IGARSS, pp. 4401–4402. IEEE, Honolulu (2010)
3. Moller, D., Farquharson, G., Esteban-Fernandez, D.: Assessment of near-nadir correlation characteristics over water bodies using interferometric SAR: Implications for the swot mission. In: Proceedings of the IEEE IGARSS, pp. 3219–3222. IEEE, Beijing (2016)
4. Kong, W., Chong, J., Tan, H.: Performance analysis of ocean surface topography altimetry by Ku-Band Near-Nadir Interferometric SAR. Remote Sens. **9**(9), 933 (2017)
5. Ren, L., Yang, J., Zhang, Y., Zhai, W., Zheng, G., Wang, J.: Exploring for the wind speed retrieval from Interferometric Imaging Radar Altimeter. In: Proceedings of the IEEE IGARSS, pp. 4637–4640. IEEE, Beijing (2016)
6. Fu L., Alsdorf D., Morrow R., Rodríguez E. and Mognard N.: SWOT: The Surface Water and Ocean Topography mission: Wide-swath altimetric measurement of water elevation on Earth. Jet Propulsion Lab., NASA Technical Reports Server, Pasadena (2012)
7. Yang, J., Ren, L., Zheng, G.: The first quantitative ocean remote sensing by using Chinese interferometric imaging radar altimeter onboard TG-2. Acta Oceanol. Sin. **36**(2), 122–123 (2017)
8. Bao, Q., Lin, M., Zhang, Y., Jia, Y., Lang, S.: Wind speed inversion for imaging microwave altimeter. J. Remote Sens. **21**, 835–841 (2017)

9. Fu, L.-L., Rodriguez, E.: High-resolution measurement of ocean surface topography by Radar Interferometry for Oceanographic and Geophysical Applications. In: The State of the Planet: Frontiers and Challenges in Geophysics, 1st edn. American Geophysical Union, Washington, D. C. (2004)
10. Durand, M., Fu, L.L., Lettenmaier, D.P., Alsdorf, D.E., Rodriguez, E., Esteban-Fernandez, D.: The surface water and ocean topography mission: observing terrestrial surface water and oceanic submesoscale eddies. In: Proceedings of the IEEE IGARSS, pp. 766–779. IEEE, Honolulu (2010)
11. Li, S., Tan, H., Liu, Z., Zhou, Z., Liu, Y., Zhang, W., Liu, K., Qin, B.: Mapping high mountain lakes using space-borne near-nadir SAR observations. Remote Sens. **10**(9), 1418 (2018)
12. Goldstein, R.M., Werner, C.L.: Radar interferogram filtering for geophysical applications. Geophys. Res. Lett. **25**(21), 4035–4038 (1998)
13. Jin, G., Xu, Q., Zhang, H.: Synthetic Aperture Radar Interferometry, 1st edn. National Defense Industry Press, Beijing, China (2014)
14. Goldstein, R.M., Zebker, H.A., Werner, C.L.: Satellite radar interferometry: two-dimensional phase unwrapping. Radio Sci. **23**(4), 713–720 (1988)
15. Ghiglia, D.C., Pritt, M.D.: Two-dimensional phase unwrapping: theory, algorithms, and software, 1st edn. Wiley, New York (1998)

High-Precision Geometric Correction of Tiangong-2 Interferometric Imaging Radar Altimeter

Zhiwen Liu(✉), Yunfei Liu, and Bo Wang

Key Laboratory of Space Utilization, Technology and Engineering Center for Space Utilization, Chinese Academy of Sciences, Beijing, China
zwliu@csu.ac.cn

Abstract. The interferometric imaging radar altimeter (InIRA) of Tiangong-2 is a new type of altimeter, which combines with traditional altimeter height tracking measurement technology, SAR synthetic aperture technology and SAR interference measurement technology. According to its imaging characteristics, a geometric correction algorithm for InIRA is proposed, and some positioning error sources included the virtual ellipsoid model, elevation errors and atmosphere transmission delay are corrected. Firstly, a geometric correction model is established based on slant Range-Doppler equations. Secondly, we correct the original virtual ellipsoid model of the geometric correction model to a new one for quite well describing its actual geometric relationship. Thirdly, we achieve elevation values from corresponding elevation data of InIRA as an input parameter of the geometric correction model. Fourthly, an atmosphere transmission delay is calculated by a simplified static model associated with elevation parameters. Finally, the geometric correction model is verified to be valid through actual images of InIRA. The result shows that the positioning accuracy improved by 3 pixels.

Keywords: Geometric correction · Tiangong-2
The interferometric imaging radar altimeter

1 Introduction

Different from any other spaceborne synthetic aperture radar(SAR), the interferometric imaging radar altimeter (InIRA) of Tiangong-2 chooses a near-nadir incidence of 1–8° for imaging in order to increase its altimetric precision and swath width [1]. So InIRA is not only a new type of altimeter, but also a special microware interference radar. It can be used in marine dynamic environment detection, marine forecasting, terrain topographic survey, glaciers and large water monitoring and other fields.

There are some altimeters similar to InIRA. One is the Shuttle Radar Topography Mission (SRTM) [2]. It consists of a specially modified radar system that flew on the Space Shuttle Endeavour during the 11-day STS-99 mission in February 2000. It uses

Y. Gu et al. (Eds.): *Proceedings of the Tiangong-2 Remote Sensing Application Conference*, LNEE 541, pp. 104–111, 2019.
https://doi.org/10.1007/978-981-13-3501-3_10

dual radar antennas to acquire interferometric radar data, processing to digital topographic data at 1 arc sec resolution. The other is the Synthetic Aperture Interferometric Radar Altimeter (SIRAL) [3], which is the CryoSat-2's primary instrument. It is designed to meet measurement requirements for ice-sheet elevation and sea-ice freeboard.

In recent years, the international spaceborne SAR has developed rapidly, and the quality of its image products has continued to increase, such as the Japanese ALOS PALSAR launched in 2006, the Italian COSMO-SkyMed series SAR launched in 2007 and 2008, Canada. The RadarSat-2 system and the German TerraSAR-X system [4]. Geometric correction of InIRA's image is an important processing procedure. At present, the system-level geometric positioning accuracy of these on-board SAR systems can reach 10 m or less [5, 6], especially the German TerraSAR-X standard products. After refined system calibration and processing, the positioning accuracy can reach even decimeters. Magnitude [7]. But now, there is no relevant literature to describe the direct geometric localization algorithm of those altimeters.

In this paper,we propose a geometric correction algorithm for InIRA based on the Range-Doppler Equations; in Sect. 2, the geometric correction processing is briefly introduced, and some main positioning error sources are analyzed and corrected, which include the virtual ellipsoid model, elevation error and atmosphere transmission delay; in Sect. 3, some actual images of InIRA is corrected with the geometric correction algorithm proposed and the geometric correction results are compared with Google images by some control points; some conclusions is given based on the geometric correction of InIRA in Sect. 4.

2 Geometric Correction Processing

Using platform measurement data, radar parameters and imaging processing parameters, the geographic latitude and longitude of the image is usually obtained based on the slant range Doppler equations [8]. The processing strategy we proposed is as follows (Fig. 1).

2.1 Geometric Correction Model

The geometric correction model is established based on the slant Range-Doppler equations by using platform measurement data, payload parameters and imaging parameters. It includes three nonlinear equations, which is the range equation, Doppler center frequency equation and virtual ellipsoid equation [9].

Range Equation. The range equation of SAR slant range image is the following formula

$$R_{p=}\left|C_s - C_p\right| \tag{1}$$

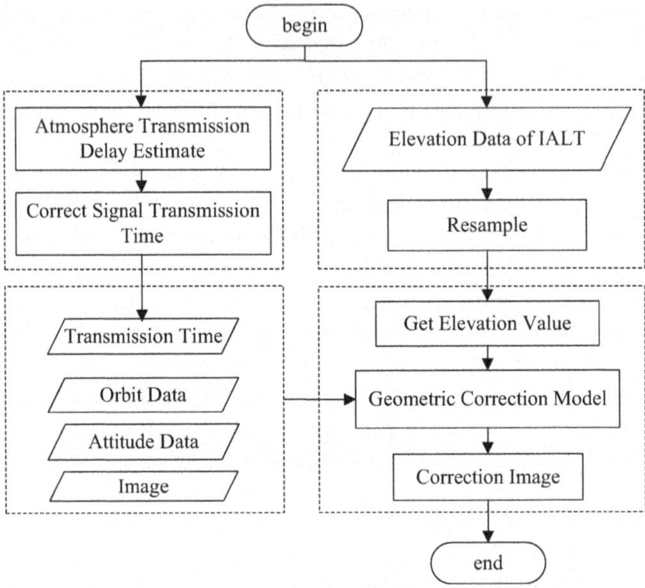

Fig. 1. Processing flowchart of geometric positioning

Where, R_p is the distance between the antenna center and the ground point, C_s is the position vector of the antenna center, $C_s = [x_s, y_s, z_s]^T$. C_p is the position vector of the ground point, $C_p = [x_p, y_p, z_p]^T$.

Doppler Center Frequency Equation. According to SAR Doppler frequency shift imaging principle, the Doppler center frequency equation is the following formula.

$$f_{dc} = \frac{-2}{\lambda R}\left(R_s - R_p\right) \cdot \left(V_s - V_p\right) \tag{2}$$

Where, V_s is the platform speed vector, $V_s = [v_{xs}, v_{ys}, v_{zs}]^T$. V_p is the speed vector of the ground point, $V_p = [v_{xp}, v_{yp}, v_{zp}]^T$. f_{dc} represents the Doppler center frequency.

Virtual Ellipsoid Equation. The virtual ellipsoid equation usually achieved by equivalently extending the long axis and short axis of the earth ellipsoid equation. Assuming that the elevation of the target point is h, the formula of the virtual ellipsoid is usually expressed as Eq. 3.

$$\frac{x^2 + y^2}{(R_e + h)^2} + \frac{z^2}{(R_p + h)^2} = 1 \tag{3}$$

Where, R_e, R_p represent the long axis and short axis of the earth ellipsoid model, respectively. h represents the elevation of the target point.

2.2 Virtual Ellipsoid Correction

Actually, the elevation of the target point at the tangent is not equal to h according to this virtual ellipsoid model. So the positioning error will be introduced by this. The point P is the target point, and the point P' is the project of the target point to the earth's surface, as shown in Fig. 2.

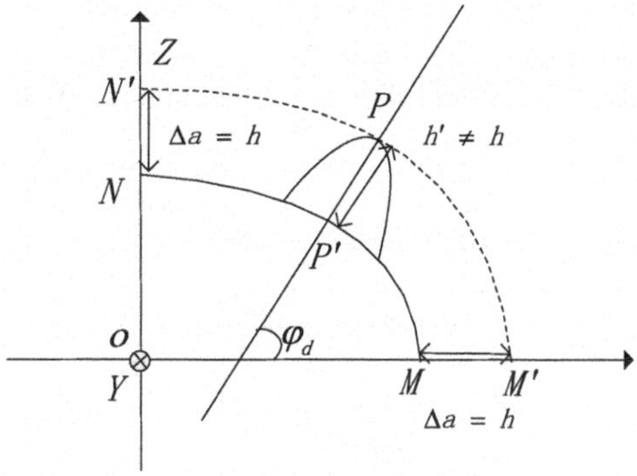

Fig. 2. A geometric map of the virtual ellipsoid model and the target point

In order to reduce this kind of errors, a new virtual ellipsoid model is established. Firstly, the earth ellipsoid is expanded outward according to its flatness. When the elevation at the target point P is equal to h, the expansion is stopped and the ellipsoid model at this moment is the new one we need, as shown in Fig. 3.

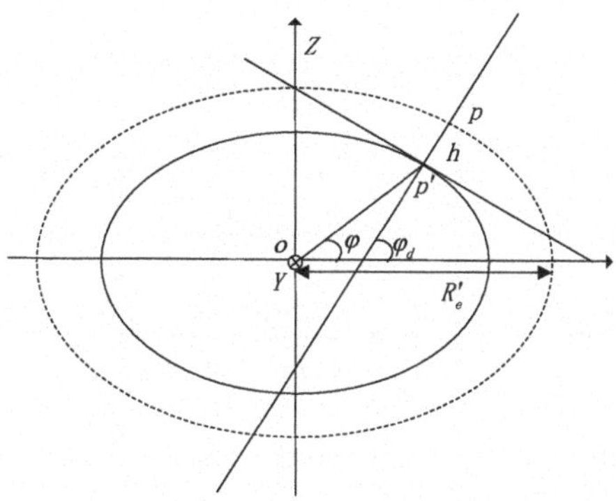

Fig. 3. A geometric schematic of elevation and new ellipsoid model

The expression of the new ellipsoid is:

$$\frac{x^2+y^2}{(R_e')^2}+\frac{z^2}{((1-e)R_e')^2}=1 \tag{4}$$

Where, R_e' represents the long semi-axis of the new ellipsoid, and e represents the flatness rate of the earth ellipsoid. So, the long semi-axis R_e' is the only parameter that need to be solved, and the calculation process is as follows.

The latitude and longitude coordinates (θ, φ) and center rectangular coordinates (x_a, y_a, z_a) of the point P' have the following relational expressions.

$$\frac{x_a^2+y_a^2}{(R_e)^2}+\frac{z_a^2}{[(1-e)R_e]^2}=1 \tag{5}$$

$$\tan(\varphi)=\frac{z_a}{\sqrt{x_a^2+y_a^2}} \tag{6}$$

$$\tan(\theta)=\frac{y_a}{x_a} \tag{7}$$

The geodetic latitude φ_d and center latitude φ have the following transform relationship.

$$\tan(\varphi_d)=\frac{\tan(\varphi)}{1-e^2} \tag{8}$$

Then, the center rectangular coordinates $(x_{a'}, y_{a'}, z_{a'})$ of the point P is:

$$x_{a'}=x_a+\frac{h}{\sqrt{1+\tan^2(\varphi_d)}}\cos\theta \tag{9}$$

$$y_{a'}=y_a+\frac{h}{\sqrt{1+\tan^2(\varphi_d)}}\sin\theta \tag{10}$$

$$z_{a'}=z_a+\frac{h\tan(\varphi_d)}{\sqrt{1+\tan^2(\varphi_d)}} \tag{11}$$

Combining Eq. 4 through Eq. 11, the long semi-axis of the new ellipsoid model is:

$$R_e'=\frac{\sqrt{(1-e)^2(x_{a'}^2+y_{a'}^2)+z_{a'}^2}}{1-e} \tag{12}$$

2.3 Elevation Value Calculation

High-precision elevation data is a very important parameter in geometric location of InIRA. Elevation error would bring positioning deviate to its actual position. The positioning deviate and elevation error have the following relationship.

$$\Delta r = \frac{\Delta h}{\tan \theta_{inc}} \tag{13}$$

Where, Δr is the positioning deviate, Δh is the elevation error, θ_{inc} is the local incident angle.

InIRA has both radiance images and interferometric data. Interference data is processed to form elevation data, whose spatial resolution and product precision are about 200 m and 10 m, respectively. We choose the elevation data of InIRA as the input parameter of the geometric correction model. The detailed processing is as follows.

At first, elevation data are resampled to images with the same resolution as radiance images. Then, the pixels of the radiance data and elevation data are in one-to-one match. We achieve the pixel value of the corresponding position of the resampled elevation data as the elevation value.

2.4 Atmosphere Transmission Delay Estimate

The propagation delay error of microwave in atmosphere will affect the slant accuracy of SAR, finally affecting the positioning accuracy. The atmospheric transmission is mainly the troposphere transmission delay.

The troposphere transmission delay is mainly related to the height, as well as the pressure, temperature and water vapor. Generally, a simplified static model associated with a height parameter can be used in the production of the data product [10].

$$\psi_{\text{tropo,zenith}} = \frac{h^2}{8.55 \cdot 10^7} - \frac{h}{3411} + 2.41 \tag{14}$$

Where, ψ_{zenith} is the tropospheric delay in the zenith direction, h is the elevation parameter in the range of 0–9000 m.

We estimate the troposphere delay in the zenith direction according to the target elevation by Eq. (14). We can obtain the troposphere delay along the line of sight of the antenna by calculating the firing angle of the signal emitted by the satellite and the Eq. (15).

$$\psi_{\text{tropo}} = \frac{\psi_{\text{zenith}}}{\cos \alpha_{\text{inc}}} \tag{15}$$

Where ψ_{tropo} is the tropospheric delay along the line of sight of the antenna, α_{inc} is the incident angle of the signal emitted by InIRA.

3 Experiment

Two images of InIRA, collected in February 6th, 2018 and January 30th, 2018, are corrected by the geometric correction method proposed in this paper. The correction results are compared with Google images by some control points. Table 1 shows the positioning accuracy of different control points, and the geometric correction accuracy is about 3 pixels. Figure 4 shows the corresponding geometric corrected images.

Table 1. The positioning accuracy of the corner reflectors

	Corner reflector	Measured latitude and longitude	Calculate latitude and longitude	Positioning accuracy/pixel
Left Image	P1	(38.9171, 121.7207)	(38.9180, 121.7216)	2.539
	P2	(38.8737, 121.8415)	(38.8745, 121.8423)	2.528
	P3	(39.0041, 121.7483)	(39.0051, 121.7488)	2.492
	P4	(38.9729, 121.8063)	(38.9737, 121.8055)	2.561
	P5	(38.7952, 121.6008)	(38.7961, 121.6017)	2.319
Right Image	P6	(31.5275, 120.3459)	(31.5264, 120.3471)	2.892
	P7	(31.6171, 120.2656)	(31.6160, 120.2664)	2.658
	P8	(331.6381, 120.1285)	(331.6371, 120.1298)	2.795
	P9	(31.4504, 120.0146)	(31.4495, 120.0155)	2.541
	P10	(31.4067, 120.2165)	(31.4052, 120.2179)	2.524

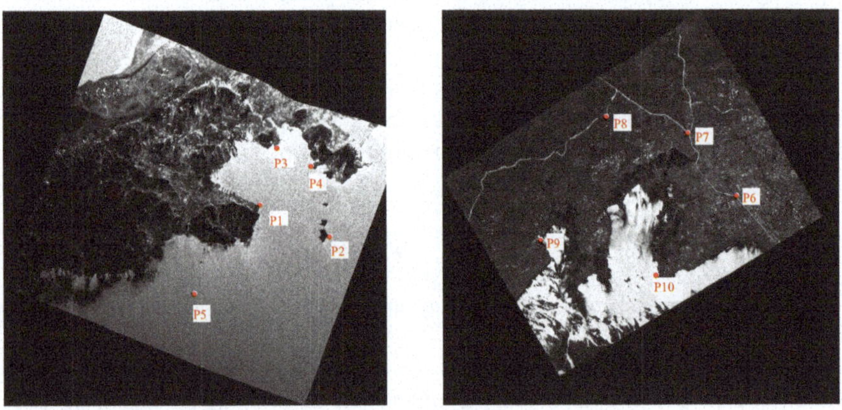

Fig. 4. The geometric corrected image of InIRA

4 Conclusion

In this paper, the geometric correction method of InIRA is proposed, and the main factors affecting positioning accuracy are analyzed and corrected. The geometric correction method is tested by actual images of InIRA. Finally, the result shows that the geometric correction method is validated for InIRA and positioning accuracy is improved by 3 pixels.

Acknowledgements. Thanks to China Manned Space Engineering for providing space science and application data products of Tiangong-2.

References

1. Kong, W., Chong, J., Tan, H.: Performance analysis of ocean surface topography altimetry by ku-band near-nadir interferometric SAR. Remote Sens. **9**(9), 933 (2017)
2. Hennig, T.A., Kretsch, J.L., Pessagno, C.J., et al.: The Shuttle Radar Topography Mission, vol. 45, no. 2, pp. 37–55. Springer, Berlin (2001)
3. Drinkwater, M.R., Francis, R., Ratier, G., et al.: The European Space Agency's earth explorer mission CryoSat: measuring variability in the cryosphere. Ann. Glaciol. **39**(1), 313–320 (2004)
4. Hong-Jian, Y., Kun, F.: Image Precise Processing of Synthetic Aperture Radar, vol. 13. Science Press, Beijing (2011)
5. Schubert, A., Small, D., et al.: COSMO-skymed, Terra SARX, and RADARSAT-2 geolocation accuracy after compensation for earth-system effects. In: 2012 IEEE International Geoscience and Remote Sensing Symposium (IGARSS2012), Germany, pp. 3301–3304 (2012)
6. Williams, D., LeDantec, P., et al.: RADARSAT-2 image quality and calibration update. In: Proceedings of the 10th European Conference on Synthetic Aperture Radar, EUSAR 2014, Denmark, pp. 1–4 (2014)
7. Eineder, M., Minet, C., et al.: Imaging geodesy—Toward centimeter-level ranging accuracy with TerraSAR-X. IEEE Trans. Geosci. Remote Sens. **49**(2), 667–671 (2011)
8. Curlander, J.C., McDonough, R.N.: Synthetic Aperture Radar: Systems and Signal Processing, pp. 374–377. Wiley, New York (1991)
9. Curlander, J.C.: Location of spaceborne SAR imagery. IEEE Trans. Geosci. Remote Sens. **20**(3), 359–364 (1982)
10. Eineder, M., Minet, C., Steigenberger, P., et al.: Imaging geodesy—toward centimeter-level ranging accuracy with terraSAR-X. IEEE Trans. Geosci. Remote Sens. **49**(2), 661–671 (2011)

Data Preprocessing and On-Orbit Quality Analysis of the Multi-azimuth UV Imager of Tiangong-2

Bangyong Qin[1(✉)], Xiaohu Yang[2], and Leijuan Li[1]

[1] Key Laboratory of Space Utilization, Technology and Engineering Center for Space Utilization, Chinese Academy of Sciences, Beijing 100094, China
qinby@csu.ac.cn

[2] Changchun Institute of Optics, Fine Mechanics and Physics, Chinese Academy of Sciences, Changchun 130033, China

Abstract. The Multi-Azimuth UV Imager (MAVI) is a new generation atmospheric sounder carried on Tiangong-2 Space Laboratory. It can obtain the vertical distribution information of the ultraviolet radiation of the global middle and upper atmosphere, and also can detect the ultraviolet radiation of the entire atmosphere in the range of ±5° from the subastral point at the same time. The design of the instrument is unique in the world. So it is different from the traditional optical remote sensing instrument in data processing and quality analysis. In this paper, the data characteristics and data preprocessing flow of MAVI are introduced, and the data quality on-orbit of MAVI is preliminarily analyzed by using several data quality evaluation indexes. The results show that the imaging performance of MAVI is stable and the data quality is reliable. The data products of MAVI can provide effective support for atmospheric scientific research and atmospheric environment monitoring.

Keywords: Tiangong-2 · The Multi-Azimuth UV imager · Data preprocessing
Data quality analysis

1 Introduction

As the first space laboratory of China, Tiangong-2 was launched successfully on Sep 15, 2016 [1]. There are more than ten space scientific experiments and application tests have been carried out on Tiangong-2 [2].

The Multi-Azimuth UV Imager (MAVI) is one of the advanced earth observation payloads carried on Tiangong-2, and is a new generation atmospheric sounder [3]. It has the function of quasi synchronous detection of the ultraviolet radiation characteristics of the global middle and upper atmosphere using the large field of view (FOV), and is mainly used for detection of the vertical structure and three-dimensional dynamic distribution of ozone, aerosol and other atmospheric composition on the mesosphere and lower thermosphere atmosphere. It can provide supports for atmospheric scientific research and atmospheric environmental monitoring applications [4].

Y. Gu et al. (Eds.): *Proceedings of the Tiangong-2 Remote Sensing Application Conference*, LNEE 541, pp. 112–122, 2019.
https://doi.org/10.1007/978-981-13-3501-3_11

MAVI carries out multi-azimuth measurements of the earth's atmosphere using six sets of CCD arrays, meanwhile, detects the entire atmosphere in the range of ±5° from the subastral point using one set of CCD array (see Fig. 1). The imaging mode of the MAVI is novel, and is the first time in the world. Therefore, it is different from the conventional optical remote sensing instrument in data processing and quality analysis.

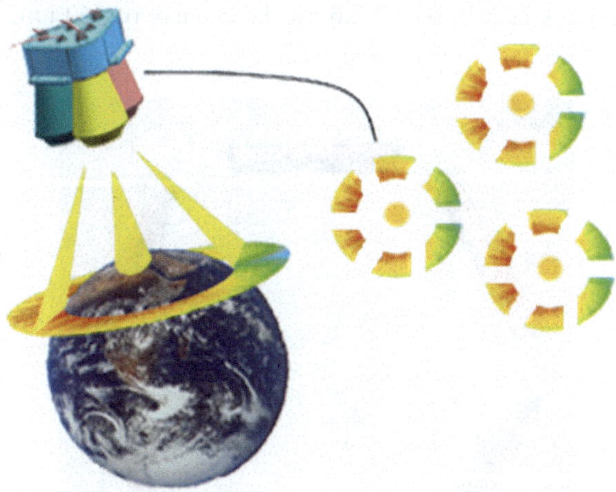

Fig. 1. Schematic diagram of MAVI observation

In this paper, the data characteristics, data preprocessing flow and on-orbit data quality of the MAVI are mainly analyzed and discussed. The results can provide useful reference for further data processing and application of the MAVI.

2 Payload and Data Indicators

MAVI has three channels in the range of ultraviolet spectrum. The center wavelength of each channels are 265 nm, 295 nm and 360 nm respective. The three channels can work together to get observation data synchronously [3]. The technical indicators of MAVI are shown in Table 1.

Table 1. Technical indicators of MAVI [4].

Indicator	MAVI
Number of channels	3
Center wavelength (nm)	265, 295, 360
Bandwidth (nm)	265 nm: 40
	295/360 nm: 20
Field of view (°)	Center FOV: ≥ 10
	Limb FOV: 141.8–146.6
Spatial resolution (km)	3
Limb height range (km)	265/295 nm: 10–80
	360 nm: 10–60

As shown is Fig. 2, the effective area in the middle of the image is the center FOV, which corresponding to the radiance of the whole atmosphere in the range of ±5° from the subastral point. The six effective areas around the center FOV are the limb FOV, which corresponding to the radiation of atmosphere in different directions and different heights. The six directions are respectively front, rear, left front, right front, left rear and right rear. The limb detection height of 265 nm channel and 295 nm channel are from 10 km to 80 km, and which is 10–60 km for the channel of 360 nm.

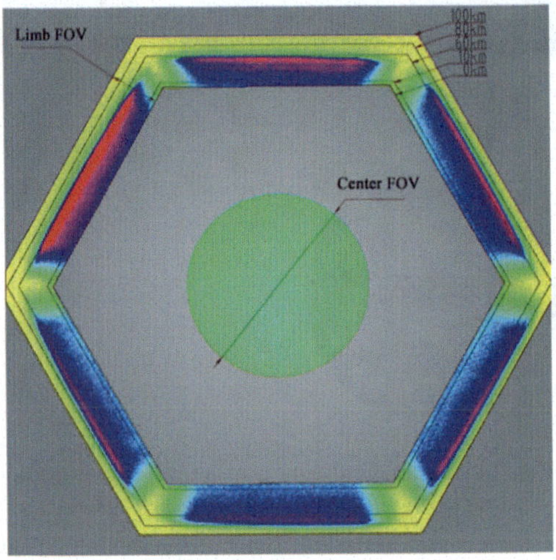

Fig. 2. Schematic diagram of MAVI image

3 Data Preprocessing

After reception of the MAVI original observation data, they will be processed with formatting disassembly and reorganization to produce the level 0 data. Then, the level 0 data will undergo a series of processing, such as data conversion, brightness processing and geometric processing, to generate the radiation data products with geometric information, which named level 1 data products [4].

The specific preprocessing flow is shown in Fig. 3.

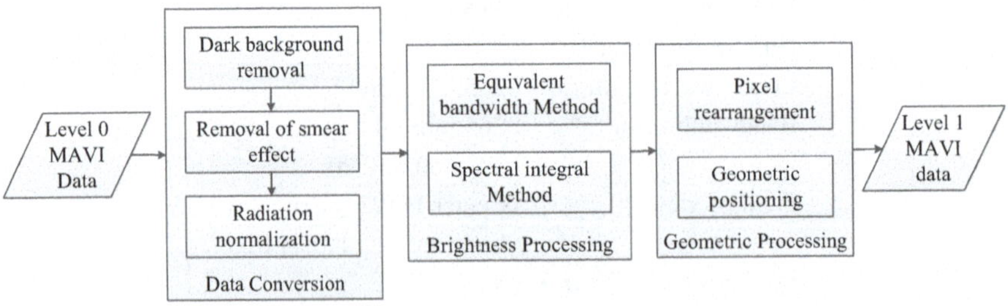

Fig. 3. Data preprocessing flow of MAVI

3.1 Data Conversion

The DN value in level 0 data of MAVI is the response value of each detecting unit. In order to get the valid data of each pixel with physical meaning, it is necessary to convert these data with special methods, including dark background removal, removal of smear effect and radiation normalization.

Dark Background Removal. Dark background is the electronic background noise of CCD under working temperature. It can be removed by subtracting the value of dark reference area from the DN value.

Smear Effect Removal. Smear effect is the trailing phenomenon formed in the charge transfer process of CCD. It will not only affect the data quality, but also bring detection bias into the data. It can be effectively removed by subtracting the value of virtual reference lines from DN value.

Radiation Normalization. Because of the difference of atmosphere radiation in different wavelength, the three channels of MAVI have been set different integrating times and gains. The DN value of each channels need to be normalization processing to gain the comparable results.

3.2 Brightness Processing

According to the calibration parameters of MAVI before launched, the DN value in MAVI images can be converted to the radiation received by instrument. There are two effective methods to calculate the radiation, which are equivalent bandwidth method and spectral integral method. The first method is use to get the radiation value at default wavelength. The second method can obtain the radiation value at specific wavelength.

Equivalent Bandwidth Method. According to the channel bandwidth and responsibility of each pixel, the pixel-by-pixel radiation can be calculated with the following formula.

$$L_{\lambda}(i,j) = \frac{F(i,j)}{Band(i,j) \times R(i,j)} \quad (1)$$

Where $L_{\lambda}(i, j)$ is the radiation of pixel (i, j), $F(i, j)$ is the image DN value, and $Band(i, j)$ are the channel bandwidth and responsibility respectively.

Spectral Integral Method. According to the channel constant, response function of each pixels, and the external relative spectrum of observation object, the pixel-by-pixel radiation can be calculated with the following formula.

$$L_m(i,j,\lambda) = \frac{F(i,j) \times L_R(i,j,\lambda)}{c_1(i,j) \times \int_{\lambda_1}^{\lambda_2} L_R(i,j,\lambda) \times R_r(i,j,\lambda) \times d\lambda} \quad (2)$$

Where $L_m(i, j, \lambda)$ is the radiation of pixel (i, j) at the wavelength of λ, $F(i, j)$ is the image DN value, $L_R(i, j, \lambda)$ is the external relative spectrum, $c_1(i, j)$ is the channel constant, $R_r(i, j, \lambda)$ is the channel relative response function.

3.3 Geometric Processing

Pixel Rearrangement. The six limb FOVs of MAVI corresponding to different observation directions, and the installation directions of the 3 channels are different too. Therefore, it is necessary to rearrange the pixel position so that the pixel distribution of each channel can correspond one by one. The array matrix is determined by the optical system of the instrument and the installation position relationship.

Geometric Positioning. On the basis of the geometric calibration parameters and the platform attitude parameters at the imaging time, the geographic longitude, latitude and the limb height of each pixel are calculated and added into the data product [5].

Image of each channels of MAVI are shown in Fig. 4.

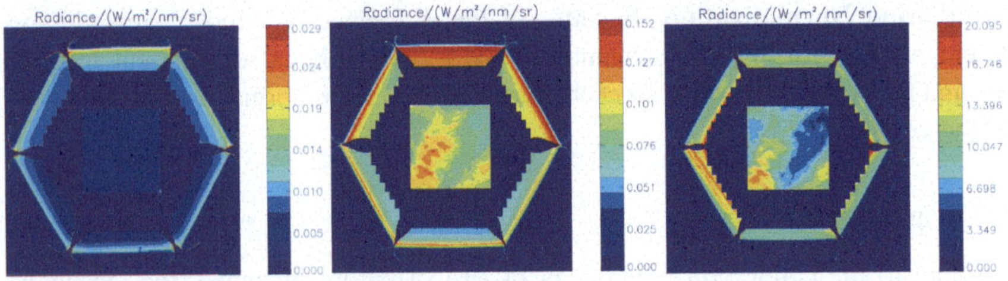

a) Image of 265nm b) Image of 295nm c) Image of 360nm

Fig. 4. MAVI images of each channels.

As is shown in Fig. 4, the ring structure is obvious in all the channels, which reflects the stratification of the atmosphere. The radiant brightness of the three channels images are obvious different. The radiance value of 360 nm channel is the highest, followed by 295 nm channel, and the radiance value of 265 nm channel is the lowest.

4 Data Quality Analysis and Evaluation

Since launched on September 15, 2016, MAVI has been operating steadily on-orbit and has obtained a large number of earth observation data. In order to understand the data quality and its changes on-orbit, several image quality evaluation methods are adopted to analyze the data quality on-orbit. A preliminary quality evaluation about the MAVI data is made based on the analysis results.

4.1 Quality Evaluation Method

The radiation brightness data products of MAVI are quite different from the traditional remote sensing images. Therefore, the quality evaluation methods of the MAVI data products need to be improved.

1) The image obtained by the MAVI is not a regular two-dimensional matrix, but consists of a center FOV and six limb FOVs, and the other regions of the image are filled with invalid values.

2) The physical meaning of DN value of center FOV is different from that of the limb FOVs, and the dynamic range of DN value is also quite different between them, which cannot be measured with the same evaluation standard.
3) Each pixel of the limb FOVs corresponds to different atmospheric limb height, so the radiation distribution has obvious stratification phenomenon, which cannot be measured with the same evaluation standard.

According to the common quality problems in optical remote sensing images, such as noise, bad pixel and response non-uniformity, this paper adopted the indexes of signal-to-noise ratio, information entropy and contrast to evaluate the on-orbit imaging quality of MAVI.

Signal-to-noise Ratio. Signal-to-noise ratio (SNR) is one of the most commonly used indexes in image quality evaluation. It can reflect the radiation quality of the image data by calculating the ratio of the average signal and noise level in the image [6].

SNR evaluation usually needs to find a relatively homogeneous area in image to statistic the noise level. Because the observation objects of MAVI are atmosphere, it is hard to confirm the homogeneous area in images. In the paper, the local mean and local variance method based on pixel statistic is adopted to calculation the SNR of the center FOV of MAVI image data.

$$SNR = 201g(\frac{M}{\sigma}) \tag{3}$$

Where, *SNR* Is the Image Signal-to-Noise Ratio, *M* Is Image Signal Mean, σ Is the Image Noise Standard Deviation.

Information Entropy. Information entropy (ENT) is used to measure the average amount of information in an image. The greater information entropy illustrate the more information contained in the image.

By comparing the information entropy of multi-frame continuous observation images, the bad pixel in images can be detected. In this paper, information entropy index is used to evaluate the image quality both of center FOV and limb FOVs [7].

$$ENT = -\sum_{i=\min}^{\max} p_i \log_2 p_i \tag{4}$$

Where, *ENT* is the information entropy of image, *i* is the gray value after normalized of image pixel, *min* and *max* are the minimum and maximum of gray value after normalized in image, p_i is pixel probability in an image which gray value equal *i*.

Contrast. Contrast reflects the brightness difference degree between target and background in an image. It can be used to evaluate the radiation uniformity of image. Owing to the obvious stratification characteristics of the atmospheric radiation, the radiation value of the limb FOVs of MAVI is symmetrical and with staircase distribution from the inside to the outside. In this paper, the contrast of MAVI images are evaluated with the contrast method based on Gray-level Co-occurrence Matrix (GLCM) [8].

$$CON = \sum_{i=0}^{L-1} \sum_{j=0}^{L-1} (i-j)^2 p(i,j) \tag{5}$$

Where, *CON* is the image contrast, *i and j* are the image gray value after normalized, *L* is the gray levels of normalized image, *p*(*i*, *j*) is the GLCM element located in *(i, j)*.

4.2 Data Quality Analysis

In order to verify the applicability of the evaluation methods mentioned above, the level 1 data product of 265 nm channel acquired on October 7, 2018 by MAVI is used as experimental data. The experimental data contains 254 continuously observed images. There is a certain correlation between adjacent image frames. The calculation results of each evaluation method are shown in Figs. 5, 6 and 7.

As shown in Figs. 5, 6 and 7, the abscissa indicates the different image frames, and the ordinate indicates the evaluation results of each frame. The evaluation results of SNR (see Fig. 5), entropy (see Fig. 6) and contrast (see Fig. 7) are consistent for different image frames.

In order to realize the imaging performance and data quality of MAVI during the period of on-orbit, the images obtained on different dates from October 2016 to October 2018 were chosen as experimental data. The results of each index are show in Figs. 8, 9 and 10.

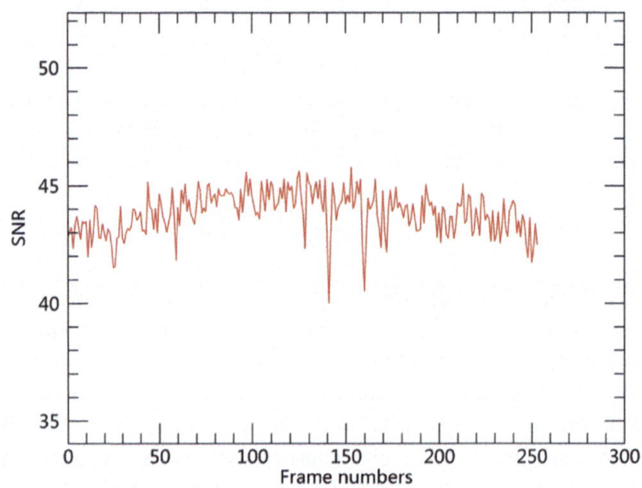

Fig. 5. The SNR of 265 nm channel image obtained on October 7, 2018.

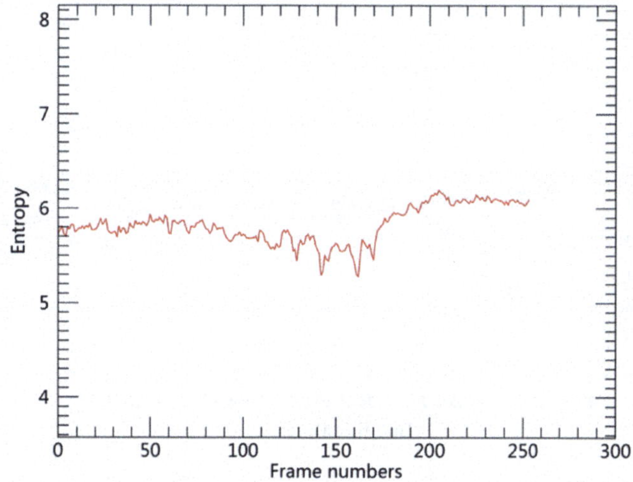

Fig. 6. The entropy of 265 nm channel image obtained on October 7, 2018

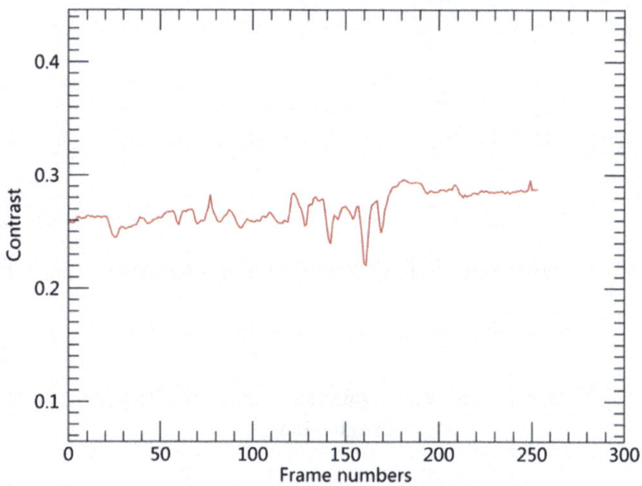

Fig. 7. The contrast of 265 nm channel image obtained on October 7, 2018.

The SNR (see Fig. 8) are very close between each channels, and there is no obvious change on different dates from October 2016 to October 2018. It can be seen that the noise level of each channel of MAVI has not increased during the period of on-orbit.

The information entropy of each channel (see Fig. 9) has not obvious change from October 2016 to October 2018. The information entropy of 295 nm channel is larger than that of 265 nm channel and 360 nm channel, which illustrates that the 295 nm channel contains more atmosphere information than the other two channels.

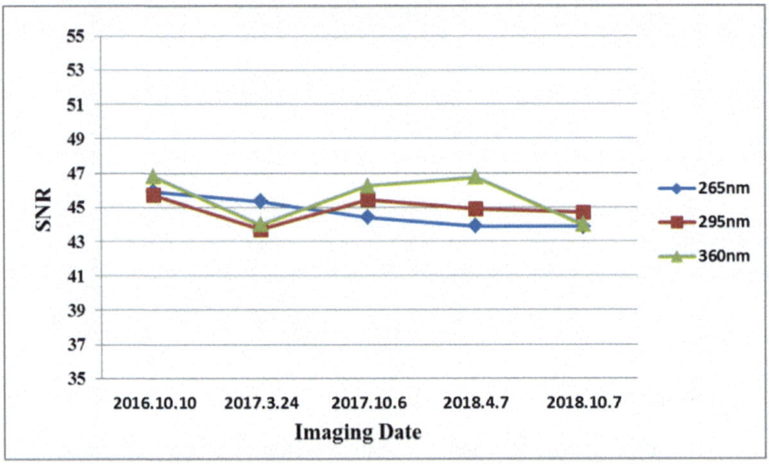

Fig. 8. The SNR evaluation results of MAVI data on different dates.

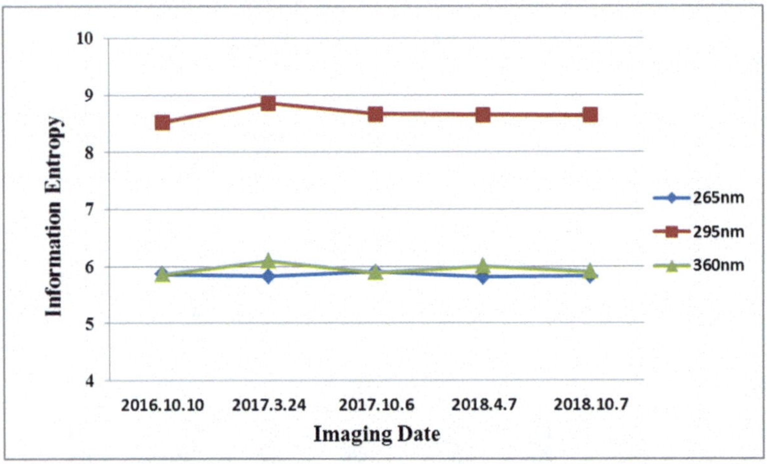

Fig. 9. The information entropy evaluation results of MAVI data on different dates.

As shown in Fig. 10, the contrast between different channels is quite different, but the contrast of all channels is relatively stable from October 2016 to October 2018, which illustrate that the data quality of MAVI has not obviously change.

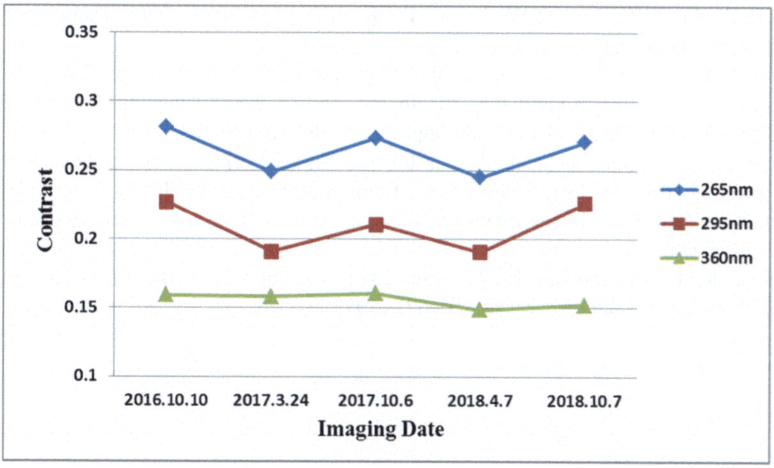

Fig. 10. The contrast evaluation results of MAVI data on different dates.

5 Conclusion

MAVI is a new atmospheric detector carried on Tiangong-2 Space Laboratory. The unique imaging mode makes it quite different from previous atmospheric detector in data processing and quality analysis. The level 0 data of MAVI needs to be preprocessed with several steps, including data conversion, radiation processing and geometric processing, to produce the level 1 data product. Based on pre-processing, a number of efficient data products of MAVI have been produced. By data quality analysis, the quality of these products has shown a stable trend since MAVI was launched, which illustrates that the imaging performance of MAVI is reliable. In the future, MAVI will continue obtaining earth observation data and providing effective data supports for atmospheric scientific research and atmospheric environmental monitoring application.

Acknowledgments. We would like to thank China Manned Space Engineering for providing data products of MAVI. Thank the National R&D Infrastructure and Facility Development Program of China, "Fundamental Science Data Sharing Platform" (DKA2018-12-02-23) who funded the project.

References

1. Wei, J.H.: China's first space laboratory Tiangong 2 was successfully launched. China Aerosp. **10**, 9 (2016)
2. Manned Space Space Application Data Promotion Service Platform. http://www.msadc.cn/en/gk/t2task/. Accessed 21 Sept 2018
3. Yang, X.H., Wang, S.R., Huang, Y., et al.: Pre-launch calibration of the multi-azimuth UV imager on Tiangong-2. In: Proceedings of the Tiangong-2 Remote Sensing Application, Technology, method and application, Beijing (2018)

4. Qin, B.Y., Li, S.Y., Hao, Z.W., et al.: Earth observation data of the Tiangong-2 Space Laboratory (2016–2017). China Sci. Data **3**(1) (2018)
5. Zhang, W.F., Liu, Z.W., Li, S.Y., et al.: Image geometric correction parallelization of the multi-azimuth UV imager based on GPU. In: Proceedings of the Tiangong-2 Remote Sensing Application, Technology, method and application, Beijing (2018)
6. Gao, B.C.: An operational method for estimating signal to noise ratios from data acquired with imaging spectrometers. Remote Sens. Environ. **43**(1), 23–33 (1993)
7. Zhou, Y.J., Tian, Q.J.: Image quality evaluation of EO-1 hyperion sensor. Geo-Inf. Sci. **5**, 678–683 (2008)
8. Haralick, R.M., Shanmugam, K., Dinstein, I.H.: Textural features for image classification. IEEE Trans. Syst. Man Cybern. **6**, 610–621 (1973)

Image Geometric Correction Parallelization of the Multi-azimuth UV Imager Based on GPU

Wanfeng Zhang(✉), Zhiwen Liu, Shengyang Li, Yuyang Shao, and Zhuang Zhou

Key Laboratory of Space Utilization, Technology and Engineering Center for Space Utilization, Chinese Academy of Sciences, Beijing, People's Republic of China
{wfzhang, zwliu, shyli, shaoyy, zhouzhuang}@csu.ac.cn

Abstract. The Multi-Azimuth UV Imager (MAVI) is designed and built by Chinese scientific institute independently. The MAVI is comprised of the ultraviolet ring imager and the ultraviolet forward imager. One track of image of ultraviolet ring imager contains hundreds of standard image frames arranged along the track direction and possesses relatively higher data volume. To produce the standard level-one product rapidly, the entire procedure of geometric correction was decomposed into parameter parsing, data reading into memory, model of geometric correction, and data exporting to hard disk. Among these four steps of ultraviolet ring imager image processing, the construction and calculating of geometric correction model is the most time consuming. There exist several for loops in steps of geometric positioning and vector calculating. Simultaneously, the majority of C code of geometric positioning and vector calculating should be rewritten to adapt the CUDA GPU in this practice. Experimental results show that parallelized geometric correction could reach a speedup of 3.2 through image partition along the column direction on GPU. A better speedup result of URI image processing would be achieved through parallelizing all for loops in every step.

Keywords: Tiangong-2 · Ultraviolet ring imager · GPU · Geometric correction Vector calculating

1 Introduction

Tiangong-2 space laboratory is a manned spaceflight vehicle for carrying out space experimental activities. It is the prototype of China Space Station [1]. The Multi-Azimuth UV Imager (MAVI) is one of the three earth observation payloads. The ultraviolet ring imager (URI) and the ultraviolet forward imager (UFI) constitute an integrated MAVI. Remote Sensing image taken by the MAVI would be applied to monitor atmospheric environmental dynamic changes between 10 and 60 km from the ground, such as PM 2.5, PM 10, ozone, and sulfur dioxide. This type of MAVI image

Y. Gu et al. (Eds.): *Proceedings of the Tiangong-2 Remote Sensing Application Conference*, LNEE 541, pp. 123–129, 2019.
https://doi.org/10.1007/978-981-13-3501-3_12

is valuable for image fusion from different sources [2]. The Ultraviolet Limb Imager had been applied to observe atmosphere for acquire the ionospheric and neutral density profiler in world [3, 4]. Like ordinary remote sensing data centres existing in at present [5, 6], there would be space science data centre to process and manage the MAVI product.

The geometric correction of the MAVI image is the most time-consuming step among these processes [7]. While timely processing of standardizing MAVI products is a key requirement for retrieving atmospheric parameters aimed at the product of URI and UFI image. Therefore, parallel processing using one type of accelerating facility such as Graphics Processing Unit (GPU) is implemented for many kinds of remote sensing image processing [8–11].

2 Related Work and Background

The MAVI was assembled on flying platform of Tiangong-II space laboratory. Therefore, this spectrometer would be flying along a non-sun synchronous orbit. The observation range of MAVI covers between 42 N and 42S.

The MAVI would take photographing from the tangential direction and sub-satellite point when the Tiangong-2 space laboratory is flying along the orbit, as illustrated in Fig. 1. So the two kinds of ultraviolet limb image could be obtained simultaneously. Imaging interval of the MAVI is about one second. So one track of URI and UFI images were usually comprised of hundreds of frames and scanned hundreds of seconds towards the tangential direction or sub-satellite point.

(1) The UFI image obtained by photographing from the coordinates of Tangent Points
(2) The URI image obtained by photographing from the coordinates of Central Points

Similar to other types of remote sensing image, the MAVI image had the characteristic of image frames regular arrangement which is convenient for parallel process on

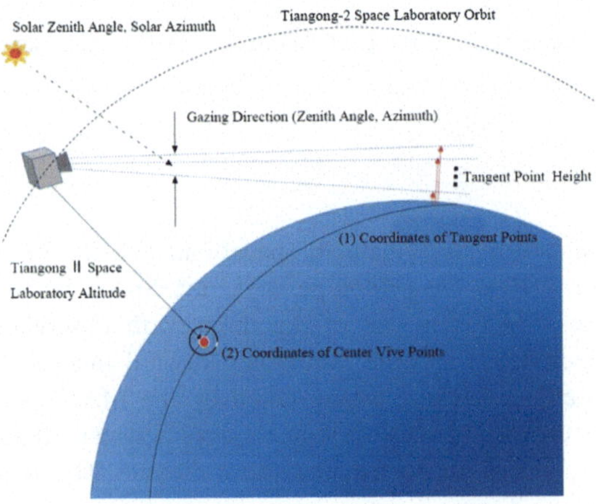

Fig. 1. Field of the MAVI on Tiangong-2 space laboratory.

GPU. The processing efficiency of GPU such as remote sensing image registration and geometric correction were growing with a speed beyond the Moore law [12–14].

Taken by the MAVI within one time interval, hundreds of standard image frames (1024 pixels × 1024 pixels) would be arranged in accordance with the order of column. So the MAVI data volume within one time interval would reach several gigabytes in general. Multi-core computing mode on GPU would be appropriate for parallel processing so many standard frames respectively. As mentioned above, the parallel processing of the MAVI image had not been carried out.

3 Problem Definition

3.1 The Arrangement of URI Image

The URI image consists of hundreds of standard image frames which are arranged along the column direction. All the standard images possess uniform width and height. So many smaller image frames formed a very big image. We call every standard image as one frame. The amount of the one track of URI image with about 250 frames will reach one gigabyte approximately. Furthermore, because the scanning time of the MAVI is variable, the quantity of URI image track frames would be likely to exceed 500 or more.

The arrangement of URI image frames is corresponding to time interval. Once the URI begin to scan at one time point, a series of frames would be photographed in accordance with the column direction per second. This feature makes the parallelization of URI image processing on GPU more convenient.

3.2 The Processing Flow of URI Image

The URI image processing comprised of parameter parsing, data reading into memory, geometric correction, and data exporting to hard disk. These four steps are indispensable for producing a standard URI product.

Parameter parsing is used to get engineering parameters and telemetry parameters from auxiliary files. URI image reading into memory was invoked by CPU to load image frames into memory for processing every frame at a high rate of speed. Among these four steps, geometric correction is the most time consuming because this process involved with vector transformation for calculating the longitude and the latitude of every pixel. So geometric correction should be parallelized on GPU to reduce the processing time greatly.

In order to prove the time consuming of URI image geometric correction, we take one track of URI image with 253 frames as an example. The experimental results are as shown in Table 1. The total processing time of one track of URI image would reach up to a quarter approximately. Actually, the geometric correction takes up over 98% of total time.

Table 1. Time consuming of URI image processing steps.

No.	Step of data processing	Time consuming(Second)
1	Parameter parsing	0.064
2	Data reading into memory	12.44
3	Geometric correction	**823.94**
4	Data exporting to hard disk	0.004
	Total	**826.448**

As shown in Table 1, the parallelization of URI image geometric correction based on GPU has become an important way of shortening the consuming time of URI image processing. It takes time to make geometric correction is nearly seventy times than the time consuming of the other three steps. Therefore, the time consuming of the other three steps could be negligible in this paper.

4 The Parallel Partition and Implement and Implementation of URI Image

4.1 Steps of Geometric Correction Algorithm

We have found that several steps were comprised of some for loops which are the most time consuming in the course of URI image rectifying. As a consequence, it is the key issue that how to implement this parallelization of this for loops during geometric correction.

In order to elaborate the geometric correction procedure at length, we have decomposed the algorithm into nine steps and located the two time-consuming steps involved with for loops. Through the matrix manipulation types of step6 and step8, we can rewrite the geometric correction code to run on CUDA GPU.

Among the entire processing procedures, we could divide them into URI image processing on CPU and *for loops* parallelization of URI image on GPU, as shown in Fig. 2. Specifically speaking, step 6 and step 8 were the most time-consuming steps which involved with two *for loops* respectively. So the parallelization of URI image is mainly concerned with vector calculating of *for loops*.

4.2 The Parallel Partition of URI Image on GPU

In the step 6, the time consuming of formula (1) was brought about by the matrix multiplication. Furthermore, different URI frames along the column direction could be divided to process concurrently. In order to allocate the URI image to workgroup on GPU conveniently, we choose one standard image frame onto a workgroup for parallel processing. One track of URI image could be divided into many work-index along the column direction. Meanwhile, GPU processors are subdivided into multiple work-groups accordingly.

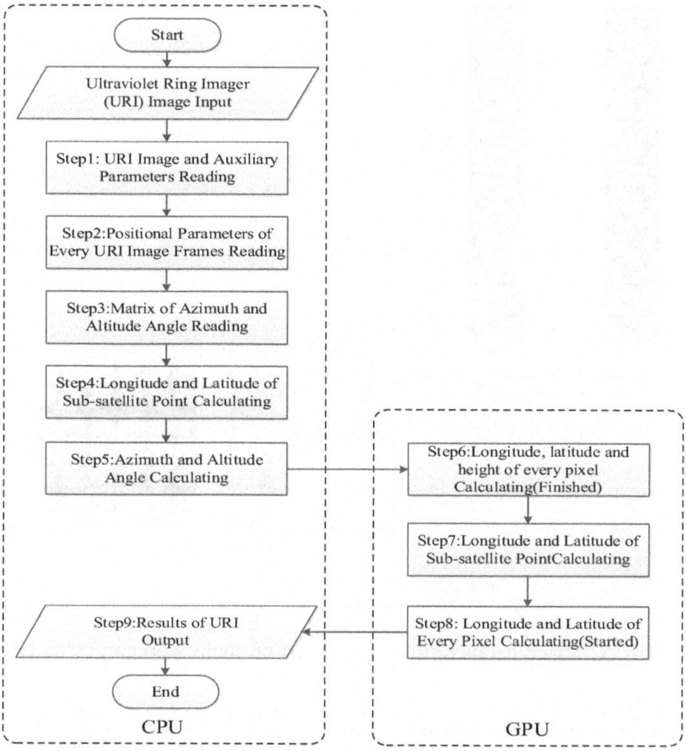

Fig. 2. Geometric correction of URI image on CPU and GPU.

4.3 Parallel Processing of URI Image Based on Workgroup

One workgroup (usually called block) is composed of many GPU processor units to improve computation efficiency. Within one workgroup, the whole cores would transmit communications information by shared memory. So the global variable within one frame of URI image such as auxiliary parameters, positional parameters, and the matrix of azimuth and altitude angle could be transferred from one work-index to another by the high bandwidth.

Under normal circumstances, the quantity of URI image frames are more than the GPU cores. So more than one work-index could be submitted to a workgroup to realize parallel geometric correction. When quite some URI image frames would be executed on one workgroup, a scheduling module could decide on what is the appropriate execution sequence on the basis of the state of thread. We are going to select the default scheduling strategy provided by the GPU in this paper.

One workgroup (block) consists of multiple threads, and there would be three steps to be parallelized on each thread. All parts processed by each thread could be exported to the global shared memory because one frame of URI image is integrity. One thread would be used to execute the three steps of one frame, as is shown in Fig. 3.

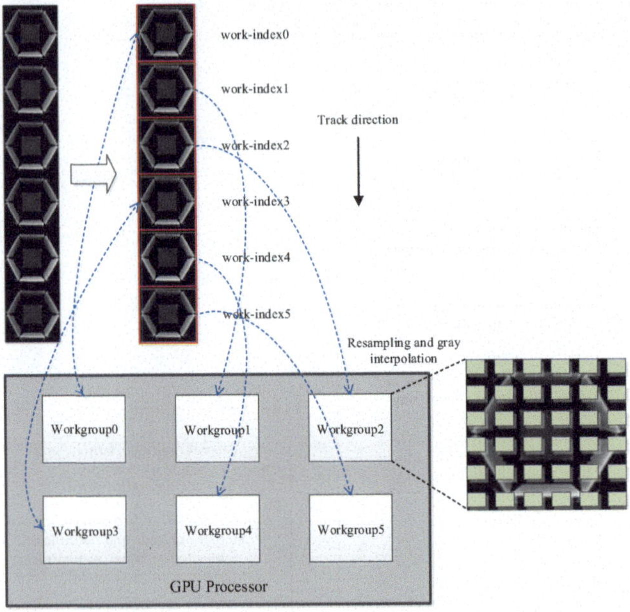

Fig. 3. URI image parallel processing based on multi-workgroups on GPU.

5 Experiments and Results

To test parallel performance of URI image processing on GPU, the CPU and GPU were the two ways of different frames URI image experiment. Table 2 shows the results of two ways of experiment using CPU and GPU while using different URI image frames. The amount of URI image frames could be determined by a clipping programme to meet the testing requirements of various quantities. In the actual URI image processing system, the clipping programme is also a parallelized executable module because some parts of URI image is sufficient for atmospheric research.

Table 2. URI image processing experimental.

URI image frames	10	20	40	100	200	300
CPU(s)	34	68	132	328	653	987
GPU(s)	16	31	60	145	237	316
Speed up	2.125	2.19	2.2	2.26	2.75	3.12

Experimental results show that the speedup of the URI image processing time is gradually increasing on GPU along with the image frames number becoming greater. Especially, when the number of the URI image frames reached 100 or more, the efficiency of URI image parallel processing had been significantly improved. Because in the process of URI image geometric positioning, vector calculating was one of the key problems in this practice. Thus, the efficiency of parallelized image processing is not necessarily ideal and optimized. Practice has proved that geometric correction of URI image was suitable for workgroup partition on GPU.

Acknowledgements. This research was supported by the National Natural Science Foundation of China (Grant No. 41701468), and the CAS Key Laboratory of Lunar and Deep Space Exploration (Grant No. 16120452). Thanks to China Manned Space Engineering for providing space science and application data products of Tiangong-2.

References

1. China Manned Space. http://en.cmse.gov.cn. Accessed 14 Apr 2017
2. Wei, J., Wang, L., Liu, P., Song, W.: Spatiotemporal fusion of remote sensing images with structural sparsity and semi-coupled dictionary learning. Remote Sens. **9**(1), 21 (2017)
3. McCoy, R.P., et al.: Special sensor ultraviolet limb imager: an ionospheric and neutral density profiler for the Defense meteorological satellite program satellites. Opt. Eng. **33**(2), 423–429 (1994)
4. Sandel, B.R., Broadfoot, A.L., Curtis, C.C., et al.: The extreme ultraviolet imager investigation for the image mission. Space Sci. Rev. **91**(1, 2), 197–242 (2000)
5. Zhang, Wanfeng, Wang, Lizhe, Ma, Yan, Liu, Dingsheng: Design and implementation of task scheduling strategies for massive remote sensing data processing across multiple data centers. Softw. Pract. Exper. **44**(7), 873–886 (2014)
6. Ranjan, R., Wang, L., Zomaya, A.Y., Tao, J., Jayaraman, P.P.: Dimitrios Georgakopoulos.: advances in methods and techniques for processing streaming big data in datacentre clouds. IEEE Trans. Emerg. Topics Comput. **4**(2), 262–265 (2016)
7. Jiang, Y., Yang, X., Yi, H.: Parallel algorithm of geometrical correction for satellite image. Chinese J. Comput. **27**(7), 944–951 (2004)
8. Chen, D., Li, D., Xiong, M., Bao, H., Li, X.: GPU-aided ensemble empirical mode decomposition for EEG analysis during anesthesia. IEEE Trans. Inf Technol. Biomed. **14**, 1417–1427 (2010)
9. Harish, P., Narayanan, P.J.: Accelerating large graph algorithms on the GPU Using CUDA. In: Aluru, S., Parashar, M., Badrinath, R., Prasanna, V.K. (eds.) High Performance Computing—HiPC 2007. HiPC 2007. Lecture Notes in Computer Science, vol 4873. Springer, Berlin, Heidelberg (2007)
10. Belleman, R.G., Bédorf, J., Portegies Zwart, S.F.: High performance direct gravitational N-body simulations on graphics processing units II: an implementation in CUDA. New Astron. **13**, 103–112 (2008)
11. Kniaz, V.V.: Real-time optical flow estimation on a GPU for a skied-steered mobile robot. In: Proceedings SPIE 9897, Real-Time Image and Video Processing 2016, 989706 (2016/04/29)
12. Owens, J., Luebke, D., Govindaraju, N., et al.: A survey of general-purpose computation on graphics hardware. Comput. Graphics Forum **26**(1), 80–113 (2007)
13. Deren, L., Shugen, W., Yueqin, Z.: An Introduction to Photogrammetry and Remote Sensing. 2nd ed. Surveying and Mapping Press, Beijing (2008)
14. Fengshun, Lu, Junqiang, Song, Fukang, Yin, et al.: Survey of CPU/GPU synergetic parallel computing. Comput. Sci. **38**(3), 5–10 (2011)

Cross-Calibration of S-NPP/VIIRS and Tiangong-2/MAI Visible Channels Using the SNO Method

Jun Jiang[1], Zhigang Han[1(✉)], Zhigang Yao[1], Zengliang Zhao[1], and Junjie Guo[1,2]

[1] Beijing Institute of Applied Meteorology, Beijing 100029, China
hanzg@pku.org.cn

[2] College of Meteorology and Oceanology, National University of Defense Technology, Nanjing 211101, China

Abstract. The space laboratory "Tiangong-2" (TG-2) was launched on 15 September 2016. The Multi-Angle Polarization Imager (MAI), onboard TG-2, is the first visible and near-infrared multi-angle polarization imager developed in China. MAI can accomplish polarization observation from up to 14 different viewing angles, which is of great significance for relevant research on the microphysical characteristics of clouds and aerosols. Aimed at the problem of a lack of an onboard calibration device for MAI, to correct TG-2's calibration coefficients in time, the Simultaneous Nadir Overpass (SNO) cross-calibration method is proposed as a reasonable and feasible method for MAI 565- and 670-nm channels. Data from the M4 and M5 channels of the Visible Infrared Imaging Radiometer Suite (VIIRS), with good calibration accuracy, are used as reference data, and a series of matching conditions and a spectral adjustment algorithm are specified. To apply the SNO method, eight groups of examples from December 2016 to February 2017 are selected, and a large number of matching pixel samples covering land, ocean surface, and cloud layers are obtained. The resulting cross-calibration curves between MAI 565-nm and VIIRS M4 channels, and between MAI 670-nm and VIIRS M5 channels, show correlation coefficients of reflectivity of 0.986 and 0.994, mean biases of 2.48% and 5.90%, and RMSEs of 3.56% and 6.95%, respectively. Overall, the cross-calibration method achieved good results, which are of reference significance to correct the laboratory calibration coefficient of MAI. The proposed method effectively solves the problems of MAI's subsequent on-orbit monitoring and vicarious calibration, and also lays a foundation for later applications of MAI data in research on cloud and aerosol microphysical parameters.

Keywords: Tiangong-2 · Cross-calibration · SNO method · MAI VIIRS · Visible channels

Y. Gu et al. (Eds.): *Proceedings of the Tiangong-2 Remote Sensing Application Conference*, LNEE 541, pp. 130–143, 2019.
https://doi.org/10.1007/978-981-13-3501-3_13

1 Introduction

The space laboratory "Tiangong-2" (TG-2), which was successfully launched on 15 September 2016, is China's first facility of this kind. A total of 14 application loads are carried by TG-2, among which the Multi-Angle Polarization Imager (MAI) can detect the polarized light scattered by the earth–atmosphere system, and provide multi-angle (up to 14 angels) information for the same pixel. This high-quality angle and polarization information is of great significance for relevant research on the microphysical characteristics of clouds and aerosols.

Radiometric calibration of satellite remote sensors is a prerequisite for the quantitative application of observational data, and must run throughout the entire lifetime of the instrument. MAI underwent laboratory-based absolute calibration before its launch [1]. However, it is difficult for laboratory-based calibration to mimic a satellite's on-orbit operational environment. This is affected by many factors and, with respect to MAI specifically, this instrument's spectral response characteristics will inevitably decay, resulting in changes in the radiometric calibration relationship. Considering that MAI is not equipped with an onboard calibration device, in order to guarantee its expected scientific and application objectives, it is necessary to adopt vicarious calibration methods to carry out continuous calibration and data verification during its onboard operation. Compared with field-based calibration, cross-calibration, as a commonly used vicarious calibration method, has the advantages of low cost, rich scene types of targets, high frequency, and achieving radiometric calibration between multiple remote sensors [2–5]. Because of the advantages of cross-calibration, several experts and scholars have carried out a series of cross-calibration studies on different satellite remote sensors in visible channels, and have produced relatively good results [6–11]. Andrew et al. [6] used Aqua/MODIS to cross-calibrate the NOAA-9/AVHRR 640-nm channel with the Libyan Desert as the target, and the results showed that the observed reflectance deviation was less than 5%. Thome et al. [7] selected the Railroad Valley Playa as the target scene, and compared the observational data of each channel of Earth resource satellite loads such as EO-1/ALI, EO-1/Hyperion, Landsat-7/ETM+, and Ikonos/OSA with Aqua/MODIS. Their results showed that, except for OSA, whose observed reflectance deviation was greater than 7% in the 565-nm and 665-nm channels, the deviations from visible channels of other sensors were all less than 5%. Rao et al. [8] used ERS-2/ASTR-2, with high calibration accuracy (within 2%), as a standard instrument to calibrate Terra/MODIS. The comparison results of observational data from the Libyan and Sonoran deserts showed that the deviations of the 565- and 645-nm channels of MODIS were within 1% and 3% respectively. Meirink et al. [9] carried out cross-calibration analysis for the geostationary satellite load MSG/SEVIRI using Aqua/MODIS as a standard, and performed sensitivity experiments on the selection conditions of matched pixel sets. The results showed that the observed reflectance deviation was less than 8% in the 600-nm channel, and the uncertainty introduced by the calibration method itself was less than 1%. Chen et al. [10] applied the cross-calibration method to visible channels of FY-3C/VIRR, FY-3C/MERSI and Aqua/MODIS. They concluded that, when the sample target scenes were ice, snow, clouds and seawater, the relative deviations of the remaining visible channels were all less than 2%, except for the 505-nm channel of VIRR and the 551-, 678- and 748-nm channels of MERSI, whose relative deviations with MODIS were greater than 5%. Yu et al. [11] used S-NPP/VIIRS to analyze Himawari-8/AHI, and the results showed that the observed reflectance deviations of visible channels in both seawater and deep convective cloud targets were within 5%.

Aimed at the problem of a lack of an onboard calibration device, this paper proposes the Simultaneous Nadir Overpass (SNO) cross-calibration method for the 565- and 670-nm visible channels of TG-2/MAI, based on S-NPP/VIIRS data. The structure of the paper is as follows: Sect. 2 introduces the basic situation regarding MAI and the reference remote sensor, VIIRS; Sect. 3 describes the implementation process of the SNO cross-calibration method; Sect. 4 concludes the calibration results; and Sect. 5 summarizes the paper and further discusses the cross-calibration error sources.

2 Data

2.1 MAI

MAI is carried by the TG-2 space laboratory, with an orbital height of 350 km, a swath of about 770 km, a viewing field of 87.6° in both the along- and cross-track directions, and a spatial resolution of about 3 km × 3 km. The remote sensor is a multi-angle polarization imager in the visible and near-infrared regions, which can be used for analysis of the global radiation budget, cloud macro- and microphysical properties, aerosols, cloud-top pressure, and water vapor. A total of six observation channels are set in MAI, three of which are the 565-, 670- and 865-nm polarized channels, and the other three are the 763-, 765- and 910-nm unpolarized channels. When the satellite passes over a certain target, MAI can observe it from 14 different directions at most, from the target entering the viewing field of the remote sensor to it finally disappearing.

Before the launch of TG-2, laboratory-based absolute radiometric calibration established the relationship between the count value and the radiance of the integrating sphere. In addition, at the end of autumn 2015, four 17-h airborne observation tests of MAI were carried out in the east sea area of Chengshantou. The aviation flight data under clear-sky and cloud scenes and corresponding ground coordination test data were obtained, and each performance index of MAI was verified and analyzed [12]. Overall, the indexes of MAI were good and the data achieved a high level of accuracy before launch.

2.2 VIIRS

The S-NPP satellite was launched on 28 October 2011, with an orbital altitude of 834 km and an ascending node time of 1:30 PM. Combining the characteristics of NOAA/AVHRR, EOS/MODIS and DMSP/OLS, VIIRS is one of the main remote sensors onboard the satellite. It is the most comprehensive scanning imaging radiometer in the world, and can be used to measure cloud cover, ocean water color, aerosol characteristics, sea-ice movement, ocean and land surface temperatures, fires, Earth surface albedo, and so on. The scanning angle of VIIRS is about ±56°, the swath width is about 3000 km, and the spatial horizontal resolutions are better than 400 m and 800 m at the nadir point and the scanning edge, respectively. VIIRS has a total of 22 channels in the range of 0.4–12 μm, nine of which are located in the visible/near-infrared (0.4–0.9 μm), eight in the short/medium-wave infrared (1–4 μm), four in the thermal infrared (8–12 μm), and one morning faint/night visible channel in low light conditions [13].

VIIRS is a follow-on instrument to MODIS and has many similar design features as its predecessor, including their onboard calibrators. As a result, it is an ideal candidate to replace MODIS and to serve as the future GSICS (Global Space-based Inter-Calibration System) reference sensor. Since its launch, S-NPP/VIIRS has already been operational for more than seven years, and its overall performance has been extensively characterized and demonstrated to meet its overall design requirements [14–16]. Lei et al. [17] used 950 days of data observed by the Sunlit Solar Diffuser Panel onboard S-NPP to calibrate the visible/near-infrared channels of VIIRS, and the results showed that the relative standard deviation between the observed reflectance and true value in the visible channels was within 1.3%. Wolfe et al. [18] analyzed and calibrated the geolocation of VIRRS before and during its on-orbit operation, which proved that the geolocation accuracy of VIIRS is comparable to MODIS, and is expected to be further improved. Considering the good overall performance of VIIRS, this study employs VIIRS as a reference remote sensor to perform calibration analysis on the visible channels of MAI. The similarity of the spectral center wavelength and spectral width between VIIRS and MAI is the main principle for selecting the matching channels. The visible channels of VIIRS and MAI, along with the selection of calibration channels, is shown in Table 1.

Table 1. The visible channels of VIIRS and MAI, and the selection of calibration channels.

Central wavelength (nm)	Spectral width of MAI (nm)	Horizontal resolution of nadir (km)	Central wavelength (nm)	Spectral width of VIIRS (nm)	Horizontal resolution of nadir (km)
–	–	–	M1 (412)	20	0.742
–	–	–	M2 (445)	20	0.742
–	–	–	M3 (488)	20	0.742
B1 (565)	20	3	M4 (555)	20	0.742
B2 (670)	20	3	M5 (672)	20	0.742
–	–	–	M6 (746)	15	0.742

3 Method

In principle, the SNO cross-calibration method needs to ensure that the two remote sensors observe the same uniform target at the same time and under the same angles. The cross-calibration process is shown in Fig. 1.

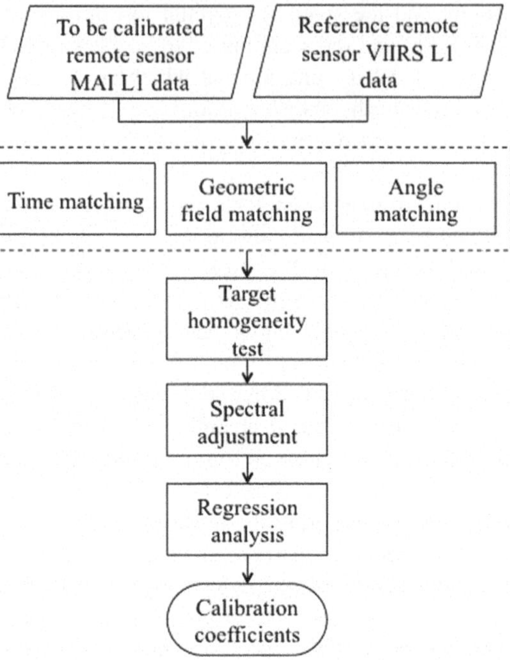

Fig. 1. Flow chart of the SNO cross-calibration method.

3.1 Time Matching

In order to ensure that the ground solar irradiance, atmospheric conditions, and cloud characteristics during the observation times of MAI and VIIRS are almost identical, the time matching condition of the cross-calibration is set as the time difference of simultaneous nadir overpass, $\Delta T < 15$ min. The orbit prediction software SGP4 is used to forecast the orbital intersections of TG-2 and NPP from December 2016 to February 2017 [19], and eight groups of cross-observation examples can therefore be screened. The concrete space–time matching situations are shown in Table 2 and Fig. 2 (using the 565-nm channel of MAI and the M4 channel of VIIRS as an example).

3.2 Geometric Field Matching

The nadir spatial resolution of MAI's visible channel is about 3 km $\times$ 3 km, and that of VIIRS is about 0.75 km $\times$ 0.75 km. In order to achieve geometrical field matching of an observational target, an MAI pixel is used as a basic pixel, and then the spatial distance D between a VIIRS pixel and a basic pixel can be calculated. The cross-calibration geometric field matching condition is set as $D < 2$ km. The matching pixel of VIIRS is taken as the central pixel, and the surrounding 5 $\times$ 5-pixel range is used as the target area matching the MAI basic pixel.

Table 2. Space–time cross-calibration matching examples.

Number	MAI observation time (UTC)	Latitude (°)	Longitude (°)	ΔT (s)	Target scene types
1	2016.12.22 15:24:17	−27.50	−26.53	312	Sea surface
2	2017.01.22 09:24:04	−0.91	63.93	852	Sea surface
3	2017.01.27 07:33:52	−30.17	94.76	451	Sea surface
4	2017.02.03 20:31:42	−42.90	−96.32	528	Sea surface, cloud layer
5	2017.02.10 16:35:04	−31.76	−41.16	225	Sea surface, cloud layer
6	2017.02.20 06:37:51	12.99	97.31	587	Sea surface, land
7	2017.02.21 11:47:41	22.67	23.63	711	Land
8	2017.02.24 05:43:20	32.21	111.49	384	Land, cloud layer

(*Note* "latitude" and "longitude" refer to the latitude and longitude of the nadir at the center of the intersection of MAI and VIIRS images).

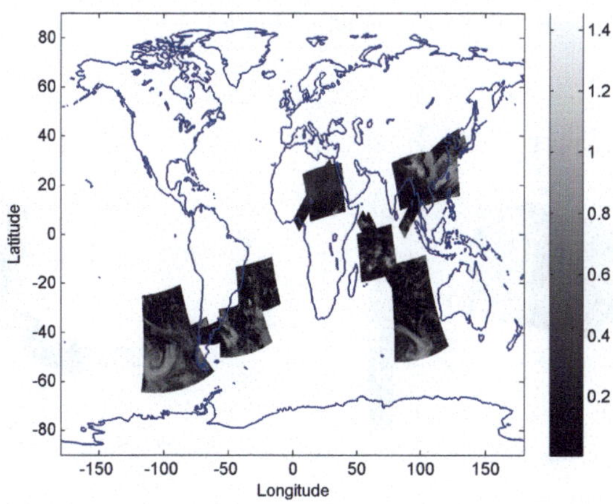

Fig. 2. Sketch map of the space-time cross-calibration matching.

3.3 Angle Matching

To reduce the cross-calibration error caused by the different angles, the pixel matching pairs of MAI and VIIRS should be principally limited to have the same angles. However, it is difficult to meet this condition in operation. The conditions of angle matching are set as follows:

$$\theta_{S\_VIIRS} < 50^{\circ} \ \& \ \theta_{S\_MAI} < 50^{\circ} \tag{1}$$

$$|\theta_{S\_MAI} - \theta_{S\_VIIRS}| < 4^{\circ} \tag{2}$$

$$\theta_{V\_VIIRS} < 30^\circ \ \&\ \theta_{V\_MAI} < 30^\circ \quad (3)$$

$$\left| \frac{\cos \theta_{V\_MAI}}{\cos \theta_{V\_VIIRS}} - 1 \right| < 0.01 \quad (4)$$

$$\left| \varphi_{V\_MAI} - \varphi_{V\_VIIRS} \right| < 10^\circ \quad (5)$$

where θ_{S_MAI}, θ_{V_MAI} and φ_{V_MAI} represent the solar zenith angle, the viewing zenith angle, and the relative azimuth angle of MAI, respectively; and θ_{S_VIIRS}, θ_{V_VIIRS} and φ_{V_VIIRS} represent those of VIIRS. It can effectively reduce the influence of the bidirectional reflection distribution function (BRDF) on the cross-calibration results by setting the solar zenith angle of MAI and VIIRS to be less than 50° and the viewing zenith angle to be less than 30° [7, 20, 21]. Setting $\left| \frac{\cos \theta_{V\_MAI}}{\cos \theta_{V\_VIIRS}} - 1 \right| < 0.01$ and $\left| \varphi_{V\_MAI} - \varphi_{V\_VIIRS} \right| < 10^\circ$ can ensure the viewing angles of MAI and VIIRS are fairly close. Figure 3 illustrates cross-calibration examples of three typical targets (land, sea surface, and cloud layer) after angle matching (using the 565-nm channel of MAI and the M4 channel of VIIRS as an example).

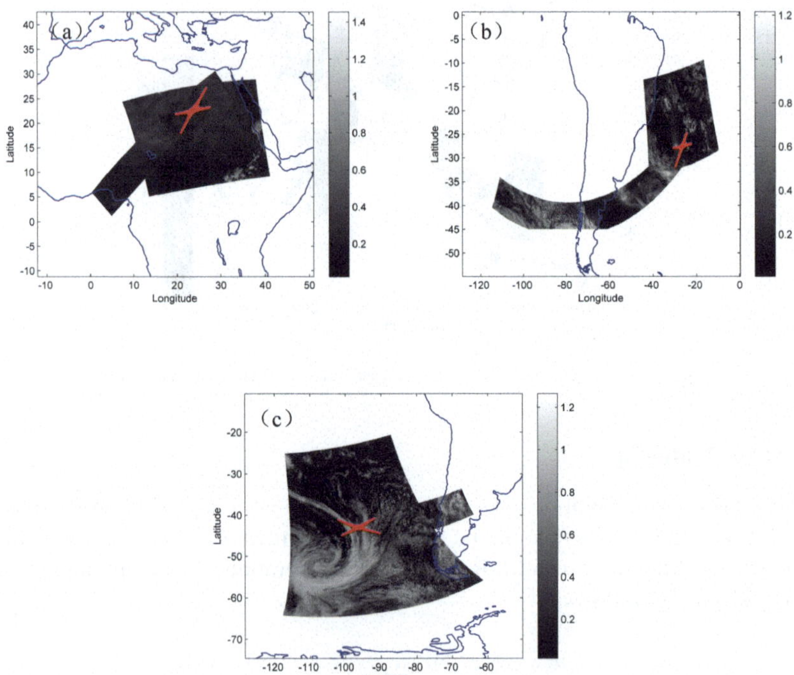

Fig. 3. Cross-calibration examples of (a) land, (b) sea surface, and (c) cloud layer, after angle matching (red coloring indicates the matching pixels).

3.4 Target Homogeneity Test

To minimize the influence of geometric field matching error, the target homogeneity test condition should be set to guarantee that the spectral properties of the target area are relatively uniform. The target homogeneity test condition is set as follows:

$$\frac{\sigma_{MAI\_3\times3}}{\bar{\rho}_{MAI\_3\times3}} < 0.03 \tag{6}$$

$$\frac{\sigma_{VIIRS\_5\times5}}{\bar{\rho}_{VIIRS\_5\times5}} < 0.03 \tag{7}$$

where $\sigma_{MAI\_3\times3}$ and $\bar{\rho}_{MAI\_3\times3}$ represent the standard deviation and average value of the observed reflectance values for the MAI 3 × 3-pixel region, respectively; and $\sigma_{VIIRS\_5\times5}$ and $\bar{\rho}_{VIIRS\_5\times5}$ represent those for the VIIRS 5 × 5-pixel region. Figure 4 shows the target homogeneity distribution for the typical cross-calibration examples of land, sea surface, and cloud layer (using the 565-nm channel of MAI and the M4 channel of VIIRS as an example).

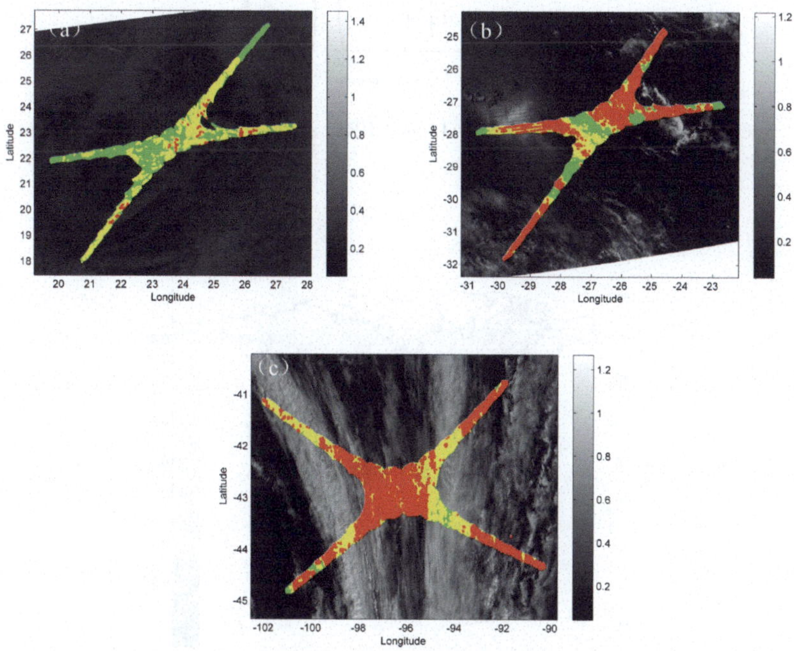

Fig. 4. Target homogeneity distribution of (a) land, (b) sea surface, and (c) cloud layer (green: *homogeneity* < 0.03; yellow: 0.03 ≤ *homogeneity* < 0.1; red: *homogeneity* ≥ 0.1).

3.5 Spectral Adjustment

MAI and VIIRS have similar spectral response functions near the 565- and 670-nm channels, as shown in Fig. 5. To reduce the cross-calibration error caused by the difference in spectral response functions [22–24], this paper employs Metop-B/GOME-2 hyperspectral data (wavelength range: 240–790 nm; spectral resolution: 0.2–0.51 nm) to adjust the spectral response functions of MAI and VIIRS [25]. The selected GOME-2 hyperspectral data totals 24 tracks, and each track contains 32 × 24

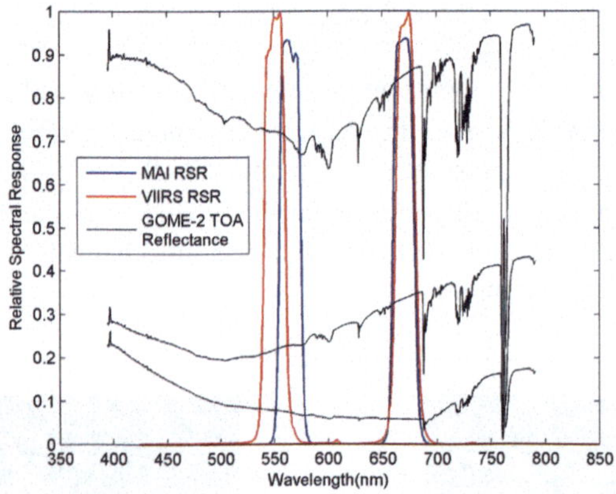

Fig. 5. Spectral response functions of MAI and VIIRS, along with the observed reflectances of GOME-2.

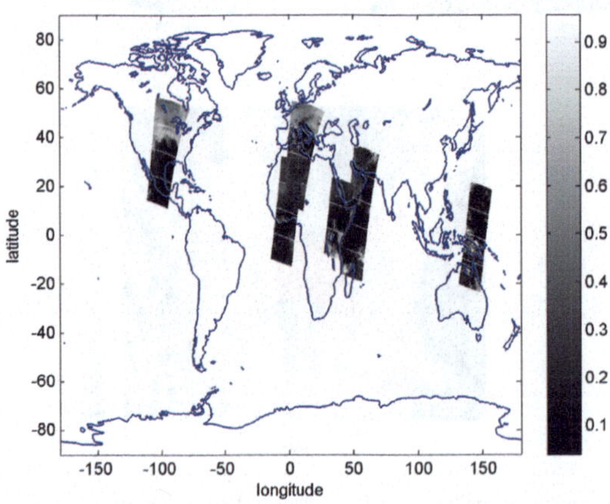

Fig. 6. Spatial distribution of the GOME-2 hyperspectral data.

pixels, as shown in Fig. 6. The timespan is from December 2016 to February 2017, which can fully cover the land, sea surface, and cloud layer, ensuring that the spectral adjustment results are well represented in the low-, medium, and high-reflectivity ranges.

The related calculation formulas for the spectral adjustment are as follows:

$$\bar{\rho}_{MAI} = \frac{\sum_{i=1}^{n} \int \rho_i(\lambda) f_{MAI}(\lambda) d\lambda}{\mathrm{n} \int f_{MAI}(\lambda) d\lambda} \tag{8}$$

$$\bar{\rho}_{VIIRS} = \frac{\sum_{i=1}^{n} \int \rho_i(\lambda) f_{VIIRS}(\lambda) d\lambda}{\mathrm{n} \int f_{VIIRS}(\lambda) d\lambda} \tag{9}$$

$$SBAF = \bar{\rho}_{MAI} / \bar{\rho}_{VIIRS} \tag{10}$$

$$\rho^*_{MAI} = \rho_{MAI} / SBAF \tag{11}$$

where $\rho_i(\lambda)$ represents the observed reflectance of GOME-2 at the wavelength λ; n is the total number of GOME-2 pixels; f_{MAI} and f_{VIIRS} represent the channel spectral response functions of MAI and VIIRS, respectively; and $\bar{\rho}_{MAI}$ and $\bar{\rho}_{VIIRS}$ represent the reference reflectance of GOME-2 relative to the spectral response function of the MAI and VIIRS channels, respectively. *SBAF* is the spectral band adjustment factor, ρ_{MAI} is the observed reflectance of MAI channels, and ρ^*_{MAI} is the corrected reflectance of MAI channels after the spectral adjustment. Since *SBAF* uses an average value, it will cause certain errors in the calibration results for specific pixels and atmospheric conditions.

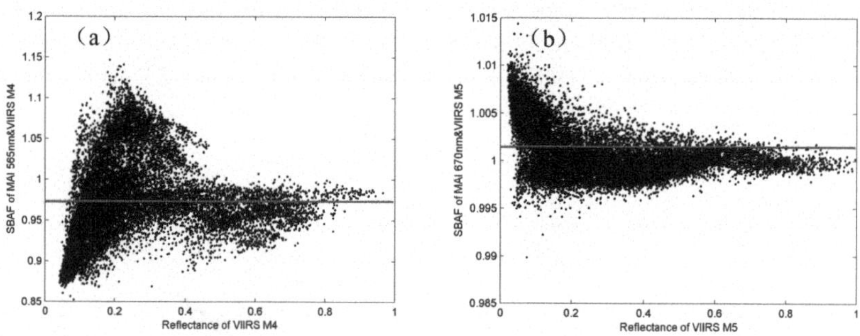

Fig. 7. *SBAF* calculation results (the ordinate value of red solid line): (a) MAI 565-nm and VIIRS M4; (b) MAI 670-nm and VIIRS M5.

SBAF can be calculated using the above formulas. The *SBAF* of the 565-nm channel of MAI and the M4 channel of VIIRS, and the 670-nm channel of MAI and the M5 channel of VIIRS, are 0.973 and 1.002, respectively. The results are shown in Fig. 7.

3.6 Regression Analysis

The final matching pixels can be filtered by completing the time matching, geometric field matching, angle matching, target homogeneity test, and spectral adjustment. On this basis, the calibration coefficients can be calculated through linear regression analysis on the reflectance of the final matching pixels.

4 Results

For the eight groups of cross-calibration examples from December 2016 to February 2017, the linear regression results of observed reflectance for the 565-nm channel of MAI and the M4 channel of VIIRS, as well as the 670-nm channel of MAI and the M5 channel of VIIRS, can be respectively obtained by performing a series of matching conditions and spectral adjustment with the above cross-calibration method.

As seen from Fig. 8, the final matching pixel numbers of the 565-nm channel of MAI and the M4 channel of VIIRS, and the 670-nm channel of MAI and the M5 channel of VIIRS, are 5330 and 4078, respectively. The observed reflectance of matching pixels basically covers the low-, medium- and high-reflectivity ranges. The number of matching pixels is large enough, and the target scene types are rich, indicating that the cross-calibration results are well represented. The correlation coefficients of the reflectivity of the 565-nm channel of MAI and the M4 channel of VIIRS, and the 670-nm channel of MAI and the M5 channel of VIIRS, are 0.986 and 0.994, respectively, presenting a significant linear relationship. The mean bias and root-mean-square error (RMSE) of the reflectivity for the 565-nm channel of MAI and the M4 channel of VIIRS are 2.48% and 3.56%, respectively, while those for the 670-nm channel of MAI and the M5 channel of VIIRS are 5.90% and 6.95%. The linear regression equations of the cross-calibration are $y = 0.993x + 0.026$ and $y = 1.11x + 0.023$. The calibration

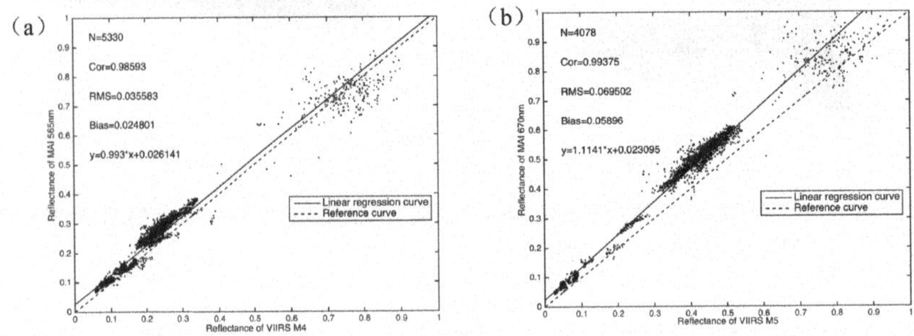

Fig. 8. Linear regression results of cross-calibration: (a) MAI 565-nm and VIIRS M4; (b) MAI 670-nm and VIIRS M5.

slope and intercept are 0.993 and 0.026, respectively, for the 565-nm channel of MAI and the M4 channel of VIIRS, and 1.11 and 0.023 for the 670-nm channel of MAI and the M5 channel of VIIRS. The above results indicate that, after 3–5 months of on-orbit operation, different degrees of offsets occurred for the laboratory-based calibration coefficients of the 565- and 670-nm channels of MAI. While the calibration coefficient offset of the 565-nm channel is relatively small, that of the 670-nm channel is quite obvious. The linear regression equations can be used to correct the system offsets of the MAI observed reflectance.

5 Summary and Discussion

Because TG-2/MAI is not equipped with an on-board calibration system, its laboratory-based calibration coefficients will inevitably shift during on-orbit operation. In order to correct the laboratory-based calibration coefficients in time, SNO cross-calibration is proposed as a feasible method for the 565- and 670-nm channels of MAI. In this method, data from the M4 and M5 channels of VIIRS, with good calibration accuracy, are used as reference data, and a series of matching conditions and a spectral adjustment algorithm are specified. To carry out the cross-calibration method, eight groups of examples from December 2016 to February 2017 are selected, and a large number of matching pixel samples covering land, ocean surface, and cloud layers, are obtained. The resulting cross-calibration curves of the 565-nm channel of MAI and the M4 channel of VIIRS, and the 670-nm channel of MAI and the M5 channel of VIIRS, show correlation coefficients of reflectivity of 0.986 and 0.994, mean biases of 2.48% and 5.90%, and RMSEs of 3.56% and 6.95%, respectively.

The error sources of the cross-calibration method proposed in this paper mainly come from four parts: First, there is the uncertainty of the calibration results caused by different matching conditions; for example, if the time matching threshold ΔT is too large, the position of the clouds will move obviously, resulting in inconsistent observational targets between MAI and VIIRS. Second, there is a certain deviation in the reference data of VIIRS itself, which is indirectly transmitted to the MAI visible channels through the cross-calibration process. Third, MAI's CCD currently has non-uniformity problems, which affects the quality of MAI data and thus affects the results of cross-calibration. And fourth, the accuracy of MAI's geolocation needs to be further improved, and its impact on cross-calibration also requires a follow-up assessment.

Overall, the cross-calibration results obtained here are encouraging, but we still have a lot of work to do to address the problems that existed. The proposed method effectively solves the problems of MAI's subsequent on-orbit monitoring and vicarious calibration, and also lays a foundation for later applications of MAI data in research on cloud and aerosol microphysical parameters [26, 27].

Acknowledgments. Thanks to China Manned Space Engineering for providing the MAI data products of Tiangong-2. This work was co-supported by the TG-2 Mission of the Manned Space Flight Project, the National Natural Science Foundation of China (NSFC41575031), and the China Postdoctoral Science Foundation (2015M580124).

References

1. Wenbin, X., Xiaobing, Z., Weining, Y.: Cross-calibration method based on hyperspectral imager hyperion. Acta Opt. Sin. **33**(5), 279–286 (2013)
2. Guanhua, Z., He, J., Huijie, Z., et al.: Imaging spectrometry radiometric cross-calibration based on precise spectral response matching. Spectrosc. Spectr. Anal. **32**(12), 3416–3421 (2012)
3. Chander, G., Hewison, T.J., Fox, N., et al.: Overview of intercalibration of satellite instruments. IEEE Trans. Geosci. Remote Sens. **51**(3), 1056–1080 (2013)
4. Xiong, X., Cao, C., Chander, G.: An overview of sensor calibration inter-comparison and applications. Front Earth Sci. China **4**(2), 237–252 (2010)
5. Boming, X.: Meteorological Satellite Payload Technology. China Astronautic Publishing House, Beijing (2005)
6. Heidinger, A.K., Cao, C., Sullivan, J.T.: Using moderate resolution imaging spectrometer (MODIS) to calibrate advanced very high resolution radiometer reflectance channels. J. Geophys. Res. Atmos. **107**(D23), AAC-1-AAC 11-10 (2002)
7. Thome, K.J., Biggar, S.F., Wisniewski, W.: Cross comparison of EO-1 sensors and other earth resources sensors to Landsat-7 ETM+ using Railroad Valley Playa. IEEE Trans. Geosci. Remote Sens. **41**(6), 1180–1188 (2003)
8. Rao, C.R.N., Cao, C., Zhang, N.: Inter-calibration of the moderate-resolution imaging spectro radiometer and the along track scanning Radiometer-2. Int. J. Remote Sens. **24**(9), 1913–1924 (2003)
9. Meirink, J.F., Roebeling, R.A., Stammes, P.: Inter-calibration of polar imager solar channels using SEVIRI. Atmos. Meas. Tech. **6**(9), 2495–2508 (2013)
10. Xu, N., Chen, L., Wu, R., et al.: In-flight intercalibration of FY-3C visible channels with AQUA MODIS. In: SPIE Asia-Pacific Remote Sensing, International Society for Optics and Photonics, pp. 163–176 (2014)
11. Yu, F., Wu, X.: Radiometric inter-calibration between Himawari-8 AHI and S-NPP VIIRS for the solar reflective bands. Remote Sens. **8**(3), 165 (2016)
12. Junjie, G., Yaozhigang, H.Z., et al.: Airborne experiment of TG-2 multi-angle polarization imager for cloud phase identification. Remote Sens. Technol. Appl. **33**(3), 439–448 (2018)
13. Lee, T.E., Miller, S.D., Turk, F.J., et al.: The NPOESS VIIRS day/night visible sensor. Bull. Am. Meteor. Soc. **87**(2), 191–199 (2006)
14. Xiong, X., Angal, A., Butler, J., et al.: Global space-based inter-calibration system reflective solar calibration reference: from Aqua MODIS to S-NPP VIIRS. In: SPIE Asia-Pacific Remote Sensing, p. 98811D (2016)
15. Goldberg, M., Ohring, G., Butler, J., et al.: The global space-based inter-calibration system. Bull. Am. Meteor. Soc. **92**(4), 467–475 (2011)
16. Xiong, X., Salomonson, V., Barnes, W., et al.: An overview of terra MODIS reflective solar bands on-orbit calibration. In: IEEE International Conference on Geoscience and Remote Sensing Symposium, pp. 1111–1114 (2006)
17. Lei, N., Wang, Z., Xiong, X.: On-orbit radiometric calibration of Suomi NPP VIIRS reflective solar bands through observations of a sunlit solar diffuser panel. IEEE Trans. Geosci. Remote Sens. **53**(11), 5983–5990 (2015)
18. Wolfe, R.E., Lin, G., Nishihama, M., et al.: Suomi NPP VIIRS prelaunch and on-orbit geometric calibration and characterization. J. Geophys. Res. Atmos. **118**(20), 11508–11521 (2013)
19. Vallado, D., Crawford, P.: SGP4 orbit determination. In: AIAA/AAS Astrodynamics Specialist Conference and Exhibit, pp. 558–570 (2008)

20. O'Brien, D.M., Mitchell, R.M.: An error budget for cross-calibration of AVHRR shortwave channels against ATSR-2. Remote Sens. Environ. **75**(2), 216–229 (2001)
21. Chang, T., Xiong, X., Angal, A., et al.: Aqua and Terra MODIS RSB calibration comparison using BRDF modeled reflectance. IEEE Trans. Geosci. Remote Sens. **55**(4), 2288–2298 (2017)
22. Teillet, P.M., Fedosejevs, G., Thome, K.J., et al.: Impacts of spectral band difference effects on radiometric cross-calibration between satellite sensors in the solar-reflective spectral domain. Remote Sens. Environ. **110**(3), 393–409 (2007)
23. Chander, G., Mishra, N., Helder, D.L., et al.: Applications of spectral band adjustment factors (SBAF) for cross-calibration. IEEE Trans. Geosci. Remote Sens. **51**(3), 1267–1281 (2013)
24. Doelling, D.R., Lukashin, C., Minnis, P., et al.: Spectral reflectance corrections for satellite intercalibrations using SCIAMACHY data. IEEE Geosci. Remote Sens. Lett. **9**(1), 119–123 (2011)
25. Munro, R., Lang, R., Klaes, D., et al.: The GOME-2 instrument on the Metop series of satellites: instrument design, calibration, and level 1 data processing–an overview. Polym. Int. **8**(8), 8645–8700 (2016)
26. Hagolle, O., Goloub, P., Deschamps, P.Y., et al.: Results of POLDER in-flight calibration. IEEE Trans Geosci. Remote. Sens. **37**(3), 1550–1566 (1999)
27. Fougnie, B., Bracco, G., Lafrance, B., et al.: PARASOL in-flight calibration and performance. Appl. Opt. **46**(22), 5435–5451 (2007)

On-Orbit Polarization Calibration of the Multi-angle Polarization Imager Based on Sunglint Over the Ocean

Junjie Guo[1,2,3], Zhigang Yao[2,3(✉)], Zhigang Han[2,3], Zengliang Zhao[2,3], Jun Jiang[2,3], and Wei Yan[1]

[1] College of Meteorology and Oceanology, National University of Defense Technology, Nanjing 211101, China
[2] State Key Laboratory of Geo-Information Engineering, Xi'an 710054, China
yzg_biam@163.com
[3] Beijing Institute of Applied Meteorology, Beijing 100029, China

Abstract. The Multi-angle Polarization Imager (MAI) is the first multi-angle polarization imaging instrument operating in orbit in China. It is important to the development of the polarized remote sensing technology. This study uses on-orbit polarization calibration of the MAI based on sunglint over the ocean to determine the accuracy of the MAI data and to lay the foundations for the study of clouds and aerosols using this instrument. The retrieved degree of polarization at the top of atmosphere (TOA) is in good agreement with the degree of polarization simulated by the Second Simulation of a Satellite Signal in the Solar Spectrum – Vector (6SV) radiation transfer mode with a correlation coefficient of 0.992 and a method uncertainty of <1%. The calibration pixels are selected by combining the theoretical analysis and 6SV simulation, and the MAI on-orbit polarization calibration test is carried out using the selected pixels. There is good agreement between the retrieved degree of polarization at the TOA and the measured degree of polarization by the MAI, with a correlation coefficient of 0.9576, an average absolute deviation of 6.06%, and a standard deviation of the difference between the retrieved degree of polarization and the measured degree of polarization of 1.29%. The method of on-orbit polarization calibration based on sunglint over the ocean can realize on-orbit polarization monitoring and calibration of the MAI.

Keywords: Tiangong-2 · Multi-angle polarization remote sensing
On-orbit calibration · The sunglint over the ocean · 6SV ·
Degree of polarization

1 Introduction

The Multi-angle Polarization Imager (MAI) has an important role in studies of the global radiation budget, the macroscopic and microscopic physical properties of clouds, and the properties of aerosols. Polarized remote sensing has two main advantages over traditional non-polarized remote sensing [1–9]. First, traditional

Y. Gu et al. (Eds.): *Proceedings of the Tiangong-2 Remote Sensing Application Conference*, LNEE 541, pp. 144–159, 2019.
https://doi.org/10.1007/978-981-13-3501-3_14

remote sensing only considers the intensity of electromagnetic waves, whereas polarized remote sensing also records the spectral characteristics of electromagnetic radiation, such as the direction, phase, and polarization state. The polarized remote sensing can obtian the polarization information while obtaining the radiance, which helps to improve the accuracy of target detection and feature recognition. Second, polarized remote sensing is superior to traditional remote sensing in retrieving the size and distribution of particles in clouds and aerosols.

To carry out quantitative studies on clouds and aerosols using MAI polarization data, it is first necessary to calibrate the MAI on-orbit. Goloub et al. [10] performed an on-orbit polarization calibration for the Polarization and Directionality of the Earth's Reflectances (POLDER) based on the non-polarization characteristics of water clouds near the 100° scattering angle and the strong polarization characteristics of sunglint over the ocean. Toubbé et al. [11] used the polarization calibration method for sunglint over the ocean to analysis POLDER data.

The ideal ocean surface can be regarded as a specular surface. According to the laws of reflection, sunglint can be observed in the direction of reflected light. In practice, the real ocean surface is prone to appear the phenomenon of sunglint as result of the wide distribution of waves. In traditional non-polarized remote sensing, sunglint is often regarded as noise and needs to be suppressed. However, in polarized remote sensing, when light is incident at Brewster's angle, the reflected light is a fully linearly polarized light with a degree of polarization equal to 1. Sunglint has both strong reflection and polarization characteristics. Studies showed that the intrinsic degree of polarization of sunglint only depends on the geometry of the observation and is not restricted by other factors. Therefore, sunglint over the ocean can be used as a standard alternative light source with strong polarization for the calibration of multi-angle polarization instruments. Compared with the 565 nm and 670 nm bands, the 865 nm band is least affected by atmospheric molecules, and the water-leaving radiance at 865 nm band can be ignored.

Combining the method proposed by Toubbé et al. [11] and the characteristics of observation data of the MAI, this paper uses the 865 nm band as a representative to study the on-orbit polarization monitoring and calibration method of MAI based on sunglint over the ocean.

2 MAI Instrument Concept

2.1 MAI Mission

The main objectives of developing MAI are as follows. (1) To deepen the research and the application of the vector radiation transfer model. The MAI is used as a test instrument to study multi-angle polarized remote sensing technology, to develop a theoretical system for multi-angle polarized remote sensing, and to lay the technical foundations for developing the business loads of subsequent meteorological and marine satellite and on-orbit applications. (2) To determine the macroscopic and microscopic physical properties of clouds, such as the cloud phase, the cloud top height, the cloud top pressure, and the cloud droplet effective particle radius. In addition, it can also be used to retrieve the dominant shape and orientation of ice crystals. (3) To determine the distribution of global atmospheric aerosols (especially over land) and to study their physical and optical properties based on multi-angle, multi-wavelength and multiband imaging information from the same target, and to consider its effects on global radiation. (4) To determine the water vapor content in the atmosphere using the multi-angle observation advantage of MAI. (5) To study the bidirectional reflection and polarization characteristics of both the land surface and vegetation.

2.2 MAI Characteristics

The MAI was launched on the TG-2 space laboratory on September 15, 2016. The swath of the MAI is 770 km and the spatial resolution is 3 km × 3 km. The MAI has a lower platform and higher spatial resolution than POLDER, a similar international instrument. And the MAI can generally achieve 12 different angle observations for the same target.

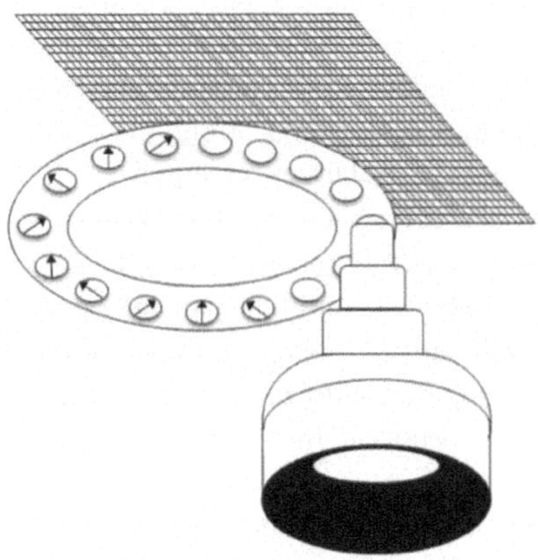

Fig. 1. Schematic diagram of the structure of the MAI.

The MAI (Fig. 1) consists of a back-illuminated charge coupled device (CCD) matrix array detector, a rotating filter wheel, and a wide field of view optical system. The CCD has a total of 264 × 268 pixels; the number of effective pixels is 256 × 256 and the remaining pixels are light-blocking pixels, which are mainly used to estimate the dark current and to eliminate the effect of image frame transfer. The pixel size is 26 μm × 26 μm. The rotating filter wheel contains 13 filters, corresponding to 13 channels [12–14] (Table 1). The MAI also includes a wide field of view optical system with an instantaneous field of view of 7.5 mRad and a full field of view of 87.6°.

Table 1. Spectral channel characteristics of the MAI

Channel	Central wavelength (nm)	Band width (nm)	Polarization	Polarization direction (°)	Main mission
1	565	20	Yes	−60	Cloud, aerosol, Earth radiation budget
2	565	20	Yes	0	
3	565	20	Yes	60	
4	670	20	Yes	−60	Cloud, aerosol, Earth radiation budget
5	670	20	Yes	0	
6	670	20	Yes	60	
7	865	40	Yes	−60	Cloud, aerosol, Earth radiation budget
8	865	40	Yes	0	
9	865	40	Yes	60	
10	763	10	No	–	Cloud top pressure
11	765	40	No	–	
12	910	20	No	–	Amount of water vapor
13	–	–	–	–	Dark current

The working mode of the MAI is shown in Fig. 2 [12]. The rotating filter wheel rotates four times in one frame period (8.625 s) and samples are only taken on the first circle. For the same polarization band, the wedge-angle compensation technique is applied to the −60° and 60° polarizer grooves to ensure accurate registration of the three-polarizer measurement. The registration accuracy is <0.2 pixel.

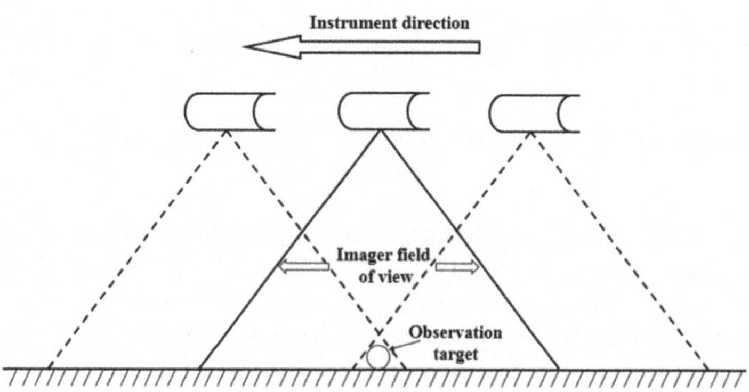

Fig. 2. Schematic diagram of the working mode of MAI

3 Principle of Polarization Calibration

This work uses the Cox and Munk model to describe the ocean surface scene [15] and the Second Simulation of a Satellite Signal in the Solar Spectrum – Vector (6SV) vector radiation transfer mode to simulate radiation transfer processes in the atmosphere–ocean system and to calculate the radiance and polarization radiance in the entrance pupil of the detector. This model only considers cloudless atmospheric conditions and uses the successive order of scattering (SOS) method to solve the scattering process with relatively high precision, which is widely used in remote sensing [16, 17].

The radiance at the TOA over the ocean includes the reflected radiance by the sunglint and white foam, the water-leaving radiance, and the scattered radiance by atmospheric molecules and aerosols.

(1) Contribution of sunglint

Based on the established statistical model of the distribution of wave slopes by Cox and Munk [13], a Gaussian distribution is used to describe the distribution of the slope of the waves. The probability density function $P(Z_x', Z_y')$ of the ocean surface wave distribution can be expressed as:

$$P\left(Z_x', Z_y'\right) = \left(\frac{1}{\pi(0.003 + 0.0512v^2)^{1/2}}\right)^2 \cdot exp\left(-\frac{tan^2\theta_n}{0.003 + 0.0512v^2}\right) \quad (1)$$

Where v is wind speed at ocean surface, and θ_n is the slope of the wave.

The reflectance ρ_g and the polarized reflectance ρ_g^{pol} of sunglint can be expressed by formula (2), (3), respectively.

$$\rho_g = \frac{\pi R(\Omega) P(Z_x', Z_y')}{4cos\theta_s cos\theta_v cos^4\theta_n} \quad (2)$$

$$\rho_g^{pol} = \frac{\pi R^{pol}(\Omega) P(Z_x', Z_y')}{4cos\theta_s cos\theta_v cos^4\theta_n} \quad (3)$$

Where θ_s, θ_v, φ_s, and φ_v represent the solar zenith angle, the observed zenith angle, the solar azimuth, and the observed azimuth, respectively. $R(\Omega)$ and $R^{pol}(\Omega)$ are the Fresnel reflection coefficient and the polarization reflection coefficient, respectively. Ω is the angle of incidence of sunlight relative to the wave surface.

The intrinsic degree of polarization of sunglint P_g is:

$$P_g = \frac{\rho_g^{pol}}{\rho_g} = \frac{R^{pol}(\Omega)}{R(\Omega)} \quad (4)$$

(2) Contribution of white foam on the ocean surface

Waves form white caps and foam on the ocean surface due to the action of wind and other factors. White caps and foam are collectively referred to as ocean surface white foam. White foam is a mixture of air and water. It consists of bubbles with a diameter of about 1.55 mm and the volume of gas can reach 94–97%. Ocean surface white foam has very different reflection and radiation characteristics from the background seawater in the visible band. Therefore, it is necessary to consider its impact on the radiance at the TOA. The reflectance of sea surface white foam ρ_f is defined as follow:

$$\rho_f = 0.22 \times 2.95 \times 10^{-6} v^{3.52} \tag{5}$$

(3) Contribution of water-leaving radiance

For Case I water (e.g. the South Indian Ocean and Southeast Pacific Ocean), no floating objects nor dust aerosol is contributed by the continental shelf. Therefore the water-leaving radiance is modeled by a Lambertian diffuse reflection ρ_w is roughly 0 at 865 nm band and is almost 0 at 565 and 670 nm bands.

Therefore it is assumed that ocean surface white foam and seawater make no polarization contribution to sunglint over the ocean at low wind speed. The reflectance, polarization reflectance, and degree of polarization of the TOA are:

$$\rho \cong (\rho_g + \rho_w + \rho_f)e^{-M\delta} + \rho_m + \rho_a \tag{6}$$

$$\rho^{pol} \cong \rho_g^{pol} e^{-M\delta} + \rho_m^{pol} + \rho_a^{pol} \tag{7}$$

$$P = \frac{\rho^{pol}}{\rho} = \frac{\rho_g^{pol} e^{-M\delta} + \rho_m^{pol} + \rho_a^{pol}}{(\rho_g + \rho_w + \rho_f)e^{-M\delta} + \rho_m + \rho_a} \tag{8}$$

Where M is the air mass factor, and δ is the total optical thickness. ρ_m and ρ_m^{pol} represent the reflectance and the polarization reflectance correspond to the scattering of atmospheric molecules, respectively. ρ_a and ρ_a^{pol} represent the reflectance and the polarization reflectance correspond to the scattering of aerosol, respectively.

According to the 6SV simulation under typical conditions, when $v < 10$ m/s, the sunglint reflectance ρ_g is between 0.2 and 0.6 and the white foam contribution $\rho_w < 0.001$. Therefore the contribution of white foam can be ignored under these conditions. In addition, the contribution of water-leaving radiance is negligible at 865 nm band. Therefore the degree of polarization at the TOA can be approximated as formula (9).

$$P = \frac{\rho_g P_g e^{-M\delta} + \rho_m P_m + \rho_a P_a}{\rho_g e^{-M\delta} + \rho_m + \rho_a} \tag{9}$$

The strict specular reflection direction is considered—that is, the detector has a relative azimuth angle of 180° with the sun and the observation zenith angle is equal to the solar zenith angle. The degree of polarization at the TOA is simulated when the solar zenith angle is set to 20–60°, the step length is 5°, the aerosol type is marine, the aerosol optical thickness is 0.025, 0.05, 0.075, and 0.1, and the sea surface wind speed is 2, 5, 10, and 15 m/s. The intrinsic degree of polarization of sunglint is calculated using these geometric conditions. Figure 3 shows the relationship between the simulated degree of polarization at the TOA in the specular direction and the intrinsic degree of polarization of the sunglint. The simulated degree of polarization at the TOA in the specular direction is always lower than the intrinsic degree of polarization of sunglint. The difference increases significantly as the degree of polarization increases. This is mainly because electromagnetic waves are depolarized by clouds, aerosols, etc. during the transmission in the atmosphere. And the depolarization effect is significantly enhanced as the degree of polarization increases.

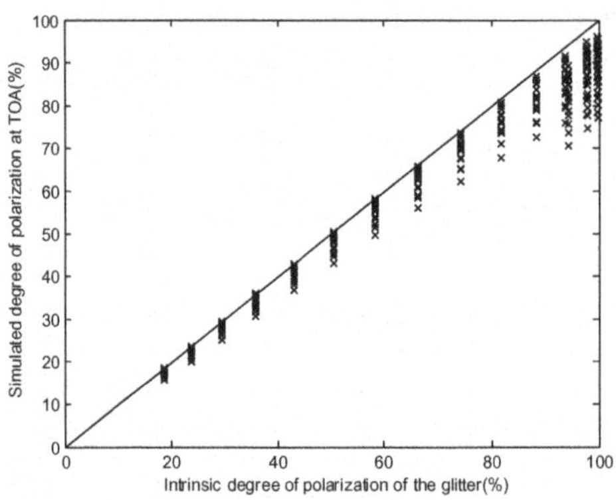

Fig. 3. Relationship between the simulated degree of polarization at the TOA in the specular direction and the intrinsic degree of polarization of sunglint.

The difference between the simulated degree of polarization at the TOA in the specular direction and the intrinsic degree of polarization of sunglint is expressed as $\Delta P = P - P_g$. According to Goloub et al. [10], for different aerosol optical thicknesses and different sea surface wind speed, the atmospheric molecular reflectance ρ_{atm}^{*865} has a significant linear relationship with $\frac{\Delta P}{P}\rho^{-\beta}$ and this relationship exists at each solar zenith angle, and $\rho_{atm}^{*865} = \rho_{atm}^{865} - \rho_{atm}^{*}$, where ρ_{atm}^{865} represents the reflectance of the calibration pixel which measured in some viewing directions that are not affected by the sunglint, to simplify the presentation, hereinafter referred to as the atmospheric direction reflectance. ρ_{atm}^{*} is the reflectance corresponding to ρ_{atm}^{865} regardless of the action of aerosols. Therefore the retrieved degree of polarization P_{ret} at the TOA can be expressed as:

$$P_{ret} = P_g + \Delta P = P_g[1 + a_0 + \left(\rho^{865}\right)^{\beta}\left(a_1 + a_2\rho_{atm}^{*865} + a_3\rho^{865} + a_4\rho^{865}\rho_{atm}^{*865} + a_5\rho_{atm}^{*865^2} + a_6\rho^{865^2}\right)] \quad (10)$$

Where $\beta = -1.1 \pm 0.1$, a_0, a_1, a_2, a_3, a_4, a_5 and a_6 are all fitted coefficients, which are constants.

Figure 4 compares the retrieved degree of polarization with the simulated degree of polarization under various zenith angle conditions. There is a significant linear relationship between the retrieved results and the simulated results, with a correlation coefficient of 0.9998, and standard deviation of the difference between the retrieved degree of polarization and simulated degree of polarization is 0.05%. Therefore the method of degree of polarization retrieval has a higher precision.

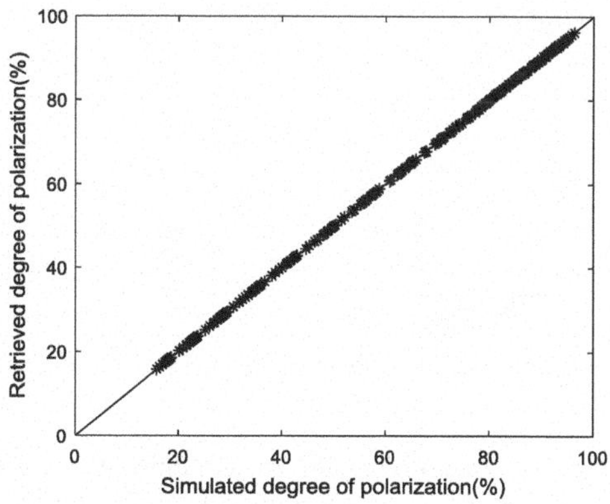

Fig. 4. Relationship between the retrieved degree of polarization and the simulated degree of polarization under various zenith angle conditions.

4 Polarization Calibration

4.1 Calibration Data

Polarization calibration needs to be carried out under stricter conditions to ensure the accuracy. This involves combining the MAI data characteristics and screening the polarization calibration data.

(1) Calibration area

The 6SV vector radiation transfer model was used to study the degree of polarization and reflectance of sunglint under different aerosol conditions and different sea surface wind field. Figure 5 shows that the characteristics of sunglint vary greatly under different aerosol conditions. In addition, the wind field has a certain influence on the degree of polarization. And when the wind speed is less than 10 m/s, the wind direction has no influence on the degree of polarization. Therefore, it is necessary to select areas as far from the coastline as possible, thereby reducing the effects of urban aerosols, dust aerosols, and floating objects. At the same time, the wind speed should generally below 10 m/s.

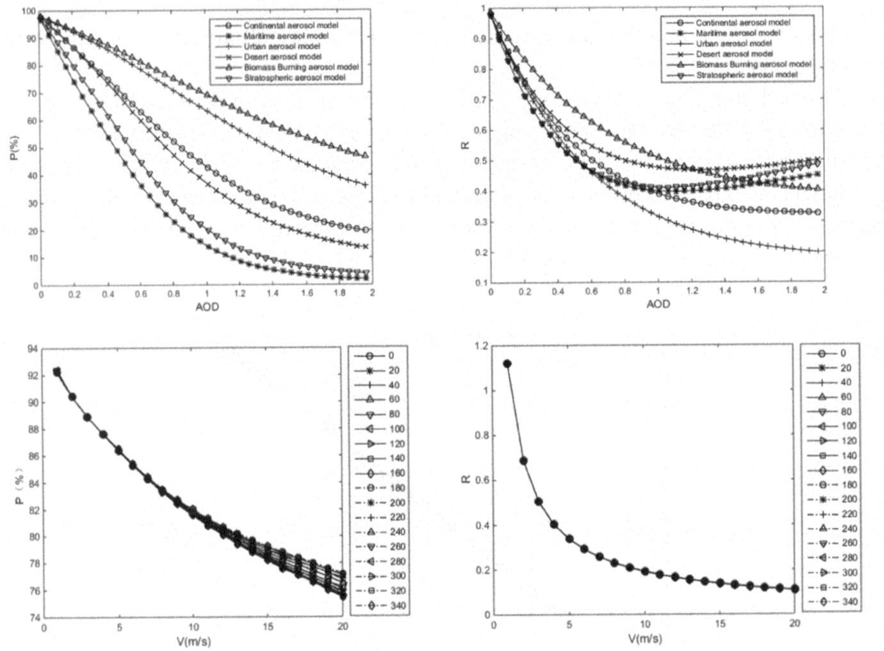

Fig. 5. The influence of different aerosol conditions and different sea surface wind field on the characteristics of sunglint. (a) Degree of polarization and (b) reflectance as a function of aerosol type and aerosol optical thickness. (c) Degree of polarization and (d) reflectance as a function of sea surface wind speed and direction.

According to the above analysis, this study selected the strong sunglint observation data of the MAI in the Gulf of Mexico on January 25, 2017 for analysis. The calibration area is shown in Fig. 6 as a black rectangle.

(2) Data matching

The synchronized Moderate Resolution Imaging Spectroradiometer (MODIS) aerosol products and the National Centers for Environmental Prediction (NCEP) ocean surface wind field data were matched according to the selected MAI calibration data. Data from MODIS and NCEP were then interpolated according to the resolution of the MAI.

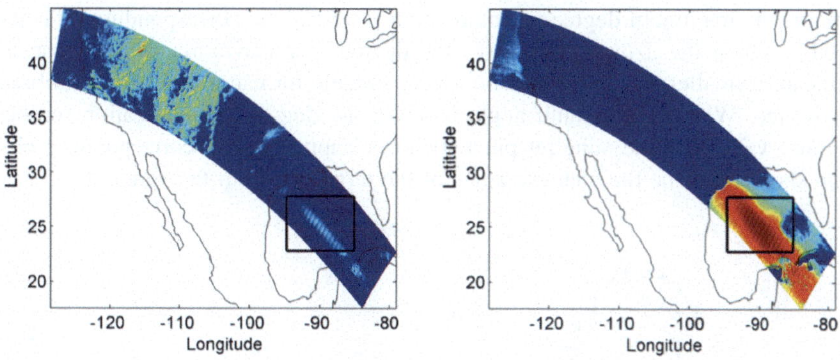

Fig. 6. Observation of sunglint over the ocean by the MAI on January 25, 2017. (a) Reflectance and (b) polarization.

(3) Data preprocessing

To minimize the impact of noise on the MAI instrument, the 3 × 3 pixel observations of the MAI were averaged.

Broken clouds are often present in the sky, which will affect the calibration of MAI polarization due to the significant depolarization effects of clouds. Therefore it is necessary to detect clouds in the sunglint area and to eliminate the cloud pixels. Sunglint can only occur under certain geometric conditions, so, for multi-angle observations of the most targets, the characteristics of high degree of polarization and reflectance will only occur in certain directions. There is a large difference in reflectance between the sunglint direction and other directions far from the sunglint. However, the reflectance of the cloud pixels in each observation direction does not vary very much. Clouds have a strong depolarization effect, which significantly lowers the degree of polarization of cloud pixels. The method of cloud detection in the sunglint region was developed based on these characteristics of the sunglint area and clouds. The uniformity of the degree of polarization was detected in the direction of sunglint, and then the standard deviation of the reflectance in each direction of the same pixel was limited to eliminate the cloud pixels. According to Toubbé et al. [11], the smaller the angle between the wave facet and the horizontal plane, say, wave angle, the higher the accuracy of the calibration. Therefore, samples only with wave angles $\theta_n < 10°$ were selected.

(4) Limiting the observation angle of the pixel

Sunglint only appears at specific angles of observation. Therefore an analysis of the influence of the observation angle on the characteristics of sunglint is important in the choice of calibration pixels. Figure 7 shows the degree of polarization and reflectance as a function of solar zenith angle, view zenith angle and the relative azimuth of the sun and the detector. Figure 7(a), (b) show that when the solar zenith angle is constant and the difference of view zenith angle and the solar zenith angle are within 20°, the degree of polarization is high, and the maximum value is >50%; Fig. 7(c), (d) show that when the azimuth angle is between 160°–180°, the degree of polarization and reflectance are high, and the maximum degree of polarization is >50%, the corresponding reflectance is >0.05. When the azimuth is <160°, the degree of polarization and reflectance is significantly smaller and sunglint is no longer suitable for using as a strong calibration light source. When the azimuth angle is <30°, the degree of polarization reaches a minimum value and the sunglint phenomenon disappears. The corresponding observation should then be the characteristics of the target far from the sunglint.

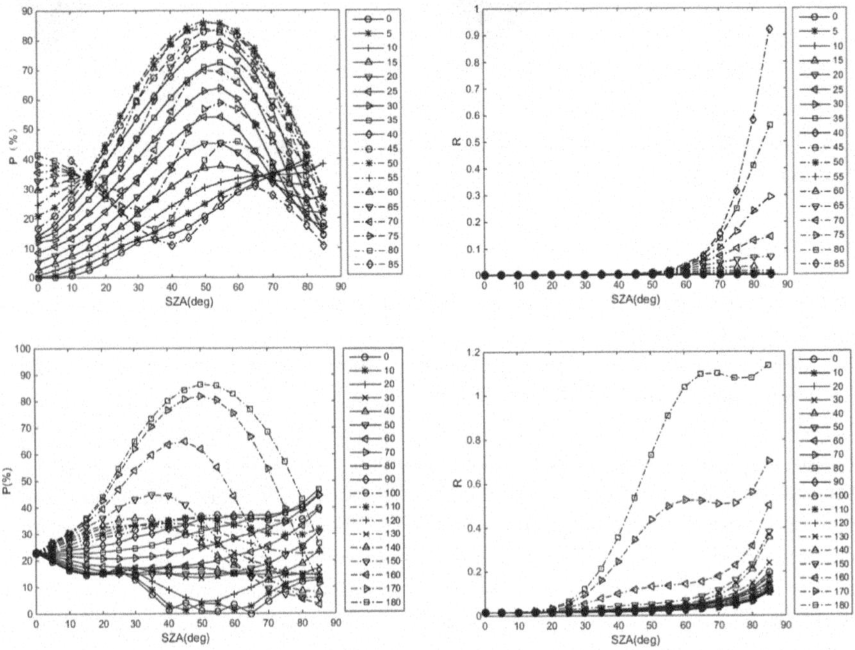

Fig. 7. (a) Degree of polarization and (b) reflectance as a function of the solar zenith angle and the view zenith angle under back-reflection, (c) Degree of polarization and (d) reflectance as a function of the solar zenith angle and the relative azimuth of the instrument and the sun under specular reflection.

To ensure the strong polarization and reflectance characteristics of the selected calibration pixel, the direction of sunglint and the directions far from the sunglint observation, say, atmospheric direction, of the same pixel need to be considered in the process of degree of polarization retrieval. It is therefore necessary to choose pixels that include both the sunglint direction and the atmospheric direction. The selected pixels generally have at least 12 observation directions.

For the sunglint direction, the absolute value of the difference between the observation zenith angle and the solar zenith angle is <20°. The absolute value of the difference between the observed azimuth angle and the solar azimuth angle is 160°–200°.

For the atmospheric direction, the smallest difference between the observed azimuth and the solar azimuth in all directions was taken as the atmospheric direction. The MAI can only obtain observations in 12 directions. It may be difficult for most targets to simultaneously observe the sunglint direction and the atmospheric direction due to the existence of the ocean surface waves, leading to a high reflectance value in the atmospheric direction of the MAI. We therefore replaced the reflectance of the atmospheric direction of the MAI with the 6SV simulation result to reduce the uncertainty in the polarization retrieval.

4.2 Calibration Results

Using the calibration pixel obtained in Sect. 4.1, the intrinsic degree of polarization of the sunglint was first calculated. Then, combined with the synchronized MODIS aerosol data and NCEP ocean surface wind speed data, the reflectance, the polarized reflectance and the degree of polarization at the TOA were simulated based on the 6SV vector radiation transfer mode. The degree of polarization at the TOA was also retrieved.

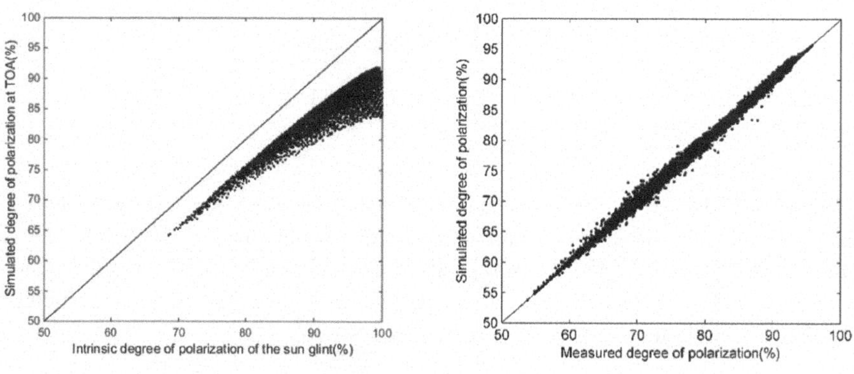

Fig. 8. 6SV simulation results. (a) Relationship between the intrinsic degree of polarization of sunglint and the simulated degree of polarization at the TOA. (b) Relationship between the retrieved degree of polarization and the simulated degree of polarization.

Figure 8(a) shows that the degree of polarization obtained by 6SV simulation is always less than the intrinsic degree of polarization of sunglint, mainly due to the depolarization of aerosols and atmospheric molecules and clouds. This is consistent with previous theoretical simulations.

ΔP strongly depends on the solar zenith angle and the azimuth angle, the observation zenith angle and the azimuth of the glint direction, and the observation zenith angle and azimuth angle of the atmospheric direction. Each MAI pixel therefore corresponds to a set of coefficients. The mode of input should be entered for each set of observation geometries and the results fitted accordingly.

Figure 8(b) shows there is a good correlation between the retrieved degree of polarization and the simulated degree of polarization with a correlation coefficient of 0.992. The linear relationship is $P_{ret} = 0.9985 \times P_{sim} + 0.0012$. The standard deviation of the difference between the retrieved degree of polarization and simulated degree of polarization is 0.26%. This further illustrates the feasibility of the polarization retrieval method.

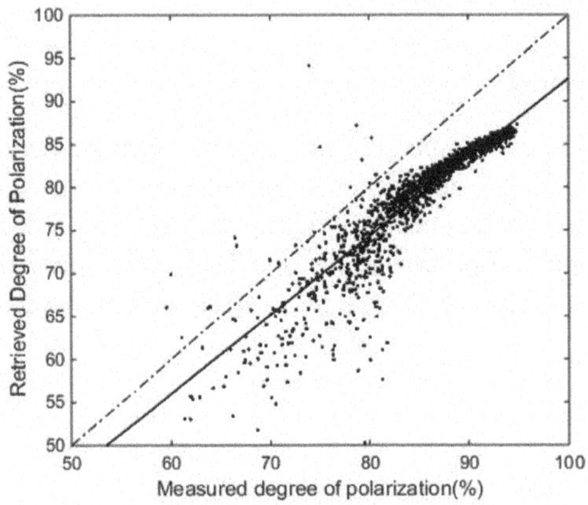

Fig. 9. Relationship between the retrieved degree of polarization and measured degree of polarization.

Figure 9 shows a significant linear relationship between the retrieved degree of polarization and the measured degree of polarization obtained by the MAI. The correlation coefficient is 0.8235, the average absolute deviation is 6.13%, and the standard deviation is 4.09%. Figure 9 shows that there are still some stray pixels, and considering that the calibration pixels are affected by some factors.

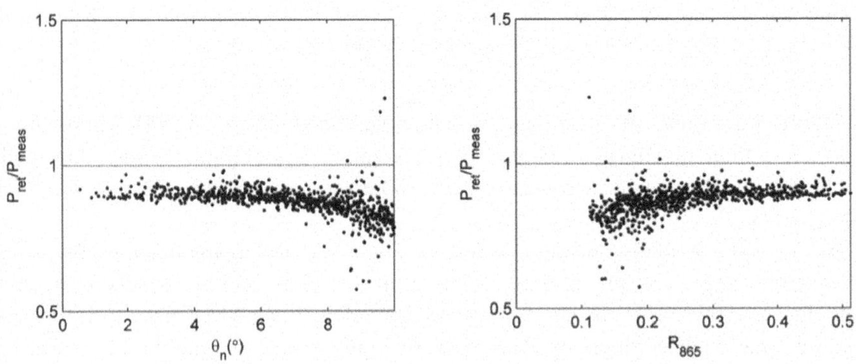

Fig. 10. Ratio of the degree of polarization of the retrieved results and the measured degree of polarization of the MAI as a function of (a) wave angle (b) reflectance of the 865 nm band.

Figure 10 shows the ratio of the retrieved degree of polarization to the measured degree of polarization of the MAI as a function of the wave angle and the reflectance of the 865 nm band. When the ocean wave angle is >6°, the retrieved results and the measured values of MAI deviated strongly. Correspondingly, the retrieved results and the measured values of MAI also deviated strongly when the 865 nm band reflectance is <0.25. The calibration result is strongly affected when the wave angle of the ocean surface was >6° or when the reflectance is <0.25.

Figure 11 shows the relationship between the retrieved degree of polarization and the measured polarization given an ocean surface wave angle threshold of 6°. The correlation coefficient is 0.9576, the average absolute deviation is 6.06%, and the standard deviation of the corresponding deviation is 1.29%.

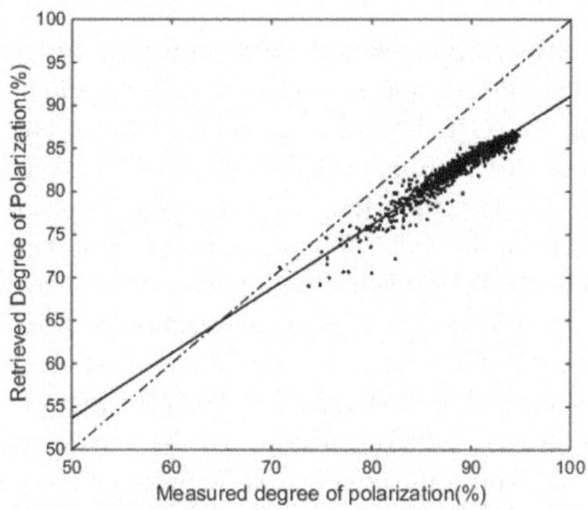

Fig. 11. Relationship between the retrieved degree of polarization and the measured polarization for the wave angle $\theta_n < 6°$.

4.3 Uncertainty Analysis

Although the degree of polarization measured by the MAI and the retrieved degree of polarization are generally consistent, there are some differences between them. This paper identify the following as the main sources of uncertainty during the calibration process.

(1) There is uncertainty in the estimation of radiance from ocean surface foam and water-leaving radiance, which are derived from the 6SV radiation transfer mode. In addition, the 6SV radiation transfer mode can't consider the influence of clouds, although it is rare to have a completely clear sky scene over the ocean. The appearance of broken clouds will affect the results.

(2) The degree of polarization is significantly affected by aerosols. According to the analysis in Sect. 4.1, the type of aerosol and the aerosol optical thickness influence the simulation of the degree of polarization. We assume that the aerosols in the study area are all marine aerosols, but other types of aerosols may be present. In addition, the aerosol optical thickness uses MODIS aerosol products, so, there

is an uncertainty in aerosol optical thickness as a result of the difference in transit time between the TG-2 and AQUA satellites.

(3) The simulated result by 6SV model is used for the atmospheric direction because the MAI observations are not far from the direction of sunglint. But, the regional average of aerosol concentration and wind speed are used in the model calculation, which may introduce uncertainties. When selecting the oceanic and oligotrophic areas in subsequent observations, a region with small ocean waves should be selected so that the atmospheric direction is not affected by sunglint and the calibration uncertainty can be further reduced.

5 Discussion and Conclusions

This paper adopt the on-orbit polarization calibration method based on sunglint over the ocean surface to study the accuracy of data of the MAI. Firstly, the reflectance and polarized reflectance at the TOA above the ocean are analyzed based on the established statistical model of the distribution of wave slopes of Cox and Munk [15]. And combining the degree of polarization retrieval method proposed by Toubbé et al. [11] with the actual results from the MAI, this paper proposes a method of retrieving the degree of polarization at the TOA when sunglint is present. There is a significant linear relationship between the retrieved degree of polarization and the simulated degree of polarization with a correlation coefficient of 0.992 and a standard deviation of 0.26%. Secondly, the polarization calibration area and the principle of selecting calibration pixels in the sunglint area are determined through theoretical analysis and 6SV simulation. Finally, The MAI on-orbit polarization calibration experiment is carried out using the selected pixels. There is good agreement between the retrieved degree of polarization and the measured degree of polarization from the MAI, the correlation coefficient is 0.9576, the average absolute deviation of 6.06%, and the standard deviation of the difference between the retrieved degree of polarization and the measured degree of polarization is 1.29%. The method reported here can realize polarization monitoring and calibration of the MAI during orbital operation with a high degree of accuracy.

Acknowledgments. Thanks to China Manned Space Engineering for providing space science and application data products of Tiangong-2. This work was co-supported by the TG-2 Mission of the Manned Space Flight Project, the National Natural Science Foundation of China (NSFC41575031), and the China Postdoctoral Science Foundation (2015M580124).

References

1. Gong, J.Q., Zhan, H.G., Liu, D.Z.: A review on polarization information in the remote sensing detection. Spectrosc. Spectr. Anal. **30**(4), 1088–1095 (2010)
2. Deschamps, P.Y., Breon, F.M., Leroy, M., Podaire, A., Bricaud, A., Buriez, J.C., Seze, G.: The POLDER mission: instrument characteristics and scientific objectives. IEEE Trans. Geosci. Remote Sens. **32**(3), 598–615 (1994)

3. Bretdibat, T., Andre, A., Laherrere, J.M.: Preflight calibration of the POLDER instrument. In: Proceedings of SPIE - The International Symposium on Optical Science, Engineering, and Instrumentation, pp. 218–231 (1995)
4. Han, Z.G.: Aerosol retrievals over steppe with POLDER data- simulation model system and preliminary tests. Doctor, The Institute of Atmospheric Physics, Chinese Academy of Sciences (1999)
5. Chen, H.B., Fan, X.H., Han, Z.G.: A review on remote sensing from POLDER multispectral, multidirectional and polarized measurements. Remote. Sens. Technol. Appl. **21**(2), 83–92 (2006)
6. Duan, M.Z., Lü, D.R.: Simultaneously retrieving aerosol optical depth and surface albedo over land from POLDER's multi-angle polarized measurement. II: a case study. Chin. J. Atmos. Sci. **31**(5), 757–765 (2008)
7. Fan, X.H., Chen, H.B., Lin, L.F., Han, Z.G., Goloub, P.: Retrieval of aerosol optical properties over the Beijing area using POLDER/PARASOL satellite polarization measurements. Adv. Atmos. Sci. **26**(6), 1099–1107 (2009)
8. Fan, X.H., Chen, H.B., Lin, L.F., Han, Z.G., Goloub, P.: Validation of POLDER-2/PARASOL aerosol products over Beijing area. J. Remote Sens. **13**(1), 137–144 (2009)
9. Yao, Z.G., Han, Z.G., Zhao, Z.L., Lin, L.F., Fan, X.H.: Synergetic use of POLDER and MODIS for multilayered cloud identification. Remote Sens. Environ. **114**(9), 1910–1923 (2010)
10. Goloub, P., Toubbe, B., Herman, M., Bailleul, T., Hagolle, O., Martinuzzi, J.M., Rouge, B.: In-flight polarization calibration of POLDER. In: Advanced and Next-Generation Satellites II. International Society for Optics and Photonics, pp. 299–310 (1997)
11. Toubbé, B., Bailleul, T., Deuzé, J.L., Goloub, P., Hagolle, O., Herman, M.: In-flight calibration of the POLDER polarized channels using the Sun's glitter. IEEE Trans. Geosci. Remote Sens. **37**, 513–524 (1999)
12. Guo, J.J., Yao, Z.G., Han, Z.G., Zhao, Z.L, Yin, D.K., Yan, W.: Airborne experiment of TG-2 multi-angle polarization imager for cloud phase identification. Remote Sens. Technol. Appl. **33**(3), 439–448 (2018)
13. Guo, J.J., Yao, Z.G., Han, Z.G., Zhao, Z.L, Yan, W.: Inter-calibration for TG-2/MAI visible band based on Metop-B/GOME-2. Spectroscopy and Spectral Analysis, Accepted
14. Guo, J.J., Yao, Z.G., Han, Z.G., Zhao, Z.L, Yan, W.: TG-2/MAI CCD dark current characteristics and channel dependence analysis and correction. Chinese Journal of Lasers, Accepted
15. Cox, C., Munk, W.: Measurement of the roughness of the sea surface from photographs of the Sun's glitter. J. Opt. Soc. Am. B **44**(11), 838–850 (1954)
16. Patel, P.N., Bhatt, H., Mathur, A.K., Prajapati, R.P., Tyagi, G.: Reflectance-based vicarious calibration of INSAT-3D using high-reflectance ground target. Remote Sens. Appl. Soc. & Environ. **3**, 20–35 (2016)
17. Lu, J.Y., He, M., Chen, L., Xu, X., Li, X.: Automated radiation calibration method based on dunhuang radiometric calibration site. Acta Opt. Sin. **37**(8), 0801003 (2017)

A Management and Service Approach for Mass Remote Sensing Data of Tiangong-2 Space Laboratory

Haijun Yu[1,2,3], Bo Wang[1,2(✉)], Wanfeng Zhang[1,2], and Tao Zhang[1]

[1] Technology and Engineering Center for Space Utilization, Chinese Academy of Sciences, Beijing 100094, China
wangbo@csu.ac.cn
[2] Key Laboratory of Space Utilization, Chinese Academy of Sciences, Beijing 100094, China
[3] University of Chinese Academy of Sciences, Beijing 100049, China

Abstract. Visualization effect and service quality is pivotal for remote sensing data managing and applying. Based on analyzing characteristics of remote sensing data, we studied on data fragmentation and parallel scheduling to improve loading and downloading efficiency of remote sensing data. In this paper, a new method of the visualized management is proposed and used in the visualization management subsystem we developed. The method which is used in distributed storage system utilizes a pyramid model with multi-resolution to make remote sensing data visible and operable. Then, data processed by the visualization management subsystem is submitted to data promoting service subsystem of china manned space engineering for distribution and utilization. In consequence, the data promoting service subsystem accumulates mass of high-precision remote sensing data received from payloads of Tiangong-1 aircraft, Tiangong-2 space laboratory and Tianzhou-1 aircraft. To ensure security and service efficiency, the subsystem enhances identity authentication, authority management and database management on the basis of some previous work. Moreover, data exchange module, GIS-based visual data retrieval module and FTP-based file download tool are increased in the subsystem to meet the diverse requirements of scientists for remote sensing data sharing.

Keywords: Tiangong-2 · Tile · Visualization · Data service
Remote sensing data

1 Introduction

With the rapid development of earth observation technologies and projects, remote sensing data growth has exploded, which makes it an urgent problem that how to manage mass of remote sensing data effectively. In recent years, studies and applications for remote sensing data management from domestic and foreign have increased significantly. Meanwhile, plenty of application platforms such as NASA's Earth Observing System Data and Information System (EOSDIS) [1], the open-type of virtual globe sharing integration platform (GeoGlobe) [2] and Remote Sensing Cloud Mart

Y. Gu et al. (Eds.): *Proceedings of the Tiangong-2 Remote Sensing Application Conference*, LNEE 541, pp. 160–169, 2019.
https://doi.org/10.1007/978-981-13-3501-3_15

have been developed and played a big role in data promotion and sharing [3, 4]. However, some limitations such as simple management, weak operability and little expression appeared with the advance of visualization technology and web service. In this paper, we studied on image visualization and online data services of remote sensing data and introduced a visual management method based on pyramid model. The method which is used in visualization management subsystem and data promoting service subsystem demonstrated to be not only high-efficiency but also stabilized.

2 System Architecture

The most important requirement of remote sensing data management and service is displaying and retrieving data instantly and accurately. While it is a key problem that how to get the results the user needed from a mass of data and transfer to the user in a remarkably short period of time. Comprehensive utilization of NoSQL technology, visualization technology and web service technology, this paper constructs a high-efficiency platform for managing and sharing remote sensing data, which contains two subsystems. The system architecture of the platform is shown in Fig. 1.

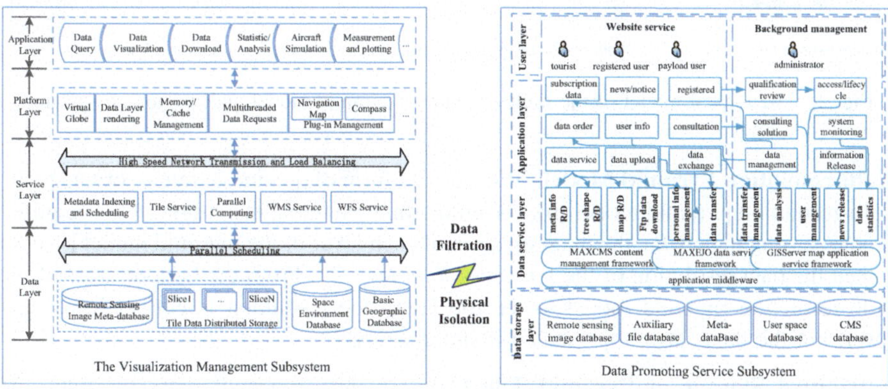

Fig. 1. System architecture of remote sensing data visualization management and online services platform

The visualization management subsystem adopts NoSQL database to store the pyramid tile model of remote sensing data. Visualization technology is used for intuitive query and analysis. Data promoting service subsystem which acquired data from the visualization management subsystem, three frameworks and heterogeneous database technology are used in it to relieve the pressure of I/O. In order to avoid confidential data leakage, equipment of the two subsystems should be physically separated. Rather than transmitting data through internet, we set up a special physical links between the two subsystems to guarantee the security and efficiency of data transmission.

3 Visualization Management for Remote Sensing Images

3.1 Multi-resolution Pyramid Model and Its Storage

The practice shows that the visual information is more helpful to improve the work efficiency. Multi-resolution pyramid model and its storage is the foundation of efficient management of remote sensing images based on visualization technology [5]. Through band extraction and transformation, we divide the remote sensing image into rectangular "tiles" of equal size according to spatial location grid, establish multiple tile layers of decreasing resolution, and store them to high performance NoSQL database. The logical model of the multi-resolution pyramid model and its storage system designed in this paper is shown in Fig. 2.

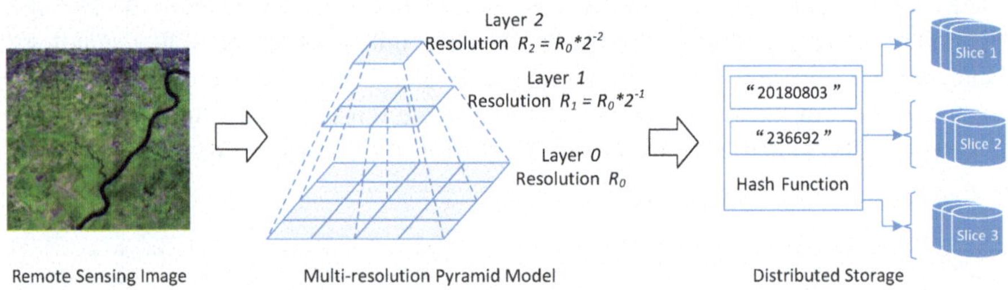

Fig. 2. The logical model of the multi-resolution pyramid model and its storage system.

Multi-resolution Pyramid Model. We take the form of quadruple tree to organize the multi-resolution pyramid model [6]. Assuming that the initial resolution of the acquired remote sensing data is R_0, the magnification of each layer of the pyramid model is set to 2, then the data resolution of the k layer is $R_k = R_0 * 2^{-k}$. From the bottom layer of the multi-resolution pyramid model to the top layer, the resolution is getting lower and lower, but the data space represented is not changed.

Distributed Storage of Pyramid Tile Data. In the storage of mass tile data, the following three issues need to be considered:

(a) With the gradual increase in the amount of tile data, the storage capacity of a single server is often insufficient. How to flexibly and conveniently expand the storage system dynamically?
(b) In the face of frequent requests of high concurrent tile data, how to ensure the I/O performance of the storage system?
(c) How to ensure the security of mass tile data and the convenience of data maintenance?

MongoDB is a new document-oriented NoSQL database using the binary as the storage format [7]. Its data structure is loose, but with high flexibility. Therefore, this paper uses MongoDB to build a distributed storage system of pyramid tile data. By designing an identifier generated by time, hardware and random number as the slice primary key and a hash function as the slice function, the tile data can be randomly and evenly distributed to each slice. With this method, its greatly increases the storage capacity and throughput of the system, which meets the demand of safe and efficient storage of massive pyramid tile data.

3.2 Tile Data Request and Parallel Scheduling

A remote sensing image may generate hundreds of tiles, and a large number of remote sensing images will generate countless tiles. So much tile data requested at the same time and loaded into the visualization management system is bound to cause delayed response of the operation, or even the collapse of the system. In order to ensure the execution efficiency of the visualization management, it is necessary to request the tile data according to the actual situation, and to load the requested tile data in parallel. The logical model of the tile data requested on-demand and parallel scheduling designed in this paper is shown in Fig. 3.

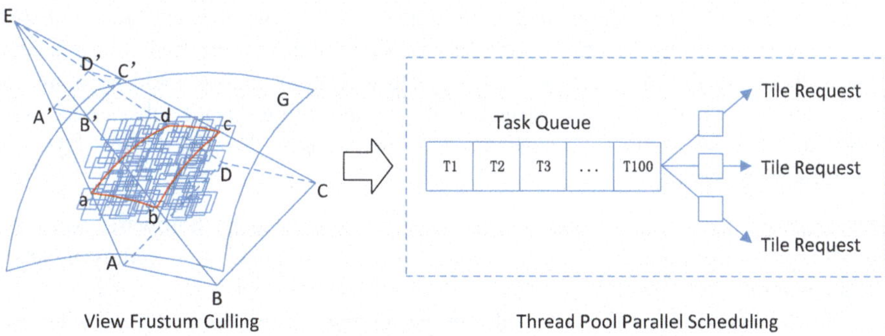

Fig. 3. The logical model of the tile data requested on-demand and parallel scheduling.

Tile Data Requested On-demand. In the process of 3D rendering, the visible range of the camera is called view frustum. The view frustum is an effective display range consisting of six surfaces calculated by the spatial transformation matrix, as shown in Fig. 3 by ABCDA'B'C'D'. According to the principle of the view frustum culling algorithm, the 3D rendering only shows the data in the range of the view frustum, and the data outside the range is not displayed [8]. Therefore, the view frustum culling can greatly reduce the amount of data to be rendered and can effectively improve the rendering efficiency.

The tile data requested on-demand in this paper is to use the view frustum culling algorithm to crop the tile data outside the view frustum. According to the real-time position and direction of the view frustum, and combined with the LOD algorithm, the tile data in the view frustum can be calculated, and the tile data that need to be loaded and rendered can be obtained.

Thread Pool Parallel Scheduling. Although the calculation of the view frustum culling is done, the amount of tile data to be loaded is still quite a lot, and concurrent requests are unavoidable. The thread pool can solve the problem of multitasking concurrency [9].

In this paper, a parallel scheduling method of thread pool for tile data is developed. There are several threads waiting to be invoked in the thread pool and remain dormant when idle. When a tile data request is received, one idle thread in the thread pool will be started to complete the loading of the requested tile. The parallel scheduling mechanism of thread pool starts multiple concurrent threads according to the number of tile data requested, and realizes synchronous and parallel tile data loading, which greatly improves the efficiency of a large number of tile data request and load.

4 Online Service for Remote Sensing Data

4.1 Subsystem Architecture

After tile-data requesting and parallel scheduling, remote sensing data with high-precision is selected from the visualization management subsystem and transmitted to the server equipment of data promoting service subsystem through high-speed special physical links. By this way, remote sensing data with high quality could be shared to the internet fleetly and safely. The structure of data promoting service subsystem can be divided into five layers for high cohesion and low coupling. We list the five layers in the following context.

Infrastructure Layer. The infrastructure layer is composed by hardware, software, and network environment of the subsystem. Specifically speaking, it mainly includes GIS application server which is rented and all equipment deployed in machine room such as database server, web server, file server, memory array, firewall, load balancers and basic software are installed in these equipment.

Data Storage Layer. This layer is designed for managing remote sensing files, remote sensing metadata, user information and web content information of the subsystem. For accessing data more secure and efficient, we designed heterogeneous databases for different data types. In addition, data access functions are encapsulated in interfaces to invoke data promptly.

Date Services Layer. Date services refer to operations involved in reading or writing actions on database. After transmitting to the data promoting service subsystem, data is displayed on website pages as images, lists or articles for public. In particular, there are two special ways of data services which made the subsystem more practical and participatory. One is that remote sensing data is presented on a fixed position of map according to latitude and longitude information stored in its metadata, the other is a client which provide the function of downloading big files from web server of the subsystem quickly and reliably through FTP.

Application Layer. In this layer, we encapsulated function module of the subsystem and developed a series of functional components for the user layer. By analyzing requirements of data promoting and sharing, application layer is composed of three parts, namely website services, background management and FTP data downloading client.

User Layer. There are four types of users in the subsystem. Excepting administrator who can login the background to update remote sensing data and manage information of the subsystem, three types of users are remained for website of the subsystem. Firstly, visitors are users who just visit the webpages or leave messages to administrator for questioning. Secondly, when a visitor completes registration and submits legitimate data requirement to subsystem, the administrator will approve the request and assign appropriate permissions to visitors. Then, the user becomes registered user and can download data within his authority. In addition, administrator can also deploy the life cycle of registered users to control the deadline for users to download data. The last type of subsystem is payloads users. This is a special role that can only be authorized to a few units who take care of payloads of aircrafts.

Demonstrating functions of data promoting service subsystem more clearly, we show the data flow of the subsystem in Fig. 4.

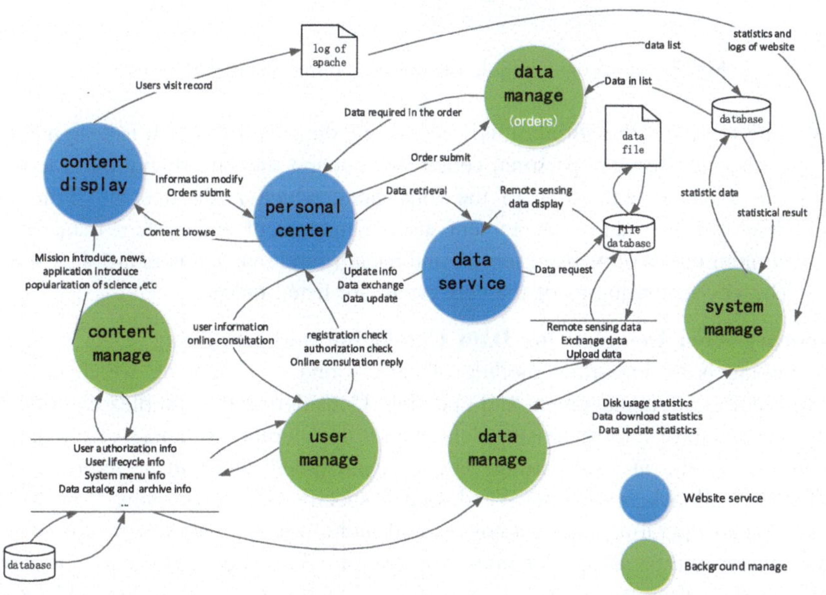

Fig. 4. Data flow of data promoting service subsystem.

4.2 Main Functions and Key Technologies

As shown in Fig. 5, data promoting service subsystem is made up by two parts-website service and background management. Besides data application results displaying, the website service also provides data retrieval and download functions. In this part, registered users can transmit their own data to other users by uploading data to user space. Furthermore, administrator will push the requested data to user space if a user submits a data order from website service. Background management refer to a website operated by administrator to manage and maintain the website service. Administrator catalogs and archived remote sensing data transferred from the visualization management subsystem. In addition, data exchange management and system data management is also designed in this part.

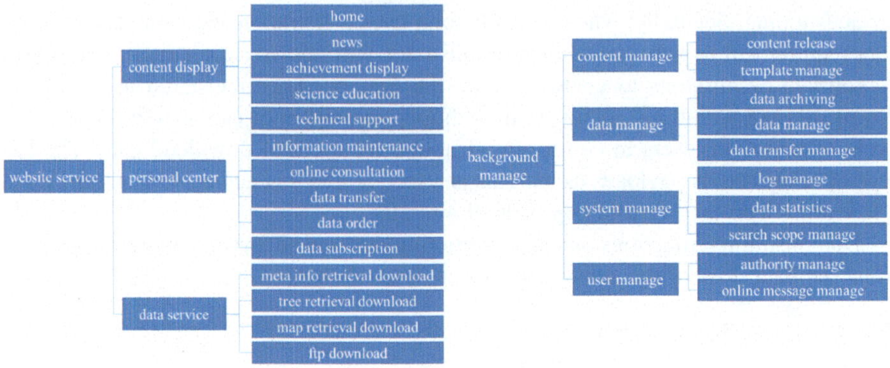

Fig. 5. Functional module of data promoting service subsystem.

Website service is the portal of user access and data acquisition. It is designed from three aspects: data service, personal center and content display. Background management is the management portal of the subsystem, which is divided into four parts: content management, data management, user management and system management. The function modules of website service and background management are shown in the Fig. 5. The key technologies of the subsystem are listed below.

Temporary Area Designed for Data Cataloging and Archiving. Data cataloging and archiving is an important module of background management, which is mainly responsible for data management and updating. Considering the complex classification and large volume of remote sensing data, we set up a temporary area to store data file without cataloguing and submitting. The archived failed data will be returned to the area for the rollback operation. We design a specific data processing flow for this module. When operating data cataloging and archiving, the subsystem will scan the temporary area and break up filenames into several fields. Then, fields and new storage path (defined by aircrafts, payloads, spectrum, production types or production levels) of each file will be stored in database, after that file will move to the new storage path from temporary area. If XML file associated with data file is existing in the temporary area, the content of XML file will be analyzed and collected into database. Before data submitting, we will check the database to ensure that no repetitive or duplicate data records exist. If the data already exists in database, we will change the field of time state. With the process of data cataloging and archiving, logs are generated and feedback to administrator to monitor operational progress. Once cataloging or archiving failed, the data will be sent back to the temporary area and wait for rollback.

Data Statistics and Analysis. The module of data statistics and analysis is designed for monitoring behavior of the subsystem. For showing statistical data visually and intuitively, we present data in histograms, pie charts, line charts or radar charts. By modifying configuration of apache tomcat and counting database by SQL, there are various statistical values included in this module such as user access statistics, user space usage statistics, data storage statistics, data downloads statistics and user uploads statistics.

Multiple Data Retrieval Techniques. Website service provides three portals for users to retrieve data. Meta-information retrieval refers to filter data according to conditions users chose. In favor of range retrieval, tree structure retrieval divides remote sensing data into a tree structure constructed according to storage directory of files. While map retrieval is a new kind of visual retrieval technology based on GIS, which applied JavaScript API 4.0 of Sky Map [10]. For data retrieving, users need to choose rectangular or circular tool provided on the webpage to define the range of latitude and longitude on map in the first. Then, other retrieval conditions such as aircraft name, sensor name, spectrum segment, product type, product level should be filled in or chose to shrink the range of data. Finally, result data of retrieval is listed on the left of map, while boxes that represent result data are drawn on the map at the same time. Users could point the boxes to download the corresponding data as well as check detailed information. It is well to be reminded that the boxes mentioned above are drawn according to the top left, bottom left, top right and bottom right latitude and longitude information of the data to ensure data accuracy and productivity level. The retrieval technology based on GIS benefits users who want images concentrated in a fixed area [11].

FTP-Based Large File Downloading Technology. As we all know, slow transmission and frequent interruptions are commonly occurred in great capacity of data transmitted through Hyper Text Transport Protocol. In order to avoid the disadvantages of HTTP, we developed a client accompanied the subsystem for large file downloads to promote the efficiency of data service. The client based on FTP will form a download queue for multiple files, and support reconnecting, resuming as well as simultaneous downloads. Before operating the client, users should export a list which contains data information that users wanted within the access permission invested by administration. If the list is imported into the client, download tasks will be created and carried out. As the server of client, the subsystem will encrypt data list by Rijndael algorithm, while the client could decrypt the list by the same secret key to protect data information of server from being compromised. For the veracity of statistics, we designed an interface to transmit downloads information of client to the subsystem. It would be shown in data statistics module as well as logs of background manage system.

5 Case Studies with Real World Remote Sensing Data

In reality, the visualization management subsystem has gathered plenty of remote sensing data acquired from multiple earth observation payloads of Tiangong-2 space laboratory. It has also achieved the purpose of retrieving and loading data visually and promptly. When the visualization management subsystem is running, response time of image loading can be controlled within 3 s while data visual rendering rate is faster than 15fps. Since coming online, on the one hand, the subsystem has managed and transmitted more than 7 TB remote sensing data to data promoting service subsystem of china manned space engineering for sharing and distributing. On the other hand, it is an efficient and intuitive mean to manage and display remote sensing images. In Fig. 6, we can see the retrieve results of the visualization management subsystem which displayed on three-dimensional digital map according to their geographical location information.

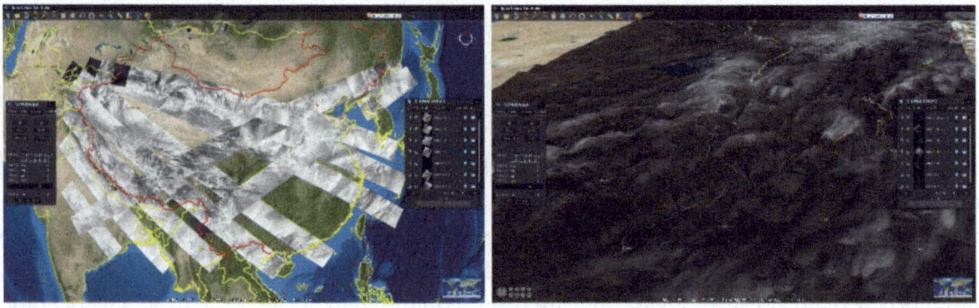

Fig. 6. Retrieve results of the visualization management subsystem.

Data promoting service subsystem of china manned space engineering [12] is a large data distribution platform in the field of remote sensing with more than 200 registered users. In terms of technical feasibility, it allows more than 1000 concurrent access while response time of data retrieving is less than 1s. The subsystem has provided more than 5.2 TB remote sensing data and customized product data to users since it is coming online. In terms of users, the subsystem has a good reputation among users according to our questionnaires (Fig. 7).

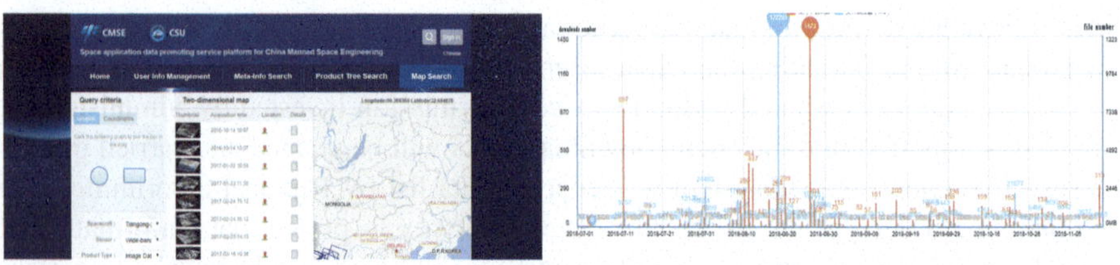

Fig. 7. Data statistics and data retrieval modules of data promoting service subsystem.

6 Conclusion

In this paper, we focused on the issue of massive remote sensing data management and presented a method on managing mass of remote sensing data visually. Subsequently, we studied the technology of multi-resolution pyramid model and storage, tile data requesting and parallel scheduling as well as data retrieval based on GIS, after which designed a platform for remote sensing data visualization management and online service. It has been proved that the platform has great improvement on efficient storage, visual management and convenient distribution.

Acknowledgments. The authors would like to thank China Manned Space Engineering for providing space science and application data products of Tiangong-2.

References

1. Esfandiari, M., Ramapriyan, H., Behnke, J., Sofinowski, E.: Earth observing system (Eos) data and information system (Eosdis)—evolution update and future. In: 2007 IEEE International Geoscience and Remote Sensing Symposium, pp. 4005–4008. IEEE, Barcelona (2007)
2. Gong, J., Chen, J., Xiang, L., Wang, Y.: GeoGlobe: geo-spatial information sharing platform as open virtual earth. Acta Geodaetica Cartogr. Sin. **39**(6), 551–553 (2010)
3. Lv, X., Cheng, C., Gong, J., Guan, L.: Review of data storage and management technologies for massive remote sensing data. Sci. China Technol. Sci. **41**(12), 1561–1573 (2011)
4. Yu, H., Li, S., Jiao, P., Hei, B.: Design and implementation of efficient sharing service platform for space utilization and science data. Manned Spacefl. **22**(4), 494–499 (2016)
5. Sample, J.T., Ioup, E.: Tile-based Geospatial Information Systems: Principles and Practices. Springer, New York City (2010)
6. Kang, J.: Technologies of Storage and Efficient Management on Cloud Computing for High Resolution Remote Sensing Image. Zhejiang University (2011)
7. Chodorow, K.: MongoDB: The Definitive Guide. O'Reilly Media, California (2013)
8. Kessenich, J., Sellers, G., Shreiner, D.: OpenGL Programming Guide, 9th edn. Addison-Wesley Professional, Boston (2016)
9. Oaks, S., Wong, H.: Java Threads. O'Reilly Media, California (2004)
10. Sky Map Web API. http://api.tianditu.com/apinew/webIndex.html. Accessed 1 Oct 2018
11. Huang, F.: Design and Implement of Massive Remote Sensing Image Management System. East China Normal University (2011)
12. Data promoting service subsystem of china manned space engineering. http://www.msadc.cn . Accessed 1 Oct 2018

Downscaling of Tiangong-2 Land Surface Temperature

Ren Luo[1], Ji Zhou[1(✉)], Jiajia Yang[1], Lijiao Ai[2], and Yilong Feng[2]

[1] School of Resources and Environment, Center for Information Geoscience, University of Electronic Science and Technology of China, Chengdu 611731, China
jzhou233@uestc.edu.cn
[2] Chongqing Landscape and Gardening Research Institute, Chongqing 401329, China

Abstract. Tiangong-2 is equipped with a new generation of earth observation instrument, i.e. Wide-band Imaging Spectrometer (WIS), which has a wide range and multiple spatial resolutions with visible and near infrared bands (VNIR, 100 m), shortwave infrared bands (SWIR, 200 m) and thermal infrared bands (TIR, 400 m). It provides the possibility to retrieve land surface temperature (LST) with a 400 m spatial resolution. However, such a spatial resolution limit the applications of TIR spectrum range of WIS in many fields. This study reports the downscaling of LST derived from Tiangong-2 WIS from 400 m to 100 m by utilizing different bands of Tiangong-2 WIS. Specifically, different downscaling methods are compared. Results shows that it is possible to downscale the 400 m LST to a 100 m resolution. Furthermore, the Random Forest (RF) method outperforms the traditional DisTrad method and the Multiple Linear Regression (MLR) method in LST downscaling. The downscaled LST at 100 m can satisfy applications such as the urban heat island monitoring, evapotranspiration calculation and environment modeling.

Keywords: Land surface temperature · Downscaling · Random Forest Wide-band Imaging Spectrometer (WIS) · Tiangong-2

1 Introduction

As a very important parameter that reflects the land surface condition, LST can be used to monitor the drought stress on vegetation and the urban heat island effect, assess soil moisture status, model the surface radiation balance and energy [1–4]. The acquisition of a wide range of LST is a key link in the study of the monitoring of terrestrial ecosystems and energy exchange between the earth's surface and the atmosphere [5]. With the development of technology, more and more satellites are equipped with TIR sensors. Thus TIR data over large area are available. However, in some situations, the spatial resolution of LST derived from satellite TIR data cannot meet the needs of a wider range of applications due to its coarser spatial resolution. Different applications have different resolution requirements for LST. For example, the needs of spatial

Y. Gu et al. (Eds.): *Proceedings of the Tiangong-2 Remote Sensing Application Conference*, LNEE 541, pp. 170–179, 2019.
https://doi.org/10.1007/978-981-13-3501-3_16

resolution for forest fire detection and urban heat island monitoring is about 100 m or higher.

The spatial resolution of TIR obtained from satellite sensors is not high as a result of the weak land surface thermal emission. For most satellite sensors, the spatial resolutions of the VNIR bands and SWIR bands are higher than the TIR bands. For example, the spatial resolutions of TIR bands of the widely-used MODIS (Moderate Resolution Imaging Spectroradiometer) and Landsat-8 are 1000 m and 100 m, respectively; the corresponding spatial resolutions of the other bands are 250–500 m and 15–30 m, respectively. For WIS of Tiangong-2, the spatial resolutions of its VNIR, SWIR and TIR bands are 100 m, 200 m and 400 m [6], respectively. In fact, the surface characteristics derived from the combination of VNIR and SWIR bands have certain correlation with LST. Thus, it is a suitable way to enhance the resolution of LST by utilizing other bands acquired from the same satellite or datasets from other sources.

In recent years, many methods have been proposed to improve the spatial resolution of LST. These methods are also known as the downscaling methods of LST. Based on the relationship between LST and other land surface parameters, Kustas et al. (2003) proposed the DisTrad method [7]. In DisTrad, the Normalized Difference Vegetation Index (NDVI) displayed as the descriptor of the LST and satisfactory downscaled LST was obtained. Subsequently, the DisTrad method has become a widely used method over vegetated surfaces such as agricultural areas. Other remote sensing indices were further added to strengthen the relationship between LST and surface parameters. When more than one parameter is considered to train the relationship, the method is regarded as the Multiple Linear Regression (MLR) method [8]. From the perspective of data fusion, many methods such as artificial neural network (ANN), kriging and Random Forest (RF) were utilized to downscale LST. Lemeshewsky et al. (2001) used data from the higher spatial resolution Landsat-7 ETM + bands to enhance the spatial resolution of TIR imageries by ANN method [9]. Pereira et al. (2018) used Geographically Weighted Regression Kriging (GWRK) method to downscale ASTER TIR images from 90 m to 15 m [10]. Hutengs et al. (2016) used RF method to downscale the LST from 1000 m to 250 m [11]. Compared with other downscaling methods, the RF method has the advantages of large sample capacity, small computational complexity and being not prone to over-fitting. The results of different methods for LST downscaling are different for different data and study areas. In this study, the RF method was adopted to downscale LST derived from Tiangong-2 WIS. For comparison purpose, two other methods, i.e. the DisTrad method and the MLR method, were also used in LST downscaling.

2 Study Area and Datasets

2.1 Study Area

In this study, the middle reaches of the Heihe River basin was selected as the study area. Heihe River is the second largest inland river in the arid regions of northwest China. The study area is distributed around Zhangye City and has an area of 36 km 36 km. The latitude and longitude ranges are from 38°43′N to 39°03′N and 100°10′E to 100°34′E, respectively. The climate in this area is relatively dry, with an average annual temperature of ~5.0 degrees Celsius. The study area is mainly composed by the Zhangye oasis, surrounded by Gobi, desert, etc. There are many croplands in the study area and a few built-up surfaces. The main crop is maize in summer. Considering the underlying surfaces in the study area, LST with a 400 m spatial resolution cannot satisfy the evapotranspiration calculation and water resource management at the field scale.

2.2 Datasets

The main datasets were obtained by Tiangong-2 WIS at 10:12 UTC on August 1, 2017, with bands of VNIR (100 m), SWIR (200 m) and TIR (400 m). Table 1 shows the information of bands of Tiangong-2 WIS. Some additional datasets were used for data processing and result validation. Atmospheric water vapor content products and near surface temperature products from the European Centre for Medium-Range Weather Forecasts were used to calculate LST. The in-situ measured LSTs provided by the Heihe Watershed Allied Telemetry Experimental Research (HiWATER) were used to validate the downscaled LST [12, 13]. HiWATER has been conducted in the Heihe River basin since 2012. There are three continuously observed ground stations in the study area, i.e. Daman, Huazhaizi and Wetland. LST was derived from the measured incoming and outgoing longwave radiation according to the Stefan-Boltzmanns' law [14]. The derived LST will be referred as in-situ LST hereinafter. The measurement interval at every station was 10 min. We used the in-situ LST from those ground stations to validate the downscaled LST. Figure 1 shows the VNIR image of the study area obtained from Tiangong-2 WIS at 10:12 UTC on August 1, 2017. Locations of three ground stations are also shown.

Table 1. Bands of Tiangong-2 WIS

Bands	VNIR band (μm)	SWIR band (μm)	TIR band (μm)
Channel range	V1: 0.970 ~ 0.990	S1: 1.230 ~ 1.250	T1: 8.125 ~ 8.825
	V2: 0.930 ~ 0.950	S2: 1.630 ~ 1.650	T2: 8.925 ~ 9.275
	V3: 0.895 ~ 0.915		
	V4: 0.845 ~ 0.885		
	V5: 0.810 ~ 0.830		
	V6: 0.740 ~ 0.760		
	V7: 0.6775 ~ 0.6875		
	V8: 0.655 ~ 0.675		
	V9: 0.610 ~ 0.630		
	V10: 0.555 ~ 0.575		
	V11: 0.510 ~ 0.530		
	V12: 0.480 ~ 0.500		
	V13: 0.433 ~ 0.453		
	V14: 0.403 ~ 0.423		
Resolution (m)	100	200	400

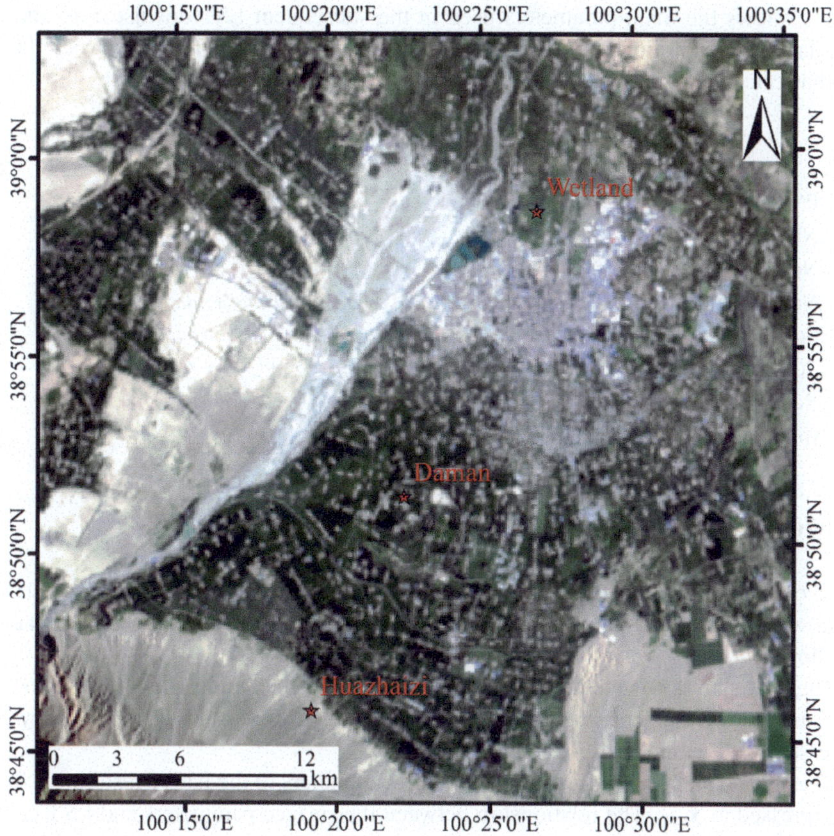

Fig. 1. Tiangong-2 VNIR images of the study area and the locations of three ground stations. The images are composited by R/G/B = V8/V10/V12.

TIR images from Tiangong-2 WIS were level 2 and the accuracy of absolute radiation calibration for TIR images was 2 K. As shown in Table 1, WIS has two bands in the TIR range, i.e. T1 and T2. The effective wavelengths of these two bands are 8.47 μm and 9.10 μm, respectively. In this study, the split window algorithm was adopted to retrieve the LST from these two bands. The split window algorithm is one of the most widely-used algorithm for LST retrieval. Its advantage over other algorithms is that the attenuation of the surface emission radiation by the atmosphere is proportional to the brightness difference of the stars on two adjacent thermal infrared channels [15]. Thus fewer input parameters are required. Zhou et al. (2018) trained 17 kinds of split window methods to find which one is the optimal method [16]. The method proposed by Wan et al. (2014) was selected as the optimal split window method in the study area [17]. The resolution of the retrieved LST was 400 m.

In addition to the LST, some other land surface parameters were calculated from the VNIR and SWIR bands. The typical features of the study area are vegetation, river, lake, building and soil. Thus, the corresponding land surface indices are considered.

Table 2 shows the main parameters used in the subsequent LST downscaling and the formulas. The resolution of indices that calculated from the VNIR bands was 100 m, and that calculated from the VNIR and SWIR bands was 200 m.

Table 2. The main land surface parameters.

Parameter	Formula
NDVI	$(\rho_{V4}-\rho_{V8})/(\rho_{V4}+\rho_{V8})$
Normalized difference water index (NDWI)	$(\rho_{V10}-\rho_{V4})/(\rho_{V10}+\rho_{V4})$
Normalized difference building index (NDBI)	$(\rho_{S2}-\rho_{V4})/(\rho_{S2}+\rho_{V4})$
Soil-adjusted vegetation index (SAVI)	$(\rho_{V4}-\rho_{V8})\times 1.5/(\rho_{V2}+\rho_{V4}+0.5)$
Soil vegetation index (SVI)	$(\rho_{S2}-\rho_{V10})/(\rho_{S2}+\rho_{V10})$
Ratio vegetation index (RVI)	ρ_{V4}/ρ_{V8}

Note ρ denotes the land surface reflectance.

3 Methodology

Based on the principle that the relationship between LST and land surface parameters is stable within a certain spatial resolution, the downscaling method can be summarized as follows.

The first step is to calculate the LST at the native resolution (i.e. 400-m) and the land surface parameters at higher resolution (i.e. 100 m and 200 m). They are expressed as T_{low} and $\mathbf{x}_{high}$, respectively;

The second step is to resample the surface parameters to the native resolution. They are expressed as $\mathbf{x}_{low}$. The relationship between the surface parameters and LST can be expressed as f, which can be linear or nonlinear functions.

$$T_{pre-low} = f_{low}(\mathbf{x}_{low}) \tag{1}$$

Where $T_{pre\text{-}low}$ is the predicted LST at the native resolution based on the surface parameters; $\mathbf{x}_{low}$ is a vector.

The error caused by the scale effect and the lack of interpretation of these parameters on the LST would affect the result of downscaling. Zhou et al. (2016) analyzed scale effect of LST downscaling and the result showed that the scale effect can affect the accuracy [18]. Thus, the residual temperature ΔT should be considered. The residual ΔT is the difference between T_{low} and the predicted $T_{pre\text{-}low}$ (Eq. 2). ΔT at the target resolution can be obtained by resampling ΔT image at the native resolution. After resampling, we get ΔT_{high}.

$$\Delta T_{low} = T_{low} - T_{pre-low} \tag{2}$$

The third step is to estimate the LST at the target resolution by using the high-resolution parameters and the relationships established at the native resolution. f_{high} is the same as f_{low} within a certain spatial resolution.

$$T_{\text{high}} = f_{\text{high}}(\mathbf{x}_{\text{high}}) + (T_{\text{low}} - T_{\text{pre-low}}) \tag{3}$$

In the DisTrad method, the utilized index was NDVI and f was a linear function. In the MLR method, the utilized indices were NDVI, NDBI, NDWI, SAVI, RVI and SVI, and f was a multiple linear function. Meanwhile, the moving window method was adopted to train the regression model for a better training purpose. Ding et al. (2018) used moving window method to downscale surface air temperature and achieved good results [19]. An optimal window was selected to maximize the correlation coefficient of the estimated LST and the true LST. The window ranged from 3×3 pixels to the maximum extent of the area.

The aforementioned single-factor and multi-factors linear regression relationship may not fully express the complex relationship between the parameters and LST. For the RF method, f is a nonlinear regression function. LST, NDVI, NDBI, NDWI, SAVI and RVI were used to train the RF model. RF uses the bootstrap resampling technique to extract multiple samples from the original training samples and combines them to generate training sample collections. Then, multiple decision trees are generated according to the training sample collections and the RF model is constructed. The classification or regression model results are determined by voting score of decision trees. In the process of LST downscaling, the relationship between LST and various parameters may not be nonlinear, and the RF method is not sensitive to multivariate collinearity, which can effectively prevent overfitting.

4 Results

Three methods were used to downscale the LST from 400 m to 100 m. The results of the LST downscaling are shown in Fig. 2. Figure 2a1 is the LST with a 400 m resolution before downscaling. Figures 2b1, f1 are the downscaled LSTs through different methods at 100 m. The spatial distribution of LST before and after downscaling is consistent, namely, the urban area is warm and surrounding suburbs and the croplands are cooler. Figure 2b1, d1 are the downscaled LSTs by the DisTrad method and the MLR method. Through comparing these two figures, the downscaled LST by the MLR method presents more details and a clearer texture, with the increases of regression factors. The DisTrad method only considers one parameter (NDVI) and it is affected by the fitting residual. Thus, the texture features are not obvious. Adding more remote sensing indices can improve the ability of the downscaling model. The moving window approach was applied in the training process of the regression model. For the DisTrad method, the optimal window was 3×3 pixels; for the MLR method, the optimal window was 7×7 pixels. The results are shown in Fig. 2c1, e1 and the moving window approach does not increase the details of images.

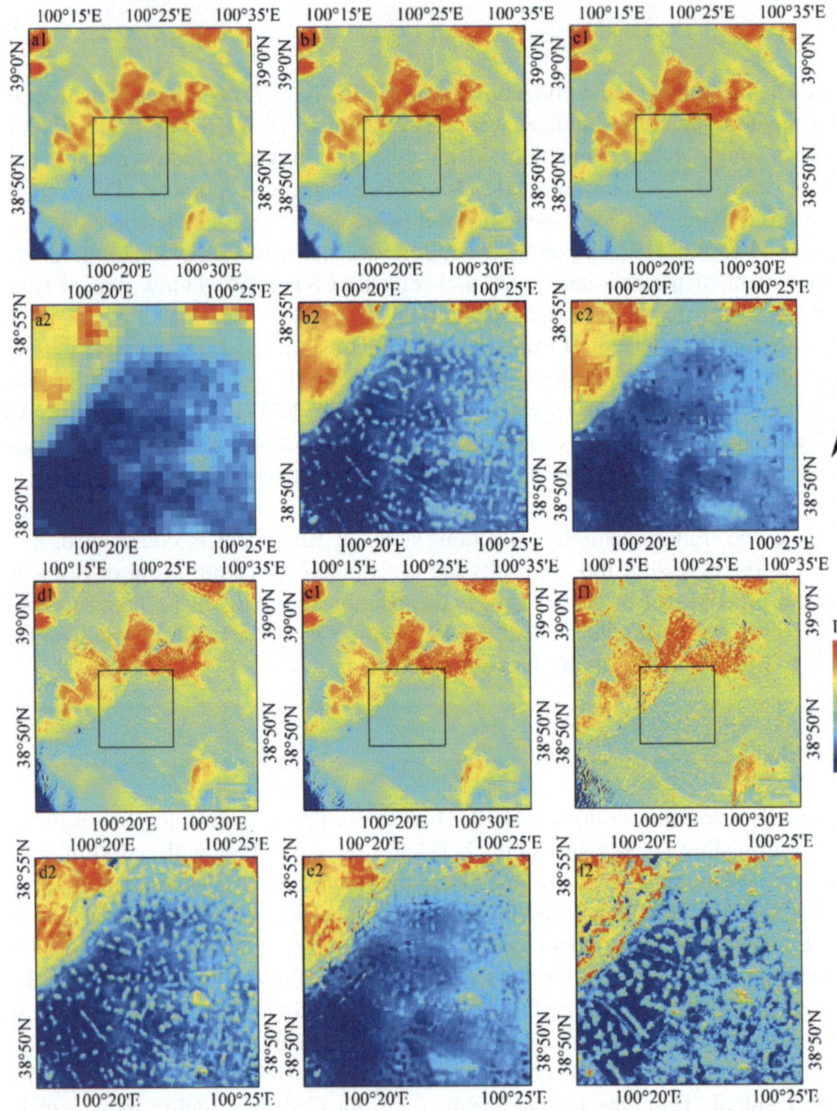

Fig. 2. Results of the LST downscaling. (a1) the LST derived from the Tiangong-2 WIS data at 400 m; (b1) the LST downscaled by the DisTrad method; (c1) the LST downscaled by the DisTrad method with the moving window approach; (d1) the LST downscaled by the MLR method; (e1) the LST downscaled by the MLR method with the moving window approach; (f1) the LST downscaled by the RF method. Figure (a2) ~ (f2) are a subset of Figure (a1) ~ (f1).

In contrast to the aforementioned downscaling method, the RF method considered the nonlinear regression between LST and other parameters. The result is shown in Fig. 2f1, when comparing the result with the other method, the downscaled LST image has more details so that it is more useful in the applications.

Figure 2a2, f2 shows the downscaled LSTs of a subset area. Comparing Fig. 2b2, d2 with Fig. 2c2 and Fig. 2e2, it is clear that the moving window approach has not improved the image quality downscaled by the DisTrad method and the MLR method. Figure 2f2 is a part of LST downscaled by the RF method. Apparently, Fig. 2f2 presents more details of the spatial distributed LST. Thus, it can better display the spatial distribution characteristics of LST. From the perspective of image accuracy, the RF method outperforms the DisTrad method and the MLR method in downscaling the LST derived from Tiangong-2 WIS.

Table 3 shows the measured LST and the estimated LST by different downscaling methods. The measured LSTs at 10:12 UTC were obtained by linear interpolation of the measured LSTs at 10:10 UTC and 10:20 UTC. Obviously, the estimated LST is closed to the measured LST. The bias and root mean squared error (RMSE) of retrieved LST is 2.72 K and 2.75 K, respectively. The bias and RMSE of the downscaled LSTs range from 2.47 K to 2.85 K and 2.50 K to 2.87 K, respectively, indicating the accuracies are acceptable. It is difficult to determine which method performs better in LST downscaling based on bias and RMSE. Nevertheless, with the increase of the archived Tiangong-2 WIS images, more datasets will be available in the future. The long-term LST derived from Tiangong-2 WIS will be benefit the comparison of LST downscaling methods.

Table 3. Validation of the downscaled LST based on in-situ LST at three ground stations.

Station	In-situ LST	Retrieved LST	Downscaling method				
			DisTrad	MLR	DisTrad-moving window	MLR-moving window	RF
Daman	297.56 K	300.02 K	299.75 K	299.47 K	300.05 K	300.05 K	299.90 K
Huazhaizi	297.66 K	300.90 K	300.71 K	300.56 K	300.92 K	300.87 K	300.42 K
Wetland	297.84 K	300.31 K	300.00 K	300.57 K	300.64 K	300.07 K	300.98 K
Mean LST	297.69 K	300.41 K	300.15 K	300.20 K	300.54 K	300.33 K	300.43 K
Bias	–	2.72 K	2.47 K	2.51 K	2.85 K	2.64 K	2.75 K
RMSE	–	2.75 K	2.50 K	2.55 K	2.87 K	2.68 K	2.77 K

5 Conclusion

In this study, three methods, including the DisTrad method, the MLR method and the RF method were applied to downscale the LST derived from WIS of Tiangong-2. We successfully improved the resolution of LST from 400 m to 100 m. From the perspective of image accuracy, the RF method outperformed the DisTrad method and the MLR method in the downscaling of the Tiangong-2 WIS LST. From the perspective of data accuracy, it is difficult to determine which method is suitable for LST downscaling because they all got an acceptable result. However, for LST downscaling, its goal is to obtain an image with better image quality. Under the similarity of data accuracy, the method making the image accuracy better is selected. Thus, we draw this conclusion that RF method is better than the traditional DisTrad method and the MLR method in downscaling of Tiangong-2 WIS LST.

Acknowledgements. Thanks to China Manned Space Engineering for providing space science and application data products of Tiangong-2. This study is supported by the National Key Research and Development Program of China under grant 2017YFB0503905.

References

1. Wan, Z., Wang, P., Li, X.: Using MODIS land surface temperature and normalized difference vegetation index products for monitoring drought in the southern great plains. USA. Int. J. Remote Sens. **25**, 61–72 (2004)
2. Norouzi, H., Bah, A., Prakash, S., Nouri, N., Blake, R.: Downscaling satellite land surface temperatures in urban regions for surface energy balance study and heat index development. In: AGU Fall Meeting (2017)
3. Han, Y., Wang, Y., Zhao, Y.: Estimating soil moisture conditions of the greater Changbai mountains by land surface temperature and NDVI. IEEE Trans. Geosci. Remote Sens. **48**, 2509–2515 (2010)
4. Voogt, J.A., Oke, T.R.: Thermal remote sensing of urban climates. Remote Sens. Environ. **86**, 370–384 (2003)
5. Quattrochi, D.A., Luvall, J.C.: Thermal infrared remote sensing for analysis of landscape ecological processes: methods and applications. Landscape Ecol. **14**, 577–598 (1999)
6. TG-2 Homepage. http://www.msadc.cn/en//gk/t2payload/mwi/. Accessed 15 Sep 2018
7. Kustas, W.P., Norman, J.M., Anderson, M.C., French, A.N.: Estimating subpixel surface temperatures and energy fluxes from the vegetation index–radiometric temperature relationship. Remote Sens. Environ. **85**, 429–440 (2003)
8. Zhan, W., Chen, Y., Zhou, J., Wang, J., Liu, W., Voogt, J., Zhu, X., Quan, J., Li, J.: Disaggregation of remotely sensed land surface temperature: Literature survey, taxonomy, issues, and caveats. Remote Sens. Environ. **131**, 119–139 (2013)
9. Lemeshewsky, G.P., Schowengerdt, R.A.: Landsat 7 thermal-IR image sharpening using an artificial neural network and sensor model. In: Aerospace/Defense Sensing, Simulation, and Controls, pp. 181–192 (2001)
10. Pereira, O., Melfi, A., Montes, C., Lucas, Y.: Downscaling of ASTER thermal images based on geographically weighted regression kriging. Remote Sens. **10**, 633 (2018)
11. Hutengs, C., Vohland, M.: Downscaling land surface temperatures at regional scales with random forest regression. Remote Sens. Environ. **178**, 127–141 (2016)
12. Li, X., Cheng, G., Liu, S., Xiao, Q., Ma, M., Jin, R., Che, T., Liu, Q., Wang, W., Qi, Y., Wen, J., Li, H., Zhu, G., Guo, J., Ran, Y., Wang, S., Zhu, Z., Zhou, J., Hu, X., Xu, Z.: Heihe watershed allied telemetry experimental research (HiWATER): scientific objectives and experimental design. Bull. Am. Meteor. Soc. **94**, 1145–1160 (2013)
13. Liu, S.M., Xu, Z.W., Wang, W.Z., Jia, Z.Z., Zhu, M.J., Bai, J., Wang, J.M.: A comparison of eddy-covariance and large aperture scintillometer measurements with respect to the energy balance closure problem. Hydrol. Earth Syst. Sci. **15**, 1291–1306 (2011)
14. Zhou, J., Li, M., Liu, S., Jia, Z., Ma, Y.: Validation and performance evaluations of methods for estimating land surface temperatures from ASTER data in the middle reach of the Heihe River Basin. Northwest China. Remote Sensing. **7**, 7126–7156 (2015)

15. Ulivieri, C., Castronuovo, M.M., Francioni, R., Cardillo, A.: A split window algorithm for estimating land surface temperature from satellites. Adv. Space Res. **14**, 59–65 (1994)
16. Zhou, J., Liang, S., Cheng, J., Wang, Y., Ma, J.: The GLASS land surface temperature product. IEEE J. Sel. Topics Appl. Earth Observations Remote Sens., 1–15 (2018)
17. Wan, Z.: New refinements and validation of the collection-6 MODIS land-surface temperature/emissivity product. Remote Sens. Environ. **140**, 36–45 (2014)
18. Zhou, J., Liu, S., Li, M., Zhan, W., Xu, Z., Xu, T.: Quantification of the scale effect in downscaling remotely sensed land surface temperature. Remote Sensing. **8**, 975 (2016)
19. Ding, L., Zhou, J., Zhang, X., Liu, S., Cao, R.: Downscaling of surface air temperature over the Tibetan Plateau based on DEM. Int. J. Appl. Earth Obs. Geoinf. **73**, 136–147 (2018)

Analysis of Slope and NDVI Effects on Land Surface Temperature Retrieval Accuracy in Mountain Area Based on WIS Data of Tiangong-2

Jingxu Wang, Bing Wang, Yangyang Liu, and Huaguo Huang(✉)

Key Laboratory for Silviculture and Conservation of Ministry of Education, Beijing Forestry University, Beijing, China
{wangjingxu_2016, huaguo_huang}@bjfu.edu.cn

Abstract. The relationship between land surface temperature (LST) and Normalized Difference Vegetation Index (NDVI) is evaluated in mountain areas using the Wide-band Imaging Spectrometer (WIS) data on Chinese Tiangong-2 (TG-2) Space Lab. The WIS is composed of three regions: Visible Near-Infrared (VNIR) (14 channels), Short-Wave Infrared (SW) (2 channels), and Thermal Infrared (TIR) (2 channels). The two TIR channels, T_1 (8.125–8.825 μm) and T_2 (8.925–9.275 μm) have not been fully exploited to extract LST. Thus, a modified split window method was proposed to retrieve LST and evaluated by comparing to the MODIS LST product. Then, the LST retrieval accuracy affected by slope and NDVI is analyzed. Results show that: (a) TG-2 LST is positively correlated to MODIS LST ($R^2 = 0.74$) but overestimated (about 3.5 K); further improvement on the split window method is still needed; (b) the TG-2 LST accuracy is different in vegetation and non-vegetation area, where non-vegetation area with small NDVI standard deviation has better agreement between MODIS LST and TG-2 than vegetation area; (c) the LST retrieval accuracy in vegetation area is independent on slopes; while increasing with the slope in non-vegetation area.

Keywords: LST retrieval · Split-window algorithm · NDVI · Slope TIR · WIS data of Tiangong-2

1 Introduction

Land surface temperature (LST) is an important geophysical variable affecting the land-surface processes at regional and global scales [1]. The long-wave radiation derived from LST affects latent heat and sensible heat flux [2], which can be used to assess surface energy and hydrological balance, thermal inertia and soil moisture. Accurate LST from remote sensing has contributed a lot to global surface temperatures and long-term dynamics [3–11]. Two main steps, including atmospheric correction and land surface emissivity correction, are usually used to extract LST. The split-window algorithm [12] is the most widely used method to remove atmospheric effect.

Y. Gu et al. (Eds.): *Proceedings of the Tiangong-2 Remote Sensing Application Conference*, LNEE 541, pp. 180–191, 2019.
https://doi.org/10.1007/978-981-13-3501-3_17

Traditionally, the split-window algorithm used the two channels centered at about 11 and 12 μm to remove the atmospheric effect through differential atmospheric absorption. The linear or nonlinear combination of the brightness temperatures is normally applied for LST estimation. This algorithm has been first applied for sea surface temperature inversion with NOAA-AVHRR thermal infrared remote sensing data. After more than 30 years of development, more than 20 inversion algorithms have been derived [13]. The Wide-band Imaging Spectrometer (WIS) remote sensing data of the Tiangong-2 (TG-2) contains rich spectral information, including visible band (VIS), near-infrared (NIR), short-wave infrared (SW), and thermal infrared (TIR) bands for LST inversion. The two TIR channels, T_1 (8.125–8.825 μm) and T_2 (8.925–9.275 μm), are not centered at the traditional 10–13 μm regions where the split window method can be used. Thus, T_1 and T_2 have not been fully exploited to extract LST. However, the TIR remote sensing instruments with 8–9.3 μm wavebands have advantages in performance and mature technology than those in 10–13 μm [13]. If the split window method could be used in 8–9.3 μm bands, the LST extract accuracy and stability might be improved. Zhang et al. (2015) has successfully demonstrated the possibility of using 8–9.3 μm wavebands to inverse sea surface temperature [13]. However, the feasibility of the split window method in 8–9.3 μm has not been evaluated in land surface area.

Therefore, in this study, the VNIR and TIR bands of WIS data of TG-2 were combined and utilized to test the possibility to inverse LST by adapting the traditional split window algorithm from 10–12 μm to 8–9.3 μm channels. To evaluate the TG-2 LST, the MODIS LST products are used for comparisons. Furthermore, based on the previous studies on the strong relationship between LST and NDVI [14, 15], the correlation between LST and NDVI is also used to show the spatial pattern of LST. Especially, we test the LST-NDVI relationship under different slope conditions, which will extend the application of using TG-2 thermal images in mountain areas.

2 Data and Methods

2.1 Study Area and Data

The study area (27°29′N-29°08′N and 100°18′E-101°45′E) is located in the south-western China, around the border of Yunnan province and Sichuan province. There are three major data sources.

First, the slope data is the Shuttle Radar Topography Mission (SRTM) slope product with 90 m resolution (see Fig. 1). This data is used to analyze the slope effects on LST-NDVI relationship.

Second, the TIR Level 2 product (400 m) of WIS data of TG-2 at 04:17:47 on January 23, 2018 has been applied to LST extraction. The T_1 with TIR channels of WIS data has a wavelength range of 8.125–8.825 μm and a center wavelength of 8.475 ± 0.105 μm; The T_2 channel has a wavelength range of 8.925–9.275 μm and a center wavelength of 9.100 ± 0.0525 μm. To extract NDVI, the closest VNIR images (100 m) taken at 10:11:59 on January 25, 2018 was used. The NDVI is calculated using the average reflectance values in red wavelengths (band 7 and 8) and NIR bands (1–6) to reduce noises.

Fig. 1. The false color composite image of study area (R:8th band, G:10th band; B: 12th band). The range of study area is from 27°29′N to 29°08′N, and from 100°18′E to101°45′E

Third, in order to verify the LST retrieval accuracy, the LST product (1 km resolution) of MODIS (00:57:28, January 23) was used as reference data, which is the closest image to TIR band data of WIS data.

Finally, to compare with the MODIS LST products, the NDVI, Slope and inversed TG-2 LST are all aggregated to 1 km.

2.2 The Algorithm of LST Retrieval

According to the previous study [13], we adapted Franca & Cracknell algorithm [16] originally using NOAA-AVHRR CH_4 and CH_5 channel. Because of the different channels, the calculation formula is adapted using T_1 and T_2 channel data of WIS of TG-2 to invert the LST (T_s), like:

$$T_s = T_1 + A(T_1 - T_2) + B \tag{1}$$

Where T_1, T_2 is the brightness temperature (BT) from the two channels, T_s is the LST retrieval. The parameters, *A* and *B,* are calculated using the method proposed by Franca and Cracknell [16]:

$$A = (C_1D_2 + C_2D_1)/(C_1D_2 - C_2D_1) \quad (2)$$

$$B = [(1 - \frac{1}{\varepsilon_2})(1 - 2W_2)C_2D_1L_2 - (1 - \frac{1}{\varepsilon_1})(1 - 2W_1)C_1D_2L_1]/(C_1D_2 - C_2D_1) \quad (3)$$

Where L_i is the parameter related to the BT at channel i;W_i is the relevant parameter of atmospheric absorption; Since L_i has the great correlation with TIR channels, Zhang et al. [13] modified the parameters for T_1 and T_2 channels between 260 K and 320 K using the Planck function in the same way as Sobrino et al. [17]. Finally, the linearly fitted formula (L_1 = 0.33T_1–47.49; L_2 = 0.3607T_2–50.83) is used; the parameters C and D are derived as:

$$C_i = \varepsilon_i\tau_i\cos\theta \quad (4)$$

$$D_i = W_i[1 + 2\tau_i(1 - \varepsilon_i)\cos\theta] \quad (5)$$

Where ε_i is the emissivity of channel i, τ_i is the atmospheric transmittance of channel i;θ is the zenith angle; W_i is calculated as: $W_i = a_1\omega + a_2\omega^2$;$\omega$ is the atmospheric water content (g/cm^3). W_i is calculated according to the relationship between τ_i and θ, proposed by Franca and Cracknell [16]. Thus, the parameters of W_i (a_1 and a_2) of different channels are obtained.

2.3 The Surface Emissivity

The calculation of the specific emissivity is based on the method proposed by Qin et al. [18].

$$\varepsilon_i = P_vR_v\varepsilon_{iv} + (1 - P_v)R_s\varepsilon_{is} + d_\varepsilon \quad (6)$$

Where P_v is the vegetation fraction:

$$P_v = (NDVI - NDVI_s)/(NDVI_v - NDVI_s) \quad (7)$$

If P_v = 0, the pixels is soil, and $\varepsilon = \varepsilon_s$; if P_v = 1, the pixels is vegetation, and $\varepsilon = \varepsilon_v$. In flat surface, d_ε = 0 is generally used; in a rough surface, d_ε can be simply estimated based on the proportion of vegetation.

Based on the formula proposed by Qin et al. [18], when $P_v < 0.5$, $d_\varepsilon = 0.003\,8P_v$; when $P_v > 0.5$, $d_\varepsilon = 0.003\,8(1 - P_v)$; when $P_v = 0.5$, d_ε is maximum, and $d_\varepsilon = 0.0019$. R_v and R_s is the temperature ratio of vegetation and soil: $R_v = 0.9332 + 0.0585P_v$; $R_s = 0.9902 + 0.1068P_v$.

2.4 The Atmospheric Transmittance

Water vapor content is the main consideration for estimating atmospheric transmittance in the process of LST inversion. Similar to the inversion of atmospheric water content using MODIS water vapor channels (the 2th and 19th bands used by Mao et al. [19]), the VNIR sensor mounted on TG-2 also contains the similar channels (the 2th band center at 940 nm and 4th band center at 865 nm), which can be used, for water vapor inversion (ω, g/cm^3):

$$\omega = \{[\alpha - \ln(v_2/v_4)]/\beta\}^2 \tag{8}$$

Where $\alpha = 0.0200$, $\beta = 0.6321$, and v_2 and v_4 is the surface reflectance of 2th band and 4th band of VNIR data of WIS of TG-2.

The common method to relate atmospheric transmittance to water vapor content is using atmospheric models such as MODTRAN, 6S and LOWTRAN. Because the TG-2 WIS channels are different from the traditional 10–13 μm channels, coefficients between atmospheric transmittance to water vapor content should be re-build. We used the fitted equation by Zhang et al [13] for 8–9 um:

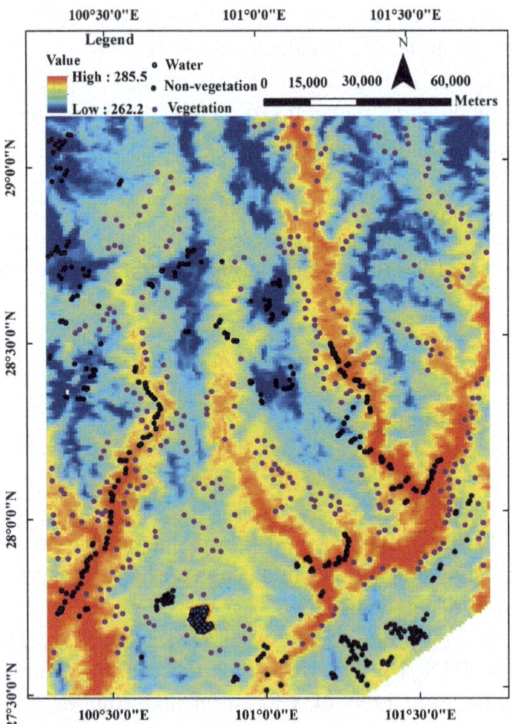

Fig. 2. MODIS LST product with 774 sample pixels in study area

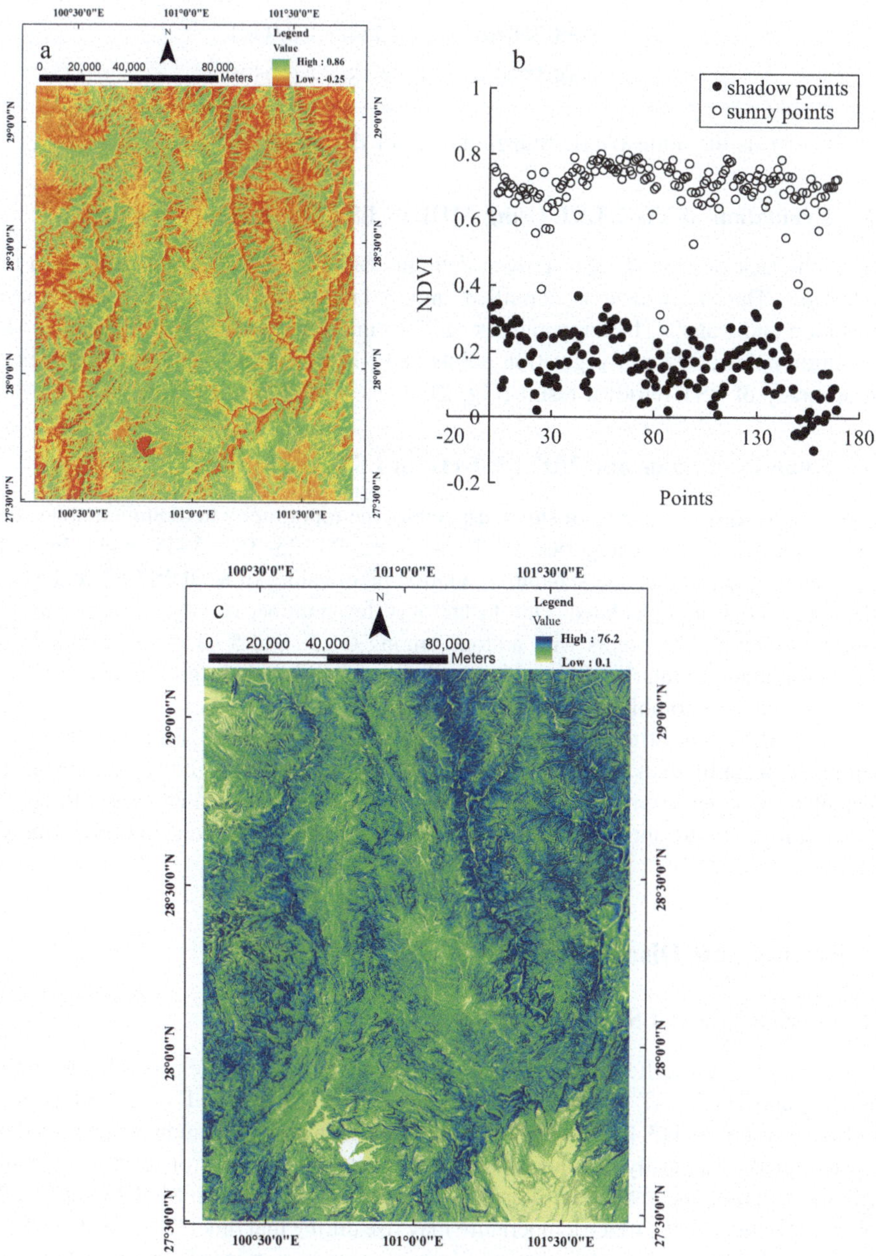

Fig. 3. (a) NDVI calculated by NIR band and Red band of WIS data of TG-2 in same area; (b) 169 NDVI points in shadow region and sunny region of vegetations; (c) SRTM slope product.

$$\tau_1 = 0.0003487\omega^2 - 0.01554\omega + 0.9301$$
$$\tau_2 = 0.0001736\omega^2 - 0.00818\omega + 0.9575 \quad (9)$$

Where τ_i is the atmospheric transmittance for the WIS band i of TG-2.

2.5 Evaluation of TG-2 LST Using MODIS LST Products

Due to the lack of ground measurement data, the MODIS LST products were used for evaluation. The study area is classified into 3 land cover types (vegetation, non-vegetation and water). The temperatures of 774 sampling points (1 km) contained 343 vegetation points, 397 non-vegetation points and 34 water points are selected to verify the accuracy of LST retrieval result (Fig. 2).

2.6 Analysis of Slope and NDVI Effects on LST Accuracy

In order to analyze the effects of slope, especially the topography in mountain areas, the slope is classified to 6 categories: (0–5, 5–15, 15–25, 25–35, 35–45, >45). In each slope category, the mean and standard deviation pixel values of MODIS LST and TG-2 LST, and NDVI are calculated. Due to the very few number of vegetation points for slope less than 5°, the points with a slope among 0°–5° and 5°–15° are combined in vegetation areas. Similarly, the non-vegetation points with slope greater than 45° are very few and also combined with slope among 35°–45°.

Due to the effect of topography, the NDVI values in shadows are generally lower than those in sunlit areas (see Fig. 3b). Considering that NDVI is the key parameter in calculation of specific emissivity, corrections are required in shadows areas. However, in this study, we do not select the pixels in shadow areas to avoid the error due to shadow.

3 Results and Discussions

3.1 Verification of LST Retrieval Accuracy

Figure 5 shows the scattering points and agreements of 774 sample pixels between the MODIS, and TG-2 LST (R^2 = 0.74, $RMSE$ = 3.5 K). The inversed TG-2 LST is generally over-estimated (3.4 K in vegetation area, 3.3 K in non-vegetation area, and 5.0 K in water areas), especially in water areas. There are two possible error sources: (1) temperature variation due to the 3 h difference of overpass time between MODIS and TG-2; for example, the water surface temperature may rise during that time; (2) we just used the center wavelengths of T_1 and T_2 instead of their spectrum response functions. Omission of the impact of spectrum response of WIS instrument may lead to the overestimation, especially bright temperature at the entrance pupil great than 280 K [13]. More convinced ground data are still needed in the future to confirm the method (Fig. 4).

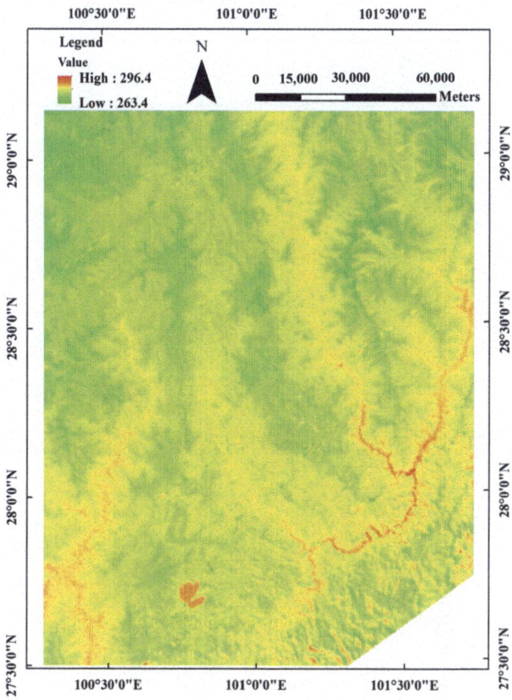

Fig. 4. Inversed TIR image of study area.

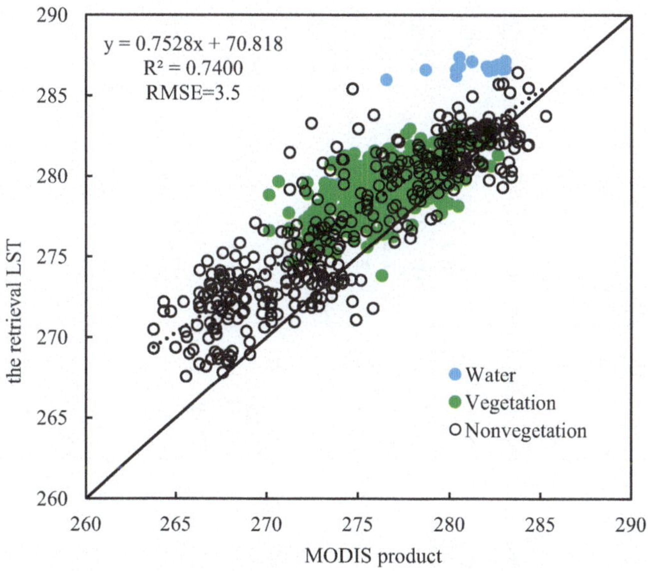

Fig. 5. Comparison of the LST retrieval result between MODIS LST products in the 774 sampling points contained vegetation, non-vegetation and water area.

Table 1. Comparison of LST retrieval results with MODIS LST product with different slope in vegetation area and non-vegetation area

Land type	Slop(degree)	Mean_MODIS (K)	Mean_LST retrieval (K)	Mean_NDVI	STD_NDVI	R	RMSE(K)
Vegetation	0–15	276.715	279.218	0.5895	0.1369	0.6543	3.0
	15–25	276.419	279.292	0.5086	0.2183	0.6055	3.3
	25–35	276.377	279.219	0.4550	0.2232	0.6081	3.5
	35–45	276.812	279.419	0.4072	0.2247	0.6142	3.2
	>45	276.089	279.394	0.5406	0.2203	0.69	3.6
Non-vegetation	0–5	272.415	274.863	0.2835	0.0558	0.6888	3.5
	5–15	271.383	274.562	0.3163	0.0871	0.8228	4.2
	15–25	275.197	277.416	0.3733	0.0866	0.8872	3.5
	25–35	277.664	279.435	0.3588	0.0742	0.9129	2.8
	>35	279.136	280.944	0.3603	0.0969	0.9121	2.6

3.2 LST Accuracy Varying with NDVI and Slope

Table 1 shows the correlation between TG-2 LST and MODIS LST product with different slopes in vegetation and non-vegetation areas. In vegetation areas with slight slopes between 0°–15°, relatively high accuracy and correlation is found with $R = 0.6543$, and $RMSE = 3.0$ K . In non-vegetation area, the accuracy of LST retrieval result increases with slope. Both the MODIS LST and TG-2 LST in vegetation area are higher than non-vegetation area when the slope less than 25°, while lower when the slope greater than 25°.

NDVI is calculated by NIR band and Red band of WIS data of TG-2 (see Fig. 3). As for NDVI, its standard deviation affects the LST accuracy. The LST accuracy in non-vegetation areas with small NDVI standard deviations is higher than those in vegetation areas with large NDVI standard deviations. The major reason is the pixel purity. For non-vegetation areas, the purity of land cover increases with slope. For example, the pixels with great slope is generally bare land; but the pixels in flat area are generally mixtures of bare land and buildings. In vegetation areas, the large variation of NDVI, which means complex pixel mixtures, results in less accurate LST.

4 Conclusion

By modifying the Sobrino's algorithm [16] and Franca [17] to retrieve the LST from WIS data of TG-2, we considered the possibility to extract LST from WIS data of Space Station Platform TG-2. Results show that the LST retrieval results using 8–9.3 μm channels are possible; but the systematic overestimation than MODIS LST around 3.5 K indicated that the split-window algorithm originally for 10–12 μm channels may not be directly applied for 8–9.3 μm channels. Because the WIS is mainly designed for marine remote sensing applications, the larger errors on LST is reasonable. But more work should be done on the algorithms to improve the LST inversion accuracy in the future.

In complex mountain areas, the accuracy of TG-2 LST is greatly affected by slope and NDVI. The major reason is the pixel purity variation caused by slope and NDVI. Generally, non-vegetation area has more stable NDVI than vegetated areas because of higher pixel purity. Since we only analyzed the effect of the slope and NDVI on LST accuracy, the relationship between NDVI and LST is still insufficient and unclear. Despite the low accuracy, this study is the first trial to inverse temperatures at complex land surface (not sea surface) using 8–9 μm channels. The inversion LST error is helpful to assess the capabilities split window methods in 8–9 μm. The improvement of parameters in the split window method, and the analysis of influencing factors on LST will be addressed in our future studies.

Acknowledgements. This work was supported by the Natural Science Foundation of China (41571332). Thanks to China Manned Space Engineering for providing Wide-band Imaging Spectrometer (WIS) data products of Tiangong-2.

References

1. Wang, K., Wan, Z., Wang, P., Sparrow, M., Liu, J., Haginoya, S.: Evaluation and improvement of the MODIS land surface temperature/emissivity products using ground-based measurements at a semi-desert site on the western Tibetan Plateau. Int. J. Remote Sens. **28**(11), 2549–2565 (2007)
2. Wang, K., et al.: Estimation of surface long wave radiation and broadband emissivity using Moderate Resolution Imaging Spectroradiometer (MODIS) land surface temperature emissivity products. J. Geophys. Res. Atmos. **110**(D11109) (2005)
3. Arnfield, A.J.: Two decades of urban climate research: a review of turbulence, exchanges of energy and water, and the urban heat island. Int. J. Climatol. **23**(1), 1–26 (2003)
4. Bastiaanssen, W.G.M., Menenti, M., Feddes, R.A., Holtslag, A.A.M.: The Surface energy balance algorithm for land (SEBAL): Part 1 formulation. J. Hydrol. **212**(98), 801–811 (1998)
5. Hansen, J., Ruedy, R., Sato, M., Lo, K.: Global surface temperature change. Rev. Geophys. **48**(4), G4004 (2010)
6. Kanani, K., Poutier, L., Nerry, F., Stoll, M.P.: Directional effects consideration to improve out-doors emissivity retrieval in the 3–13 mum domain. Opt. Express **15**(19), 12464–12482 (2007)
7. Weng, Q., Lu, D., Schubring, J.: Estimation of land surface temperature-vegetation abundance relationship for urban heat island studies. Remote Sens. Environ. **89**(4), 467–483 (2004)
8. Weng, Q.: Thermal infrared remote sensing for urban climate and environmental studies: Methods, applications, and trends. Isprs J. Photogramm. Remote Sens. **64**(4), 335–344 (2009)
9. Voogt, J.A., Oke, T.R.: Thermal remote sensing of urban climates. Remote Sens. Environ. **86**(3), 370–384 (2003)
10. Dash, P., Göttsche, F.M., Olesen, F.S., Fischer, H.: Retrieval of land surface temperature and emissivity from satellite data: Physics, theoretical limitations and current methods. J. Indian Soc. Remote Sens. **29**(1–2), 23 (2001)
11. Li, Z.L., Tang, B.H., Wu, H., Ren, H.Z., Yan, G.J., Wan, Z.M., Trigo, I.F., Sobrino, J.: Satellite-derived land surface temperature: Current status and perspectives. Remote Sens. Environ. **131**(131), 14–37 (2013)
12. Liu, X., Huang, J.X., Qin, J., Wang, P., Xu, T.: Estimating of land surface turbulent fluxes based on weak constraint variational method and GOES data. Trans. Chin. Soc. Agric. Mach. **45**(1), 236–245 (2014)
13. Zhang, X., Tang, Y.Y., Huang, X.X., Wei, J.: Performance analysis of split-window algorithms for retrieving land surface temperature using remote sensing data of 8.0–9.3μm. Remote Sens. Land Resour. **27**(2), 88–93 (2015)
14. Weng, Q.: A remote sensing-GIS evaluation of urban expansion and its impact on surface temperature in the Zhujiang Delta, China. Int. J. Remote Sens. **22**(10), 1999–2014 (2001)
15. Weng, Q.: Fractal analysis of satellite-detected urban heat island effect. Photogramm. Eng. Remote Sens. **69**(5), 555–566 (2003)
16. Franca, G.B., Cracknell, A.P.: Retrieval of land and sea surface temperature using NOAA-11 AVHRR data in north-eastern Brazil. Int. J. Remote Sens. **15**(8), 1695–1712 (1994)
17. Sobrino, J.A., Coll, C., Caselles, V.: Atmospheric correction for land surface temperature using NOAA-11 AVHRR channels 4 and 5. Remote Sens. Environ. **38**(1), 19–34 (1991)

18. Qin, H.Z., Zhang, M.H., Karnieli, A., Berliner, P.: Mono-window algorithm for retrieving land surface temperature from Landsat TM6 data. Acta Geogr. Sin. **56**(4), 456–466 (2001)
19. Mao, K.B.: A study of method for land surface temperature retrieval from MODIS data. Nanjing University, Nanjing (2004)
20. Huang, L., Huang, H., Xiang, D., Zhu, J.: The diurnal change of air temperature in four types of land cover and urban heat island effect in Nanjing, China. Ecology Environ. **16**(5), 1411–1420 (2007)

Radiometric Cross-Calibration Between the Thermal Infrared Data of MODIS and Tiangong-2 WIS for Monitoring Sea Surface Temperature

Jie Li[1,3], Qianguo Xing[1,3(✉)], Hailong Liu[1], and Lei Yu[2,3(✉)]

[1] Yantai Institute of Coastal Zone Research, Chinese Academy of Sciences, Yantai 264003, Shandong, China
xingqg@163.com

[2] Anhui Institute of Optics and Fine Mechanics, Chinese Academy of Sciences, Hefei 230031, Anhui, China
ytyu@163.com

[3] University of Chinese Academy of Sciences, Beijing 100049, China

Abstract. Following the radiometric cross-calibration procedures between the Tiangong-1 hyperspectral imaging spectrometer and the Moderate Resolution Imaging Spectrometer (MODIS) infrared bands, the thermal infrared bands of Wide-band Imaging Spectrometer (WIS) on the Tiangong-2 space station are calibrated and further used to monitor the Sea Surface Temperature (SST). Quasi-simultaneous MODIS thermal infrared images acquired in winter and spring of 2017 and 2018 are collected as the calibration reference. A linear radiance correction equation between the two data sets of MODIS and WIS is established, and the Tiangong-2 WIS thermal infrared data is calibrated by the radiance correction equation. The calibrated WIS thermal infrared data is further used to retrieve the SST of coastal waters. The results show that the thermal infrared bands of Tiangong-2 WIS are highly correlated to the 32nd band of MODIS. Credited to the high spatial resolution of WIS thermal infrared images, the SST anomalies caused by natural or human disturbance are clearly shown, e.g., the warming plume of cooling water from coastal power plant, the apparent temperature of tidal flat and the underwater tidal ridges.

Keywords: Tiangong-2 · Thermal infrared · WIS · MODIS Cross-calibration · SST

1 Introduction

With the rapidly developing economy and intensive human activities, petrochemical, nuclear and thermal power plants have been built in China's coastal zones. Large volume of seawater is used as cooling water and discharged directly to coastal waters, leading to the increase water temperature and even thermal pollution to coastal environment. In order to understand water temperature anomaly effectively, the satellite remote sensing techniques have been widely used in monitoring sea surface temperature (SST), such as, Advanced Very High Resolution Radiometer (AVHRR) and

Y. Gu et al. (Eds.): *Proceedings of the Tiangong-2 Remote Sensing Application Conference*, LNEE 541, pp. 192–200, 2019.
https://doi.org/10.1007/978-981-13-3501-3_18

Moderate Resolution Imaging Spectrometer (MODIS) infrared data used in global SST detection. However, the low spatial resolution (about 1.1 km) of AVHRR and MODIS limits their usage in small scale coastal environment. High spatial resolution remote sensing technique for the coastal SST detection is becoming more and more important with the development of coastal zone [1, 2].

Radiometric calibration is the premise of quantitative remote sensing with satellite images [3, 4]; the methods include laboratory calibration, in-orbit calibration and cross-calibration [5–7]. Laboratory calibration method obtains the sensor response coefficients by stable light sources in specific circumstance [6, 8]. In-orbit calibration method is widely used by on-board calibrators, such as the calibrators for TM and ETM+ [6, 9]. Measured surface reflectance and atmospheric parameters are also required in laboratory and in-orbit calibration methods [7, 10, 11]. Researchers proposed new methods of cross-calibration with which the precisely-calibrated sensors' data are used to replace measured data [7, 9, 12].

In cross-calibration, calibration parameters are obtained by analyzing the Digital Number (DN) values of the spatially matched regions, and the linear equation for cross-calibration is established by regression statistics [9, 12]. This method is an effective way which could realize calibration without measured data and ground calibration fields [13]. Li et al. proposed an image-based cross-calibration method for Gaofen-1 WFV cameras [14]. Xing et al. used this method to calibrate infrared data obtained by the hyperspectral imaging spectrometer on Tiangong-1: MODIS infrared data was selected as calibration reference, and then the calibrated data was obtained by the radiance correction equation [15].

The MODIS infrared data has high calibration precision and radiation resolution (12 bits), and its noise-equivalent temperature difference is up to 0.05 K (such as Band 31 and 32). Researchers used MODIS data as calibration reference to calibrate ETM+, HJ1-CCD and CBERS-CCD data [16–18].

According to measured SST data, the retrieval precision is satisfying in the East China Sea [19], the Taiwan Strait [20], the northern South China Sea [21], and some coastal water. However, the spatial resolution of MODIS infrared data (1 km) is not adequate for the applications in coastal waters. The Wide-band Imaging Spectrometer (WIS) on Tiangong-2 can produce 400 m thermal infrared images, which may have more potential use in the monitoring of coastal SST.

This paper focuses on the radiometric cross-calibration of WIS infrared data with quasi-simultaneous MODIS infrared data. The application in monitoring coastal SST is also evaluated.

2 Data and Methods

2.1 Data and Study Area

Four quasi-simultaneous thermal infrared images of WIS and MODIS acquired in winter and spring over the western Yellow Sea are used for cross-calibration and SST retrieval (see Table 1). Time differences of images acquisition are within 52 min.

Table 1. Data for infrared radiometric cross-calibration

	MODIS	Tiangong-2 WIS
Imaging time (Beijing Time)	2017.02.11 02:15	2017.02.11 01:23:09
	2018.05.04 11:00	2018.05.04 10:10:33

2.2 Radiation Correction Equation

After geometric correction and resampling to 1 km spatial resolution, Tiangong-2 WIS infrared data can be matched with MODIS infrared data. Then, through the linear regression analysis with the radiance values of MODIS infrared data, the radiation correction equation for WIS infrared data is established as Eq. (1):

$$L_{\text{calibration}} = a \cdot L_{\text{WIS}} + b \tag{1}$$

where L_{WIS} is the radiance values of WIS infrared data; $L_{\text{calibration}}$ is the radiance values of calibrated data; a and b are the scale and offset values, they can obtained by cross-calibration with MODIS infrared data.

2.3 SST Retrieval

According to the Planck formula, brightness temperature can be derived from the radiance values of WIS infrared data. Equations of brightness temperature and SST can be written as Eqs. (2) and (3):

$$T = \frac{h \cdot c}{k \cdot \lambda \cdot \ln\left(1 + \frac{2 \cdot h \cdot c^2}{L \cdot \lambda^5}\right)} \tag{2}$$

$$\text{SST} = T - 273 \tag{3}$$

where h is the Planck constant ($h = 6.6262 \times 10^{-34}$ J•s); c is the light velocity ($c = 2.9979 \times 10^{8}$ m/s); k is the Boltzmann constant ($k = 1.3807 \times 10^{-23}$ J/k); λ is central wavelength ($\lambda = 8.475 \times 10^{-6}$ m or $\lambda = 9.100 \times 10^{-6}$ m); L is incident radiance, the unit of L is W/(m^2•sr•m); T is brightness temperature, its unit is K. The unit of SST is °C.

As shown by Eqs. (2) and (3), the SST is simplified on the basis of the bright temperature (T) without consideration of sea water thermal emissivity (~0.985) and the atmospheric effects.

Table 2 shows the information of WIS and MODIS infrared data. With the similar wavelengths, four bands of MODIS infrared data (Band 29 to Band 32) can be used for cross-calibration.

Table 2. Basic factors of WIS and MODIS infrared data

Satellite	Sensor	Band	Wavelength (μm)	Central wavelength (μm)	Spatial resolution (m)
Tiangong-2	WIS	TIR Band 1	8.125–8.825	8.475	400
		TIR Band 2	8.925–9.275	9.100	400
TERRA	MODIS	Band 29	8.400–8.700	8.550	1000
		Band 30	9.580–9.880	9.730	1000
		Band 31	10.780–11.280	11.030	1000
		Band 32	11.770–12.270	12.020	1000

3 Results and Discussion

3.1 Radiometric Cross-Calibration

The wavelengths of the four thermal infrared bands of MODIS (Band 29, 30, 31 and 32) are close to the thermal infrared spectrum range of WIS. In order to select the best band for radiometric cross-calibration, this paper analyzes the correlation of radiance values for WIS and MODIS thermal infrared images. Correlation results are shown by scatter plots, and the correlation coefficients R are also calculated.

The self-correlations among the four MODIS thermal infrared bands are shown in Fig. 1. The results show that each band is highly correlated to the other bands, and all the correlation coefficients R exceed 0.90. In this case, correlation between Tiangong-2 WIS and MODIS thermal infrared data should be analyzed to determine the suitable band for radiometric cross-calibration.

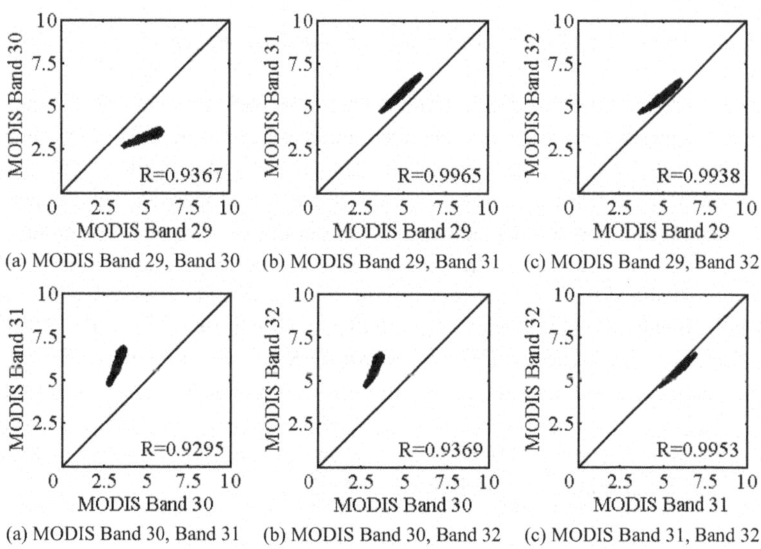

Fig. 1. Correlation of MODIS data

Figure 2 shows the correlations between MODIS and WIS resample images over the same regions. The radiance data in 2017 and 2018 images are denoted by blue and red points, respectively. As shown in Fig. 2, the thermal infrared bands of WIS are highly correlated to the Band 29, 31 and 32 of MODIS; the correlation coefficients R all exceed 0.79, and the maximum one is up to 0.9964 (Fig. 2d). The correlation coefficients R in black show the correlation of composite data, which is combined by both 2017 and 2018 data, and these correlation coefficients R (in black) of the Band 32 are higher than those of other bands. These results indicate that the Band 32 of MODIS is more suitable for radiometric cross-calibration than other bands.

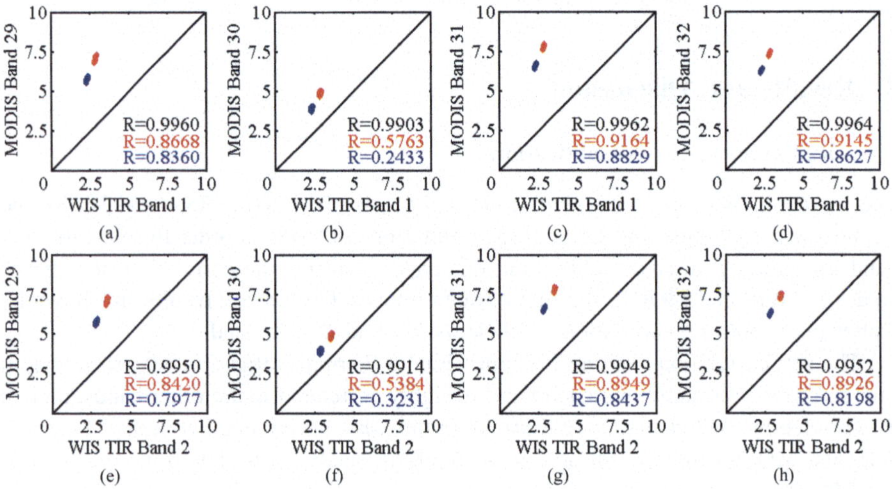

Fig. 2. Radiance scatter plots between WIS and MODIS thermal infrared data. Points in blue and red indicate the radiance in 2017 and 2018 images, respectively. R in blue, red and black are the correlation coefficients of 2017, 2018 and composite data, respectively.

In order to reduce the quasi-simultaneous error, expand dynamic range, and obtain more accurate cross-calibration results, the radiance values of WIS thermal infrared data are analyzed by linear regression statistics. Table 3 shows the linear regression results at the 0.05 significance level. a and b are the slopes and intercepts of the radiance correction equation, respectively. R^2 is goodness of fit of linear equation, and R^2 closer to 1 indicates the better fitting results. The Root Mean Square Error (RMSE) is a measure of the difference between predictive value and observed value, and the lower RMSE means the better observed values. As shown in Table 3, RMSE between the TIR Band 1 of WIS and the Band 32 of MODIS is lower than that of other bands, which indicates the two bands have the minimum error. Furthermore, the TIR Band 1 of WIS is highly correlated to the Band 32 of MODIS, and the correlation coefficient R of these two bands reaches 0.9964 (Fig. 2d). In this case, the Band 32 of MODIS is selected to cross-calibrate the TIR Band 1 of WIS.

Table 3. Linear regression of radiance between WIS and MODIS thermal infrared data

WIS TIR Band 1				
MODIS	*a*	*b*	R^2	RMSE
Band 29	2.5016	0.0908	0.9920	0.0580
Band 30	1.8019	−0.1140	0.9807	0.0653
Band 31	2.1620	1.7251	0.9924	0.0489
Band 32	1.9363	1.9380	0.9928	0.0426
WIS TIR Band 2				
MODIS	*a*	*b*	R^2	RMSE
Band 29	1.9754	0.2249	0.9901	0.0645
Band 30	1.4257	−0.0263	0.9828	0.0616
Band 31	1.7066	1.8427	0.9899	0.0563
Band 32	1.5286	2.0427	0.9905	0.0489

3.2 Observation of the Coastal SST Anomaly Caused by Natural and Human Disturbances

The positive correlation between WIS and MODIS thermal infrared data indicates that WIS thermal infrared data can be used to retrieve SST. In this paper, emissivity of seawater and water vapor absorption effect are not considered. Brightness temperature is calculated by the Planck formula (Eq. 2) and SST can be retrieved by Eq. (3).

Figure 3 shows the retrieved coastal SST retrieval results in the western Yellow Sea. Figure 3a, b are the SST retrieval results of WIS and MODIS thermal infrared data in winter of 2017, respectively. Figure 3c, d are the SST retrieval results of WIS and MODIS thermal infrared data in spring of 2018, respectively. In winter night, SST of deep water is higher than that of shallow water (Fig. 3a, b); while, in daytime of spring, the coastal SST is higher than that of the open sea (Fig. 3c, d). Figure 3a, b show that water temperature near Tianwan nuclear power plant is much higher than other regions, and the range of anomalous region can also be recognized from these figures. Figure 3c, d show seawater temperature structures at ebb tide. Due to ebb tide, tidal flat emerges and has higher apparent temperature.

Spatial resolution of infrared images affects the detail information of SST retrieval results. Compared to the retrieval results from MODIS data (Fig. 3b, d), there are more detailed information in WIS SST retrieval results. For example, the profiles of tidal flat and tidal ridges and the temperature of ebb tide in WIS SST retrieval results (Fig. 3c) are more subtle than those in MODIS SST retrieval results (Fig. 3d). This work suggests that high spatial resolution infrared images (Tiangong-2 WIS infrared images) could produce the better SST retrieval results.

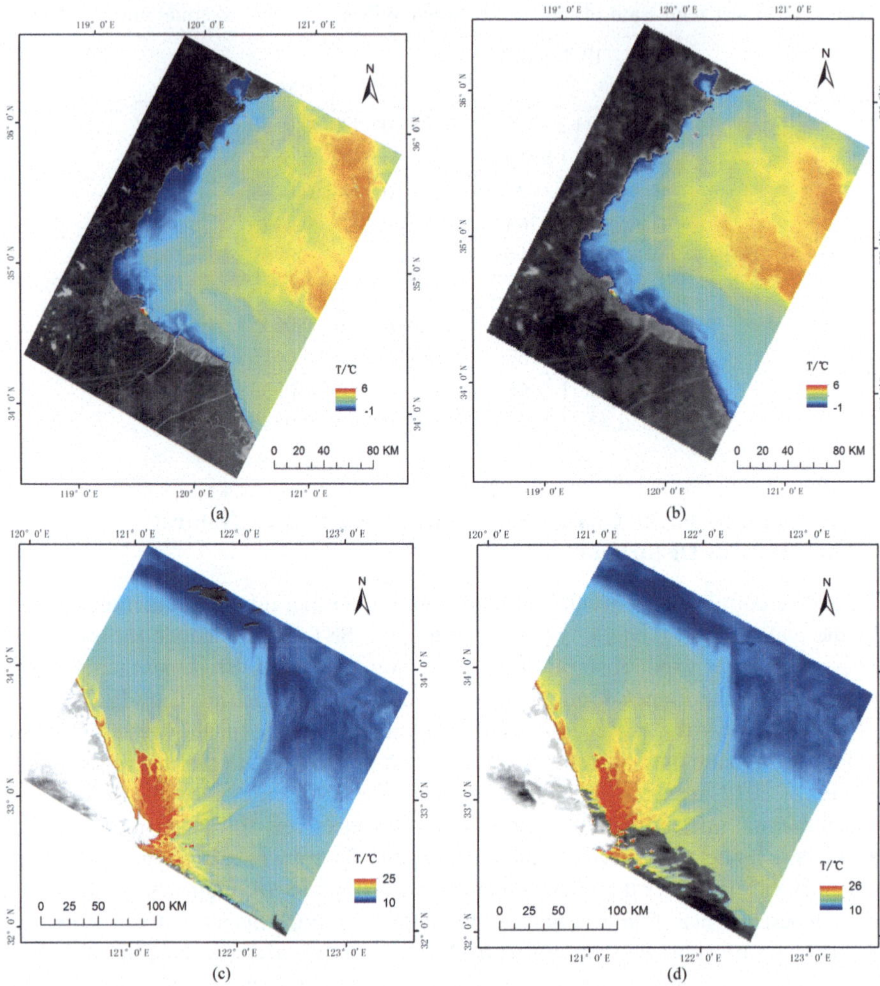

Fig. 3. Retrieval of coastal SST (bright temperature) in the western Yellow Sea based on WIS and MODIS infrared data. (a) and (b) are the 2017 winter SST retrieval results of WIS and MODIS thermal infrared data, respectively. (c) and (d) are the 2018 spring SST retrieval results of WIS and MODIS thermal infrared data, respectively.

4 Conclusions

According to WIS thermal infrared data from Tiangong-2, we use quasi-simultaneous observation method to prove the feasibility of radiometric cross-calibration based on MODIS infrared data. The applicability of SST retrieval by WIS thermal infrared data is also tested. The experimental results show that the radiometric-calibrated WIS infrared data can be used in practical quantitative evaluation. With high spatial resolution of WIS infrared data, the subtle spatial dynamics of seawater can be detected, which suggests that the thermal infrared data from Tiangong-2 has the potential in land-ocean interaction research and pollution sources monitoring and assessment.

Acknowledgment. This work is supported by Chinese Academy of Sciences key instrument project (No. YJKYYQ20170048), the National Natural Science Foundation of China (No.41676171), and the Qingdao National Laboratory for Marine Science and Technology of China (No.2016ASKJ02). Thanks to China Manned Space Engineering for providing space science and application data products of Tiangong-2.

References

1. Steissberg, T.E., Hook, S.J., Schladow, G.: High-spatial-resolution thermal infrared satellite images for lake studies. AGU. H32D-01 (2006)
2. Xing, Q., Li, L., Lou, M., Bing, L., Zhao, R., Li, Z.: Observation of oil spills through Landsat thermal infrared imagery: a case of deepwater horizon. Aquat. Procedia **3**, 151–156 (2015)
3. Chen, Q.L., Tang, J.W., Wang, X.N.: Review on the calibration for ocean color sensing. Ocean Technol. **17**(3), 13–26 (1998)
4. Zhang, Y., Gu, X.F., Yu, T., Zhang, X.Y., Qi, R.L., Li, X.W.: Radiometric cross-calibration of CBERS-02 IRMSS thermal channel. J. Infrared Milimeter Waves **25**(4), 261–266 (2006)
5. Chander, G., Markham, B.: Revised Landsat-5 TM radiometric calibration procedures and postcalibration dynamic ranges. IEEE Trans. Geosci. Remote Sens. **41**(11), 2674–2677 (2003)
6. Chander, G., Markham, B.L., Helder, D.L.: Summary of current radiometric calibration coefficients for Landsat MSS, TM, ETM+, and EO-1 ALI sensors. Remote Sens. Environ. **113**(5), 893–903 (2009)
7. Xiong, X.X., Barnes, W.: An overview of MODIS radiometric calibration and characterization. Adv. Atmos. Sci. **23**(1), 69–79 (2006)
8. Gregg, W.W., Casey, N.W., O'Reilly, J.E.: An empirical approach to ocean color data: reducing bias and the need for post-launch radiometric re-calibration. Remote Sens. Environ. **113**(8), 1598–1612 (2009)
9. Teillet, P.M., Barker, J.L., Markham, B.L., Irish, R.R., Fedosejevs, G., Storey, J.C.: Radiometric cross-calibration of the Landsat-7 ETM+ and Landsat-5 TM sensors based on tandem data sets. Remote Sens. Environ. **78**(1–2), 39–54 (2001)
10. Feng, L., Li, J., Gong, W., Zhao, X., Chen, X., Pang, X.: Radiometric cross-calibration of Gaofen-1 WFV cameras using Landsat-8 OLI images: a solution for large view angle associated problems. Remote Sens. Environ. **174**, 56–68 (2016)
11. Pahlevan, N., Lee, Z., Wei, J., Schaaf, C.B., Schott, J.R., Berk, A.: On-orbit radiometric characterization of OLI (Landsat-8) for applications in aquatic remote sensing. Remote Sens. Environ. **154**, 272–284 (2014)
12. Teillet, P.M., Markham, B.L., Irish, R.R.: Landsat cross-calibration based on near simultaneous imaging of common ground targets. Remote Sens. Environ. **102**(3–4), 264–270 (2006)
13. Pan, D.L., He, X.Q., Zhu, Q.K.: On orbit radiometric cross calibration of HY-1A COCTS. Chinese Sci. Bull. **49**(21), 2239–2244 (2004)

14. Li, J., Feng, L., Pang, X., Gong, W., Zhao, X.: Radiometric cross calibration of Gaofen-1 WFV cameras using Landsat-8 OLI Images: a simple image-based method. Remote Sens. **8** (5), 411 (2016)
15. Xing, Q.G., Lou, M.J., Zhang, J.X., Li, X.Z., Ning, J.C., Gao, Z.Q.: Radiometric cross-calibration of Tiangong-1 and MODIS infrared bands and applications in monitoring coastal sea surface temperature. J. Remote Sens. **18**(Sup), 68–73 (2014)
16. Li, X., Gu, X., Min X., Yu, T., Fu, Q., Zhang, Y., Li, X.: Radiometric cross-calibration of the CBERS-02 CCD camera with the TERRA MODIS. Science in China Ser. E Inf. Sci. **35** (S I), 41–58 (2005)
17. Xu, L., Ma, L., Hu, J., Tang, L.: Cross-calibration of HJ-1B/CCD1 against Terra/MODIS. Remote Sens. Inf. **2**, 26–31 (2011)
18. Lee, K.S., Park, S.M., Kim, S.H., Lee, H.S., Shin, J.I.: Radiometric characteristics of Geostationary Ocean Color Imager (GOCI) for land applications. Korean J. Remote Sens. **28** (3), 277–285 (2012)
19. Li, N., Sun, F.Q., Zhang, C.Y., Shang, S.L., Hong, H.S.: A primary validation of remote AVHRR SST in Taiwan Strait and its vicinity. J. Xiamen Univ. (Nat. Sci.) **45**(3), 383–387 (2006)
20. Sakuno, Y., Tsushima, K., Matsunaga, T.: Validation of ASTER water surface temperature algorithm using MODIS SST data. Proc. Conf. Remote Sens. Soc. Jpn **39**, 241–242 (2005)
21. Xing, Q., Chen, C., Shi, P., Yang, J.: Atmospheric correction of Landsat data for the retrieval of sea surface temperature in coastal waters. Acta Oceanol. Sin. **25**(3), 25–34 (2006)

Cross-Comparison of Ocean Color Products Derived from Tiangong-2/WIS and GOCI in the Yangtze Estuary, China

Rugang Tang[1], Fang Shen[1(✉)], Yanqun Pan[1], Kang Liu[2], Mengyu Li[1], Wenli Gao[1], and Chuankai Zang[1]

[1] State Key Laboratory of Estuarine and Coastal Research, East China Normal University, Shanghai 200062, China
fshen@sklec.ecnu.edu.cn
[2] Key Laboratory of Space Utilization, Technology and Engineering Center for Space Utilization, Chinese Academy of Sciences, Beijing 100094, China

Abstract. Suspended particulate matter (SPM) is one of the key issues in estuarine and coastal waters, which is widely used in water quality assessment, river erosion and siltation, and port engineering. The Wide-band Imaging Spectrometer (WIS) as one of the payloads mounted on the Tiangong-2 (TG-2) is a successor of spaceborne ocean color sensors. In this study, data from Geostationary Ocean Color Imager (GOCI) launched by Korean ocean satellite center (KOSC) and in situ measurements are used to compare and evaluate data products acquired by the TG-2/WIS. Comparison results show that the TOA reflectances of TG-2 /WIS and GOCI agree well at visible and near-infrared bands. Atmospherically-corrected remote sensing reflectances (R_{rs}) from TG-2/WIS and GOCI have good correspondences from visible to near-infrared bands Their root mean square differences (RMSD) at 660 and 745 nm are less than 0.04 sr^{-1}, and the TG2/WIS observed R_{rs} is slightly underestimated at 865 nm with the slope of 0.77. Moreover, the derived SPM concentrations from TG-2/WIS and GOCI data are quite approximate. It implies that the accuracy of TG-2/WIS data products can meet the need of marine environmental monitoring in estuaries and coastal regions. In addition, in contrast to other ocean color sensors, the TG-2/WIS may have greater potentials of water color applications, especially for meso- and small-scale phenomena and optically complex waters, due to its multi-bands and high spatial resolution advantages.

Keywords: Tiangong-2 · Wide-band Imaging Spectrometer · GOCI Cross-comparison · Suspended particulate matter · Yangtze estuary

1 Introduction

The Yangtze estuary has third-order bifurcations and four outlets. The estuary exhibits large spatial-temporal variations in SPM distribution, which are affected by tidal currents, runoffs, continental shelf circulation as well as highly-insensitive human activities. In order to explore the characteristics and distributions of estuarine sediments, a lot of researches have been carried out, including the influence of the amount of

Y. Gu et al. (Eds.): *Proceedings of the Tiangong-2 Remote Sensing Application Conference*, LNEE 541, pp. 201–211, 2019.
https://doi.org/10.1007/978-981-13-3501-3_19

sediment from the upstream to the downstream [1], the impacts of extreme climate change on the distribution of sediment [2], the erosion and deposition of the deep-water channel [3], the changes in the SPM distribution caused by seasonal changes [4] and etc. Compared with the traditional in situ surveys, the use of satellite measurements can undoubtedly reduce the cost of observation, meeting the requirements for simultaneous observation of different areas of the Yangtze estuary. Traditional methods are subject to the instability of environmental factors and variable observation conditions, which make it difficult to conduct comprehensive assessments and systematic researches on a large scale of the SPM concentration [5]. In contrast, remote sensing technology, with the advantages of spatio-temporal distribution, is a highly effective method for timely understanding the distribution of SPM concentration and macroscopic observation of the sediment transport in the Yangtze estuary [6].

Since the launching of Nimbus-7 in 1978, satellite ocean color observations and applications have gradually matured [7]. Nowadays, ocean color sensors are already being used in a variety of researches on estuarine and coastal waters. For example, 10 years of MODIS data were used to retrieve the chlorophyll concentration to find its law of change in Tampa Bay [8]; MERIS data were used to retrieve the SPM concentration in the Yangtze estuary [9] and St. Lawrence [10]. In addition, GOCI and TG-2/WIS data are also used to study the chlorophyll and SPM concentrations in the estuary [11, 12]. In this work, we use the data and products of GOCI as well as in situ measurements to evaluate the the accuracy of TG-2/WIS data and its derived SPM concentration. The results will provide useful information for the TG-2/WIS on-orbit testing and technical improvements.

2 Study Area

In this work, we select the Yangtze estuary and the Hangzhou Bay (see Fig. 1) as the study area. Cross-comparing for TOA reflectance, R_{rs} and derived SPM concentration between TG-2/WIS and GOCI in the study area are conducted. In addition, Optical Backscattering (OBS) data from three hydrological stations in Chongxi, Baozhen and Hengsha are used for validation of satellite-derived SPM concentration data. To ensure the accuracy of the OBS data, results from turbidimeter and laboratory sampling are used to calibrate the OBS turbidity after installation or replacement. Special maintenance such as removing sediment and biological debris on the sensors is carried on every two weeks. In the study area, the turbidity maximum zone of the Yangtze estuary and Hangzhou Bay are areas with high SPM concentration, where can reach 2 g/L or higher.

The turbidity (NTU) data measured by the OBS can be converted to SPM concentration using empirical formula. In this study, we use the observation results from Zao et al. [13] in Chongxi station for the conversion of SPM, shown as follows:

$$\mathrm{SPM} = \begin{cases} 0.0021x\,(\text{dry season}) \\ 0.0013x\,(\text{flood season}) \end{cases}, \tag{1}$$

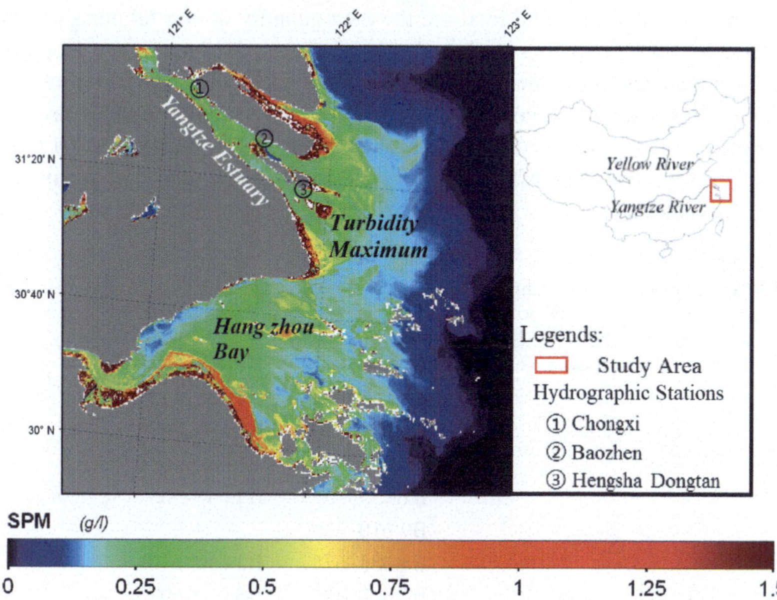

Fig. 1. Schematic diagram of the distribution of SPM concentration (g/l) in the study area. Numbers show the locations of the hydrographic stations where in situ turbidity data are obtained.

where x is turbidity (NTU) measured by the OBS, the SPM concentration is calculated under flood and dry seasons with different water and sediment characteristics, respectively.

3 Satellite Data

TG-2 was launched in Jiuquan on October 15, 2016. The Wide-band Imaging Spectrometer includes three visible and near-infrared cameras, two short-wave-infrared cameras, two thermal-infrared cameras, and one visible-band polarization camera. The visible and near-infrared wavelengths range from 403 to 990 nm, with a total of 14 bands and a bandwidth of 20–40 nm, which is narrower than traditional land observation satellites. Compared with other ocean satellites, the spatial resolution of TG-2/WIS is higher, which reaches 100 m for all the bands, thus, it is very suitable for small and medium-scale ocean observation. The data of TG-2 is released by the manned space application data promotion service platform. After registration, Level-2 products with radiation correction and system geometry correction can be obtained.

GOCI is the world's first geostationary orbit marine aqua satellite launched by the Korea Ocean Satellite Center (KOSC) in June 2010. GOCI includes 8 bands from blue to near-infrared, with a bandwidth of 20–40 nm and a spatial resolution of 500 m. GOCI visits one scene per hour from 8:30 am to 15:30 (Beijing time), in other words 8 scenes per day. Its short revisiting period can help the study of high dynamic change of estuarine and coastal waters and increase the data quantity of coastal areas affected by cloud coverage. GOCI data can be downloaded from the database of the Korea Ocean Satellite Center (KOSC, http://kosc.kiost.ac.kr/eng/), which also provides Level-1 data, i.e., radiation corrected and geo-corrected products. The satellite images used in this study are the Level-2 products of TG/WIS taken at 11:03 on January 22, 2017, and the Level-1 level products of GOCI taken at 11:30 on the same day. The detailed information of two satellite sensors is shown in Table 1.

Table 1. Comparison of technical parameters between TG-2/WIS and GOCI sensors

Satellite and sensor	TG-2/WIS	GOCI
Band range (nm)	B14:403–423	B1:402–422
	B13:433–453	B2:433–453
	B12:480–500	B3:480–500
	B11:510–530	
	B10:555–575	B4:545–565
	B9:610–630	
	B8:655–675	B5:650–670
	B7:678–688	B6:670–690
	B6:740–760	B7:735–755
	B5:810–830	
	B4:845–885	B8:845–885
	B3:895–915	
	B2:930–950	
	B1:970–990	
Spatial resolution (m)	100	500
Revisit period (days)	2–3 (minimum)	1/8
SNR range	>300	750–1170

4 Methods

4.1 TOA Reflectance

Satellite sensor-measured digital number (DN) is required to be calibrated into the TOA radiance for all bands before applications. Actually, the calibration includes pre-launch laboratory calibration and post-launch on-orbit calibration. The accuracy of the calibration can be assessed using targets features or by cross-comparing with other on-orbit routine sensors. The calibration formula can be expressed as:

$$L_{\mathrm{TOA}} = \mathrm{gain} \times \mathrm{DN} + \mathrm{offset}, \tag{2}$$

where the DN is the digital number; the gain and offset are calibrated coefficients. L_{TOA} is the TOA radiance in W $\mathrm{m}^{-2}\mu\mathrm{m}^{-1}\mathrm{sr}^{-1}$.

In order to minimize the impacts of sunlight illumination variations, we convert the radiance to the TOA reflectance (ρ_{TOA}), which is defined as the ratio of the radiance (M) at the sensor to the solar irradiance (E), calculated as follows:

$$\rho_{TOA} = \frac{M}{E} = \frac{\pi \times L_{TOA}}{E_0 \times \cos\theta / d^2}, \tag{3}$$

where E_0 is the solar irradiance; θ is the solar zenith angle; *d* is earth-sun distance, the unit is AU.

4.2 Atmospheric Correction

The objective of atmospheric correction is to remove atmospheric contributions from atmospheric molecular, particulate absorption and scattering effects such as aerosol scattering and Rayleigh scattering for obtaining water-leaving reflectance. In this work, the GOCI and TG-2/WIS derived TOA reflectance are corrected using the ESOA method of Pan et al. [14], which is to generate a lookup table by matching Aerosol Robotic Network (AERONET) data [15] and semi-empirical radiation transfer (SERT) model data [4, 9, 16]. The initial value is optimized by genetic algorithm, then the relative humidity (RH), fine mode fraction (FMF), aerosol optical thickness (AOT) and suspended particulate matter are optimized to calculate the R_{rs}.

4.3 SPM Concentration Inversion

The band settings of GOCI and TG-2/WIS are similar, only small differences exist in the position, width of the wavebands and the spectral response functions. Figure 2 shows the spectral response functions of the two sensors and the R_{rs} of water bodies with different turbidity.

It can be seen in Fig. 2 that the two sensors are similar in the position and width of the wavebands. TG-2/WIS has an extra wavelength at 620 nm compared to GOCI, which is located in the low absorption region of chlorophyll and thus it is very helpful for the inversion of sediment in the medium turbid water. In addition, TG-2/WIS has more NIR bands, which benefit to the inversion of aerosols, especially in a Case-1 water. The position of wavebands of TG-2/WIS is consistent with many other water color satellites (e.g. MERIS), which greatly provides potential application of satellites and benefits to the transplantation of algorithms.

The inversion of SPM concentration is calculated using the SERT model of Shen et al. [4, 9, 16], which is based on the Kubelka–Munk two-stream approximation of radiative transfer theory and calibrated using datasets from in situ measurements of the Yangtze estuary. The model uses a multi-conditional algorithm in a wide range to avoid saturation problems [9], as follows:

$$C_{SPM} = \frac{2\alpha R_{rs}}{\beta(\alpha - R_{rs})^2}, \tag{4}$$

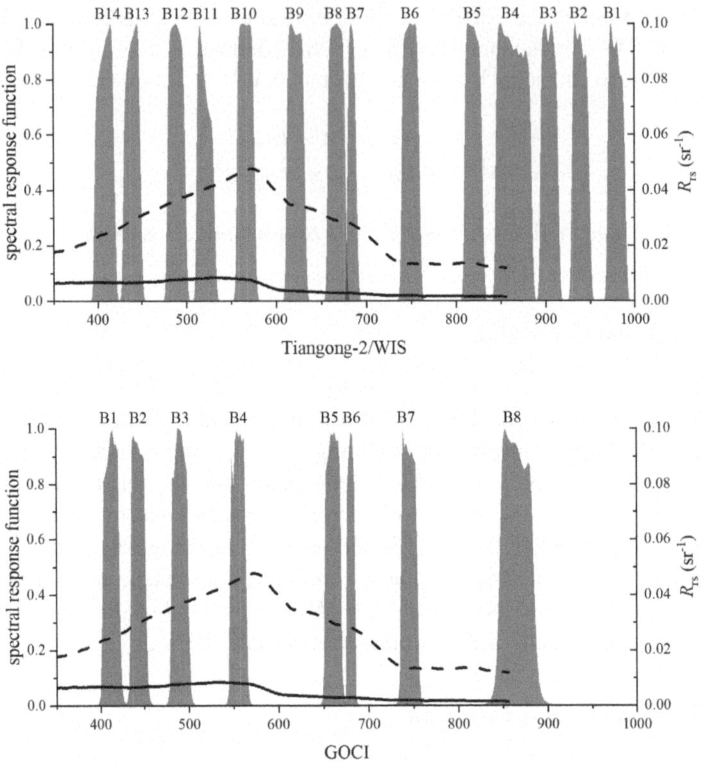

Fig. 2. Spectral response functions of GOCI and TG-2/WIS and spectral curves of R_{rs} for typical waters. Solid lines indicate the clear water and dash lines indicate the turbid water respectively.

where α is mainly affected by the illumination environment and the water case, and β is the ratio of the backscatter coefficient and the absorption coefficient. In the extremely high turbidity water of Hangzhou Bay, the green band is easy to be saturated [9, 17], thus, we use the red and near-infrared band to derive SPM (Table 2).

Table 2. The coefficients α and β at red and NIR bands calibrated by in situ measurements of the Yangtze estuary

Sensor	Band number	Central band (nm)	α	β
GOCI	5	660	0.0771	11.0158
GOCI	7	745	0.0954	2.9698
GOCI	8	865	0.1052	1.659
TG-2/WIS	8	665	0.0779	10.7085
TG-2/WIS	6	750	0.0964	2.9074
TG-2/WIS	4	865	0.1052	1.659

5 Results

5.1 Cross Comparison of TOA Reflectance

According to the calculation formula of TOA reflectance (Eq. 2) in Sect. 4.1, 660, 745 and 865 nm of GOCI and TG-2/WIS are selected for comparison. Because the high turbidity of Hangzhou Bay (SPM > 0.4 g/L), the blue and green band signals are easy to saturate [9, 17]. Therefore, only the red and near-infrared bands are used to compare and verify the sensors. The comparison of TOA reflectance of the two sensors is shown in Fig. 3.

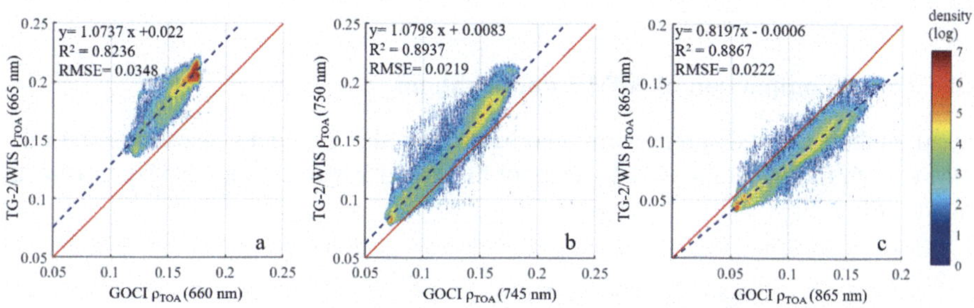

Fig. 3. Comparison of TOA reflectance of three corresponding bands (660 nm, 745 nm and 865 nm) from GOCI and TG-2/WIS in the Yangtze estuary and Hangzhou Bay.

As shown in Fig. 3, the TOA reflectance of TG-2/WIS and GOCI is in good agreements for the red and near-infrared bands ($R^2 > 0.8$). In the 865 nm band, the TOA reflectance from TG-2/WIS is slightly underestimated compared to GOCI, and the slope is 0.8197. The reason is that the calibration of the sensor is still not perfect so far. In addition, the sampling interval of GOCI and TG-2/WIS is 30 min, which can cause some errors in the comparison.

5.2 Cross Comparison of Remote Sensing Reflectance

Through the ESOA algorithm (see Sect. 4.2), the GOCI and TG-2/WIS images are atmospherically corrected, and the red to near-infrared bands are selected for comparison. The study area is consistent with that in Sect. 5.1 and the comparison of the R_{rs} of the two sensors is shown in Fig. 4.

Similar to the result of TOA radiance, TG-2/WIS and GOCI show good agreements in the red and near-infrared bands. In the 665 nm and 750 nm band, the values of R_{rs} are closer, with the slopes close to 1 and RMSE < 0.004 sr^{-1}. In the near-infrared bands, the TG-2/WIS R_{rs} values are underestimated compared to GOCI, and the slope of the scatter plot is 0.77, which is mainly caused by the difference in TOA reflectance.

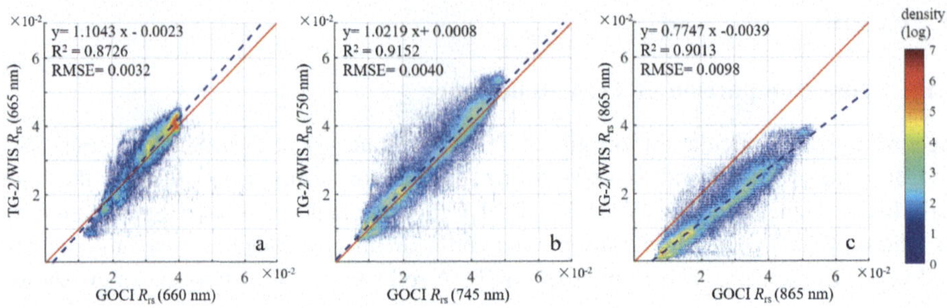

Fig. 4. Scatter plot between GOCI R_{rs} and TG-2/WIS R_{rs} for 660 nm (a), 745 nm (b) and 865 nm (c)

5.3 Cross Comparison of SPM Concentrations

Based on the R_{rs} at red and near-infrared bands, the SERT model is used to retrieve the SPM concentrations in the turbid waters of the Yangtze estuary and Hangzhou Bay, as shown in Fig. 5.

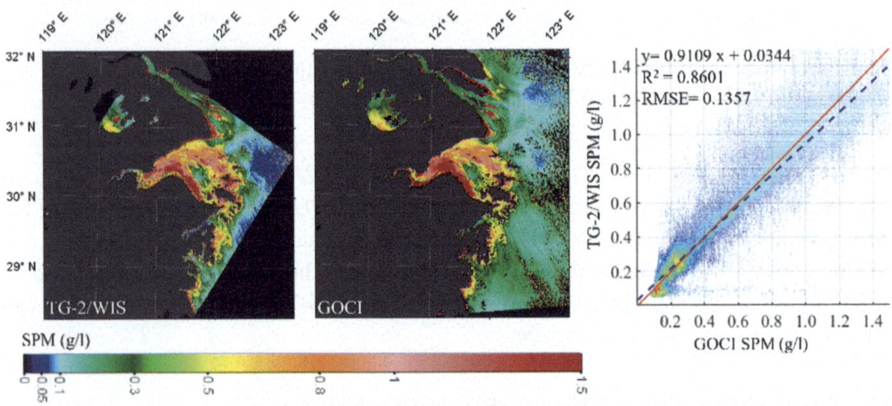

Fig. 5. Distribution of SPM concentration derived from TG-2/WIS and GOCI (left), and the corresponding scatter plot (right)

From Fig. 5, we can see that the TG-2/WIS derived SPM concentration is similar to that of GOCI. The RMSE of 0.1357 g/L seems acceptable, especially in extremely high turbid waters. Some errors can be explained by the difference between the spatial resolution and the sampling time. Generally, the SPM concentration retrieved by the two satellites has good spatial consistency in the Yangtze estuary and the Hangzhou Bay. Both sensors can be used to study the large-scale distribution of SPM.

5.4 Cross Comparison of Satellite and in Situ SPM

In this work, we used the turbidity of OBS at three hydrological stations (see Fig. 1) located in Chongxi, Baozhen and Hengsha to derive SPM (OBS-SPM), and compared with the GOCI and the TG-2/WIS derived SPM concentrations on January 22, 2017. To get rid of the influence of the land, we use the mean value of the 5×5 pixel window centered on the hydrological stations for GOCI image; for the higher resolution sensor TG-2/WIS, mean value of the 3×3 pixel window centered on the hydrological station is used. SPM concentrations derived from the OBS-SPM, GOCI and TG-2/WIS are shown in Fig. 6.

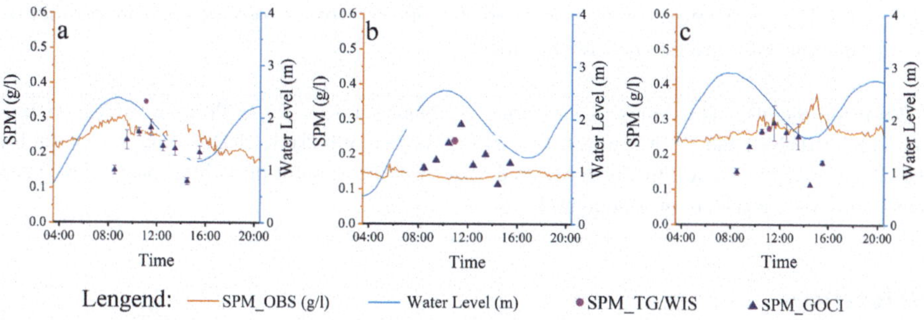

Fig. 6. Comparison of OBS-SPM, GOCI and TG-2/WIS derived SPM concentrations in Baozhen (a), Chongxi (b) and Hengsha (c) hydrological stations. Satellite-derived SPM concentration is represented by dots, OBS-SPM concentration by the yellow line, and the water level of the hydrological station by the blue line.

It is shown in Fig. 6 that the GOCI-derived SPM in Baozhen and Hengsha stations have good agreements with OBS-SPM. The underestimation before 9:30 and after 14:30 may be caused by the large values of solar zenith angle, leading to lower accuracy of the atmospheric correction. In Chongxi station, the GOCI derived SPM concentrations are overestimated, with a small difference less than 0.13 g/L. In addition, the complex SPM distribution near the stations can cause some errors in the process of averaging. Compared with the GOCI derived SPM concentration, the TG-2/WIS derived SPM concentration have good agreements, especially in Chongxi and Hengsha hydrological stations. Taking the errors caused by satellite resolution, turbidity measurements, turbidity and sediment conversion formulas into consideration, the derived-SPM is acceptable.

6 Conclusion

Through comparing TG-2/WIS with GOCI data and products, we can draw the following conclusions: the TOA reflectance of TG-2/WIS and GOCI have a good linear correlation at the red and near-infrared bands. R^2 at 660, 745 and 865 nm are higher than 0.8, although the slope of TOA reflectance at 865 nm is 0.82. The cross-comparison results of atmospherically corrected R_{rs} are similar to the cross-comparison of TOA reflectance. The TG-2/WIS R_{rs} values at 660 and 745 nm have a good consistency with GOCI, with RMSE < 0.04 sr^{-1}, and underestimated at 865 nm with the slope of 0.77. After performing the SERT model for the SPM inversion using 665, 750 and 865 nm, the TG-2/WIS derived SPM concentrations are similar to those of the GOCI derived. Compared with the in situ data, the TG-2/WIS and GOCI derived SPM concentrations have good correspondences with the OBS-SPM from 9:30 to 13:30 local time. In conclusion, the accuracy of TG-2/WIS meets the need of estuary observation. TG-2/WIS can be applied to aquatic environment observations and researches such as water quality monitoring and sediment transport study. In addition, the TG-2/WIS has advantages in the bands number and spatial resolution compared with GOCI, which show great potentials for inversions of phytoplankton pigments, sediment contents and aerosol observation.

Acknowledgments. This study is supported by National Key R&D Program of China (no. 2016YFE0103200) and NSFC projects (no. 41771378, 51739005). KORDI/KOSC is grateful for the GOCI imagery. Thanks to China Manned Space Engineering for providing space science and application data products of Tiangong-2.

References

1. Yang, S.L., Milliman, J.D., Li, P., Xu, K.: 50,000 dams later: erosion of the yangtze river and its delta. Global Planet. Change **75**, 14 (2011)
2. Chen, J., Ni, X., Mao, Z., Wang, Y., Liang, L., Gong, F.: Remote sensing and buoy based effect analysis of typhoon on hypoxia off the Changjiang (Yangtze) estuary. In: Proceedings of SPIE - The International Society for Optical Engineering. 8532. 11 (2012)
3. Du, J.L., Yang, S.L., Zhang, W.X., Li, P., Zhao, H.Y.: Study of influence on erosion and accumulation of Jiuduansha Tidal Island by deep-water channel project at north passage of the Yangtze River. Ocean Eng. **23**, 78 (2005)
4. Shen, F., Zhou, Y., Li, J., He, Q., Verhoef, W.: Remotely sensed variability of the suspended sediment concentration and its response to decreased river discharge in the Yangtze estuary and adjacent coast. Cont. Shelf Res. **69**, 52 (2013)
5. Yu, X.L.: Retrieval of Suspended Matter Concentration and Reconstruction of Missing Data Based on GOCI in Bohai and Yellow Sea. Ocean University of China (2013)
6. Peng, X.Y.: Retrivals of Surface Suspended Particulate Matter Concentration and Its Variation In Yangtze River estuary Using Multiple Polar-Orbiting and Geostationary Satellites. East China Normal University (2015)
7. Frouin, R. (ed.): In-flight Calibration of Satellite Ocean-Colour Sensors. Reports of the International Ocean-Colour Coordinating Group, No. 14, IOCCG, Dartmouth, Canada
8. Le, C., Hu, C., English, D., Cannizzaro, J., Chen, Z., Feng, L., Boler, R., Kovach, C.: Towards a long-term chlorophyll-a data record in a turbid estuary using MODIS observations. Progress Oceanogr. **109**, 90 (2013)
9. Shen, F., Verhoef, W., Zhou, Y., Salama, M.S., Liu, X.: Satellite Estimates of Wide-Range Suspended Sediment Concentrations in Changjiang (Yangtze) estuary Using MERIS Data. Estuaries and Coasts. 33. 1420 (2010)

10. Montes-Hugo, M.A., Mohammadpour, G.: Satellite-derived suspended particulates in the Saint Lawrence estuary: uncertainties due to bottom effects. Canadian J. Remote Sens. **39**, 444 (2013)
11. Sakuno, Y.: Chlorophyll data fusion in Tachibana Bay using COMS GOCI and modis data by the LCI method. Geosci. Remote Sens. Symp. **vol**, 1594 (2014)
12. Cao, Z., Duan, H., Song, Q., Shen, M., Ma, R., Liu, D.: Evaluation of the sensitivity of China's next-generation ocean satellite sensor MWI onboard the Tiangong-2 space lab over inland waters. Int. J. Appl. Earth Obs. Geoinf. **71**, 109 (2018)
13. Zhao, J., Qing, H.E., Wang, X.Y., Guo, L.C.: Field observations on the characteristics of current and sediment of the south and north branches in the yangtze estuary. Resources & Environment in the Yangtze Basin (2015)
14. Pan, Y., Shen, F., Verhoef, W.: An improved spectral optimization algorithm for atmospheric correction over turbid coastal waters: A case study from the Changjiang (Yangtze) estuary and the adjacent coast. Remote Sens. Environ. **191**, 197 (2017)
15. Ahmad, Z., et al.: New aerosol models for the retrieval of aerosol optical thickness and normalized water-leaving radiances from the SeaWiFS and MODIS sensors over coastal regions and open oceans. Appl. Optics **49**(5545) (2010)
16. Shen, F., Zhou, Y., Peng, X., Chen, Y.: Satellite multi-sensor mapping of suspended particulate matter in turbid estuarine and coastal ocean, China. Int. J. Remote Sens. **35**, 4173 (2014)
17. Luo, Y., Doxaran, D., Ruddick, K., Shen, F., Gentili, B., Yan, L., Huang, H.: Saturation of water reflectance in extremely turbid media based on field measurements, satellite data and bio-optical modelling. Opt. Express **26**, 10435 (2018)

Sharpening the VNIR-SWIR-TIR Bands of the WIS of Tiangong-2 for Mapping Land Use and Land Cover

Qingsheng Liu(✉)

Institute of Geographic Sciences and Natural Resources Research, Chinese Academy of Sciences, Beijing 100101, China
liuqs@lreis.ac.cn

Abstract. Land use and land cover maps derived from satellite remote sensing imagery are important to support landscape ecology and natural resource protecting and planning, especially over large areas. The classification information resulted from combination of the visible and near infrared bands (VNIR), shortwave infrared bands (SWIR) and thermal infrared bands (TIR) should be better than those obtained using the VNIR, SWIR and TIR images separately. Nonetheless, assessment on improvement for mapping land use and land cover based on the combination of the VNIR, SWIR and TIR imagery remain scarce. This paper evaluates the random forest classification results from four combinations from the VNIR, SWIR and TIR of the Wide-band Imaging Spectrometer onboard the Tiangong-2. The result showed that combination of fourteen VNIR bands with two TIR bands had produced the best result (overall accuracy (OA) = 84.29%), followed by the combination of fourteen VNIR bands with two SWIR bands (84.13%), then the combination of fourteen VNIR bands and two SWIR bands and two TIR bands (83.73%). Only fourteen VNIR bands had produced the worst classification result (83.5%).

Keywords: Tiangong-2 · VNIR · SWIR · TIR · Gram-Schmidt sharpening Random forest

1 Introduction

The knowledge of land use and land cover maps is important for landscape ecology and natural resource protecting and planning. In fact, with the increase of human activities, the landscape becomes fragmented, which will affect landscape ecology and habitat connectivity and further change landscape processes [1].

Satellite remote sensing has been widely used in mapping landscape features due to the extended spatial and temporal coverage, most of which can acquire free, open, routine, global and visible and infrared imagery. The visible and near infrared (VNIR) and shortwave infrared (SWIR) image observes a radiation reflected from Earth's surface over a VNIR wavelength range from 0.4 to 1.1 μm and a SWIR wavelength range from 1.1 to 2.5 μm respectively, whereas the thermal infrared (TIR) image observes radiation emitted from Earth's surface over a TIR wavelength range from 8 to 15 μm [2–4]. Compared with the VNIR imagery, the SWIR imagery can see in night and through smoke, which makes

Y. Gu et al. (Eds.): *Proceedings of the Tiangong-2 Remote Sensing Application Conference*, LNEE 541, pp. 212–221, 2019.
https://doi.org/10.1007/978-981-13-3501-3_20

the SWIR imagery detect the minerals, man-made materials, plant water status, soil moisture and fire better [5, 6]. The TIR image is sensitive to Earth's surface temperature. High spatial resolution (better than 100 m) TIR imagery has been applied in environmental monitoring such as monitoring volcanic eruptive activity, detecting forest fire, assessing urban heat island effect, and mapping ecosystem health [7]. Thus, it is often hopeful to have the high spectral and spatial resolution information from the VNIR, SWIR and TIR imagery combined in the same file [8]. However, satellite-based images currently available are often composed of VNIR bands with high spatial resolution and SWIR and TIR bands with low spatial resolution, especially the spatial resolution of TIR bands has more than 3:1 scale ratio to that of the VNIR bands, such as Sentinel-2, WorldView-3, Advanced Spaceborne Thermal Emission and Reflection Radiometer (ASTER), Landsat Thematic Mapper (Landsat TM), Landsat Enhanced Thematic Mapper Plus (Landsat ETM+), Landsat Operational Land Imager (Landsat OLI), Moderate Resolution Imaging Spectrometer (MODIS). The variance in spatial resolution among the VNIR, SWIR and TIR bands would be attributed to technical limitations on instrument design, satellite and sensor life, operational limitations such as storage and transmission, and further effects like atmospheric absorption [9], which could not be easily overcome through hardware improvements [10]. Image sharpening permits to enhance the spatial resolution of SWIR and TIR imagery based on the VNIR imagery, which broaden the baseline of physical observables so that the resulting information is better than those obtained using the VNIR, SWIR and TIR images separately [3]. Nonetheless, assessment on improvement for mapping landscape features based on the combination of the VNIR, SWIR and TIR imagery remain scarce.

Random Forest (RF) classifier is kind of data mining and machine learning techniques, which does not depend on data distribution assumption, and has good adaptive capacity. It can effectively avoid over-fitting, and requires less number of user-defined parameters than that of support vector machine often used in remote sensing image classification. In recent years, the RF has proved to be highly successful in remote sensing applications with improved classification accuracies [11–13].

In this study, we evaluate the capability of using the VNIR, SWIR and TIR imagery of the Wide-band Imaging Spectrometer (WIS) onboard the Tiangong-2 for mapping landscape features in Binzhou and Dongying City, Shandong Province, China to understand the SWIR and TIR images' effects on land use and land cover classification accuracy. For this case, the selected band scheme and Gram-Schmidt (GS) image sharpening approach embedded in ENVI software 5.1 version were adopted to fuse the VNIR, SWIR and TIR bands because the high spatial resolution VNIR bands do not spectrally overlap the low spatial resolution SWIR and TIR band, and RF classifier was used to classify land use and land cover in the study area.

2 Materials and Methods

2.1 Data and Study Area

The WIS images the Earth's surface in 300-kilometer-wide swath with 14 VNIR bands (100 m spatial resolution), two SWIR bands (200 m spatial resolution) and two TIR bands (400 m spatial resolution). The WIS Level 2 data was acquired on May 8, 2018, which was geometrically corrected and radiometric calibration product, and covered some areas of Tianjin City, Hebei Province and Shandong Province, China. However, the VNIR, SWIR and TIR images of the WIS Level 2 data need to be registered each other. Therefore, the SWIR and TIR images were registered to the VNIR images of the study area. The sub-region was trimmed for the study area from the original imagery according to the minimum bounding rectangle of the boundary of Binzhou (due to the coverage of the VNIR image, not including Zouping County) and Dongying City.

The spectral reflectance of the VNIR and SWIR bands was retrieved using the quick atmospheric correction method within ENVI software 5.1 version, which determines atmospheric compensation parameters directly from the image without ancillary information, and generally produces reflectance spectra within approximately ±15% of the physics-based approaches [14]. The Thermal Atmospheric Correction in ENVI software 5.1 version was used to approximate and remove the atmospheric contributions from the TIR bands.

2.2 Sharpening the SWIR and TIR Bands

The original VNIR and SWIR bands were resized to match the same spatial resolution (400 m) of the TIR bands. The correlation coefficient (CC) was used to select the optimal high resolution bands for sharpening the SWIR and TIR bands. The previous studies have indicated that: (1) the VNIR Band 1 has higher positive CC value with the SWIR Bands [15] and (2) the sharpened results of SWIR bands from Gram-Schmidt image sharpening approach with average low resolution bands as the low resolution pan band was acceptable [16].

Table 1 showed the CC between the VNIR, SWIR and TIR bands. The higher correlation coefficient values indicated the more similarity between the VNIR, SWIR and TIR bands. The closer wavelength of two bands is, the higher CC value between two bands has. The maximum value in CC between SWIR bands and VNIR and TIR bands was observed at the VNIR Band 1, whereas the maximum value in CC between TIR bands and VNIR and SWIR bands was observed at the SWIR Band 2. Thus, the original SWIR Band 2 was selected as the optimal high spatial resolution band for sharpening the TIR bands to 200 m. Then, the 200 m-resolution TIR was stacked with the 200 m-resolution SWIR bands. After that, the VNIR Band 1was selected as the optimal high spatial resolution band for sharpening the 200 m-resolution SWIR and 200 m-resolution TIR bands to 100 m.

2.3 Land Use and Land Cover Classification

The study area encompasses a region characterized by winter wheat, summer maize, cotton, orchard, resident, forest, grassland, river, reservoir, aquaculture, salt pan, bare soil with other land use and land cover types. According to the acquisition date and spatial resolution of the WIS imagery, land use and land cover types in this study to be classified include built-up and impervious surface, rural resident, forest land, grass land, winter wheat, jujube, ponds, salt pan, aquaculture, Yellow River, sea water, tidal flat and bare soil.

Random forest classifier in EnMap-Box 2.0.2 was used to map the land use and land cover of the study area from four combinations of VNIR, SWIR and TIR bands: (1) fourteen VNIR bands, (2) fourteen VNIR bands + two SWIR bands, (3) fourteen VNIR bands + two TIR bands, (4) fourteen VNIR bands + two SWIR bands + two TIR bands. In this study, the default values of RF were used, and the number of trees is 100, as previous studies have proved that the default values are often a good choice [12, 17].

Table 1. The correlation coefficient for the VNIR, SWIR and TIR bands with 400 m spatial resolution

Bands	SWIR Band 1	SWIR Band 2	TIR Band 1	TIR Band 2
VNIR Band 1	0.9594	0.8346	0.5269	0.5041
VNIR Band 2	0.9275	0.7872	0.4626	0.4363
VNIR Band 3	0.9235	0.7622	0.4851	0.4610
VNIR Band 4	0.9211	0.7573	0.4913	0.4685
VNIR Band 5	0.9042	0.7365	0.4780	0.4546
VNIR Band 6	0.9263	0.7822	0.5031	0.4817
VNIR Band 7	0.1014	0.3248	0.1569	0.1643
VNIR Band 8	0.0632	0.2886	0.1276	0.1356
VNIR Band 9	−0.0379	0.1866	0.0457	0.0556
VNIR Band 10	−0.1926	0.0241	−0.1310	−0.1214
VNIR Band 11	−0.0861	0.1124	−0.0928	−0.0857
VNIR Band 12	−0.0545	0.1315	−0.0815	−0.0746
VNIR Band 13	0.0589	0.2062	−0.0281	−0.0246
VNIR Band 14	0.0610	0.1831	−0.0487	−0.0473
SWIR Band 1	1.0000	0.9201	0.5823	0.5636
SWIR Band 2	0.9201	1.0000	0.6132	0.5995
TIR Band 1	0.5823	0.6132	1.0000	0.9793
TIR Band 2	0.5636	0.5995	0.9793	1.0000

2.4 Accuracy Assessment

The VNIR pseudo-color composite image, Landsat 8 OLI image and higher spatial resolution images from Google Earth were used to visually determine reference and validation samples using Function *Create Class Image from ROIs* of Region of Interesting (ROI) tool in ENVI software 5.1 version by the experts who are very familiar with land use and land cover distributed in the study area.

All four combinations were classified and validated with the same reference and validation samples. Results of land use and land cover classification were evaluated and compared based on the commonly used accuracy indices: overall accuracy (OA), Kappa value, user's accuracy (UA) and producer's accuracy (PA). In addition, band importance of RF classifier models from four combinations for feature selection was computed and analyzed.

3 Results and Discussion

Tables 2, 3, 4 and 5 summarizes the results of land use and land cover classification from four combinations as input and RF classifier. All of the OA and Kappa values were relatively high. Comparison of the overall accuracies and Kappa values created with the four combination images indicated that a combination of fourteen VNIR bands with two TIR bands had produced the best result (Fig. 1c), followed by the combination of fourteen VNIR bands with two SWIR bands (Fig. 1b), fourteen VNIR bands with two SWIR bands and two TIR bands (Fig. 1d). Only fourteen VNIR bands had produced the worst result (Fig. 1a).

Table 2. A summary of land use and land cover classification results calculated with fourteen VNIR bands by using RF classifier

Class	User's accuracy (%)	Producer's accuracy (%)
(1) Built-up and impervious surface	84.47	94.92
(2) Rural resident	71.69	41.46
(3) Forest land	40.26	66.67
(4) Jujube	80.30	96.36
(5) Grassland	96.50	81.18
(6) Tidal flat	83.45	77.59
(7) Bare soil	57.76	47.18
(8) Pond	75.64	85.71
(9) Salt pan	89.31	92.09
(10) Aquaculture	42.11	86.38
(11) Yellow River	100.00	77.50
(12) Sea water	100.00	79.20
(13) Winter wheat	94.96	86.90
	Overall accuracy = 83.5%, Kappa = 79.84%	

Comparison of PA values created with the four combination images indicated that only fourteen VNIR bands with RF classifier produced better result for only forest land, and the combination of fourteen VNIR bands with two SWIR bands produced better results for rural resident, jujube, bare soil, pond and winter wheat, and the combination of fourteen VNIR bands with two TIR bands produced better results for tidal flat, salt pan, aquaculture and Yellow River. The combination of all fourteen VNIR bands with two SWIR bands and two TIR bands produced better result for only sea water. In addition, the PA values indicated that two SWIR bands were important for mapping the built-up and impervious surface, and the PA value could not increase with injection of two TIR bands. For grass land, fourteen VNIR bands combined with either two SWIR bands or two TIR bands could produce better results. Overall, two SWIR bands is helpful for improving the classification accuracy of the built-up and impervious surface, rural resident, jujube, bare soil and pond and winter wheat, and two TIR bands is useful for improving the classification accuracy of the built-up and impervious surface, tidal flat, salt pan, aquaculture and Yellow River, which indicated that soil moisture and canopy moisture of the study area in May were obviously different, and water bodies with different types of utilization had different surface temperature.

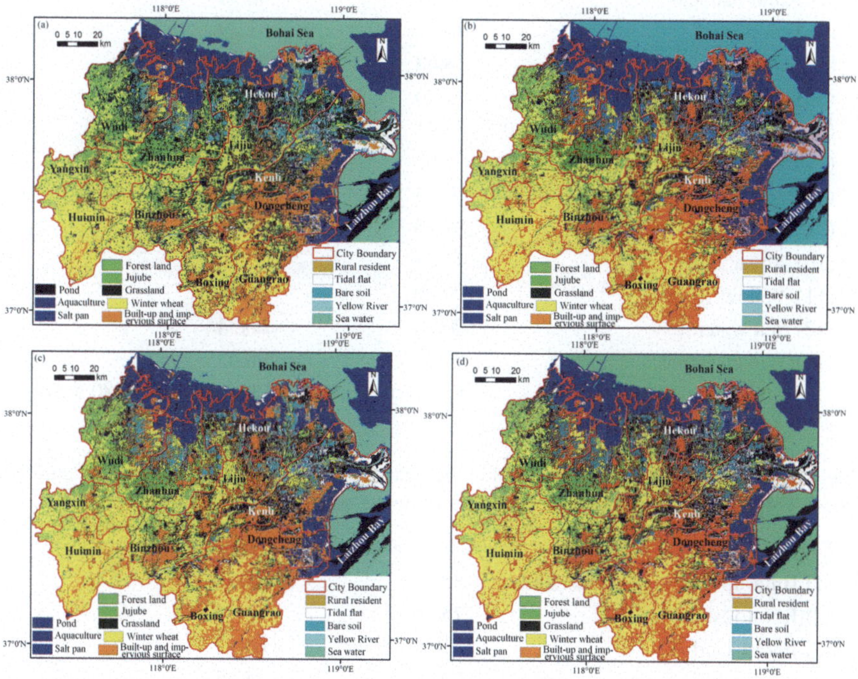

Fig. 1. Random forest classification results of land use and land cover in the study area. (a) the results from fourteen visible and near infrared bands; (b) the results from fourteen visible and near infrared bands with two shortwave infrared bands; (c) the results from fourteen visible and near infrared bands with two thermal infrared bands; (d) the results from fourteen visible and near infrared bands with two shortwave infrared bands and two thermal infrared bands.

Bands importance for classifying land use and land cover with four combinations using the RF classifier was shown in Fig. 2.

When fourteen visible and near infrared bands were used to classify land use and land cover with the RF classifier, VNIR Band 1, Band 13, Band 10 and Band 11 had greater importance (Fig. 2a). When fourteen visible and near infrared bands and two shortwave infrared bands were used together to classify land use and land cover with the RF classifier, VNIR Band 13, Band 11, Band 10 and Band 14 had greater importance, two shortwave infrared bands only had greater importance than VNIR Band 2, Band 3 and Band 4, and SWIR Band 2 had greater importance than SWIR Band 1 (Fig. 2b). When fourteen visible and near infrared bands and two thermal infrared bands were used together to classify land use and land cover with the RF classifier, VNIR Band 13, Band 11, Band 10 and Band 14 had greater importance, two

Table 3. A summary of land use and land cover classification results calculated with combined fourteen VNIR bands with two SWIR bands by using RF classifier

Class	User's accuracy (%)	Producer's accuracy (%)
(1) Built-up and impervious surface	85.27	95.30
(2) Rural resident	80.61	46.34
(3) Forest land	46.72	61.29
(4) Jujube	78.68	97.27
(5) Grassland	92.26	84.12
(6) Tidal flat	84.41	74.25
(7) Bare soil	61.02	50.70
(8) Pond	69.63	86.36
(9) Salt pan	89.30	91.11
(10) Aquaculture	43.28	85.55
(11) Yellow River	100.00	78.33
(12) Sea water	100.00	80.34
(13) Winter wheat	94.26	90.48
	Overall accuracy = 84.13%, Kappa = 80.58%	

Table 4. A summary of land use and land cover classification results calculated with combined fourteen VNIR bands with two TIR bands by using RF classifier

Class	User's accuracy (%)	Producer's accuracy (%)
(1) Built-up and impervious surface	85.43	95.22
(2) Rural resident	75.14	45.30
(3) Forest land	41.10	64.52
(4) Jujube	80.77	95.45
(5) Grassland	96.62	84.12
(6) Tidal flat	83.87	78.26
(7) Bare soil	58.26	47.18
(8) Pond	75.86	85.71
(9) Salt pan	89.62	92.26
(10) Aquaculture	43.76	87.38
(11) Yellow River	100.00	79.17
(12) Sea water	100.00	80.40
(13) Winter wheat	94.68	87.35
	Overall accuracy = 84.29%, Kappa = 80.78%	

thermal infrared bands only had greater importance than VNIR Band 2, Band 3 and Band 4, and TIR Band 1 had greater importance than TIR Band 2 (Fig. 2c). When fourteen visible and near infrared bands, two shortwave infrared bands and two thermal infrared bands were used together to classify land use and land cover with the RF classifier, VNIR Band 13, Band 10, Band 14 and Band 5 had greater importance, SWIR Band 1 and TIR Band 1 had greater importance than VNIR Band 4, Band 3, Band 2 and Band 1 (Fig. 2d).

Table 5. A summary of land use and land cover classification results calculated with combined fourteen VNIR bands with two SWIR bands and two TIR bands by using RF classifier

Class	User's accuracy (%)	Producer's accuracy (%)
(1) Built-up and impervious surface	84.92	95.30
(2) Rural resident	80.00	45.99
(3) Forest land	35.33	56.99
(4) Jujube	79.10	96.36
(5) Grassland	93.46	84.12
(6) Tidal flat	83.66	71.91
(7) Bare soil	60.17	50.00
(8) Pond	73.26	85.39
(9) Salt pan	89.22	91.22
(10) Aquaculture	43.23	87.04
(11) Yellow River	100.00	77.50
(12) Sea water	100.00	80.43
(13) Winter wheat	93.34	85.57
	Overall accuracy = 83.73%, Kappa = 80.09%	

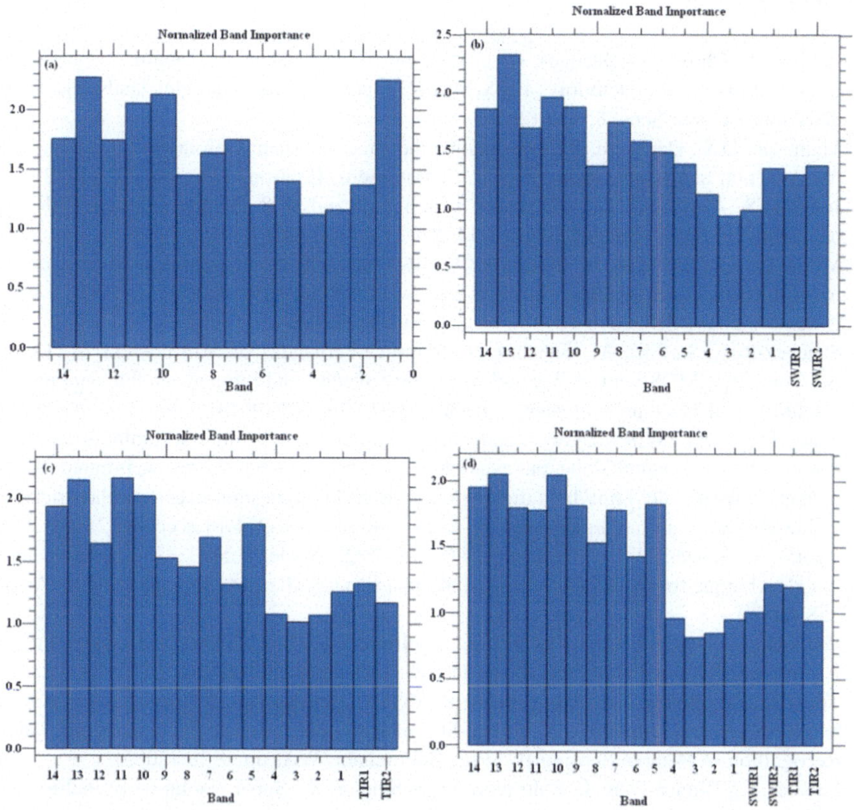

Fig. 2. Normalized band importance for classifying land use and land cover with four combinations using the random forest classifier. (a) fourteen visible and near infrared bands; (b) fourteen visible and near infrared bands with two shortwave infrared bands; (c) fourteen visible and near infrared bands with two thermal infrared bands; (d) fourteen visible and near infrared bands, two shortwave infrared bands and two thermal infrared bands.

4 Conclusions

This study analyzed the WIS imagery to evaluate four combinations of data from VNIR, SWIR and TIR imagery for mapping landscape features. The Gram-Schmidt sharpening method with the selected bands was used to fuse the VNIR, SWIR and TIR bands with 100 m spatial resolution. Random forest classifier was used to classify land use and land cover in the study area. Results in this study showed that combination of fourteen VNIR bands with two TIR bands had produced the best result (OA = 84.29%), followed by the combination of fourteen VNIR bands with two SWIR bands (84.13%), fourteen VNIR bands with two SWIR bands and two TIR bands (83.73%). Only fourteen VNIR bands had produced the worst result (83.5%). In the future, a comprehensive evaluation over more study areas should be performed.

Acknowledgment. This research work was jointly financially supported by the National Natural Science Foundation of China (Project No.41671422, 41661144030). Thanks to China Manned Space Engineering for providing the Tiangong-2 space science and application data products.

References

1. Radoux, J., Chome, G., Jacques, D.C., Waldner, F., Bellemans, N., Matton, N., Lemarche, C., dAndrimont, R., Defourny, P.: Sentinel-2's potential for sub-pixel landscape feature detection. Remote Sens. **8**, 488 (2016)
2. Matheson, D.S., Dennison, P.E.: Evaluating the effects of spatial resolution on hyperspectral fire detection and temperature retrieval. Remote Sens. Environ. **124**, 780–792 (2012)
3. Jung, H.S., Park, S.W.: Multi-sensor fusion of Landsat 8 thermal infrared (TIR) and panchromatic (PAN) images. Sensors **14**, 24425–24440 (2014)
4. Milewski, R., Chabrillat, S., Eisele, A.: VNIR-SWIR-TIR hyperspectral airborne campaign for soil and sediment mapping in semi-arid SouthAfrican environments. In: EGU General Assembly 2016, EGU2016-18438, Vienna, Austria (2016)
5. Sanchez-Ruiz, S., Piles, M., Sanchez, N., Martinez-Fernandez, J., Vall-llossera, M., Camps, A.: Combing SMOS with visible and near/shortwave/thermal infrared satellite data for high resolution soil moisture estimates. J. Hydrol. **516**, 273–283 (2014)
6. Yong, D.K.: Our SWIR imagery product is now available, at the highest resolution ever from space, only from DigitalGlobe, our eyes can play tricks on us but shortwave infrared (SWIR) imagery reveals all. http://blog.digitalglobe.com/technologies/our-eyes-can-play-tricks-on-us-but-shortwave-infrared-swir-imagery-reveals-all-part-1-of-2. Last accessed 25 Jul 2018
7. Agam, N., Kustas, W.P., Anderson, M.C., Li, F.Q., Neale, C.M.U.: A vegetation index based technique for spatial sharpening of thermal imagery. Remote Sens. Environ. **107**, 545–558 (2007)
8. Padwick, C., Deskevich, M., Pacifici, F., Smallwood, S.: WorldView-2 pan-sharpening. In: ASPRS 2010 Annual Conference, San Diego, California, USA (2010)
9. Tonooka, H.: Resolution enhancement of ASTER shortwave and thermal infrared bands based on spectral similarity. In: Proceeding of SPIE 5657, Image Processing and Pattern Recognition in Remote Sensing II, SPIE, Bellingham, WA, pp. 9–19 (2005)
10. Lanaras, C., Biouca-Dias, J., Baltsavias, E., Schindler, K.: Super-resolution of multispectral multiresolution images from a single sensor. In: IEEE Conference on Computer Vision and Pattern Recognition Workshops, Honolulu, USA, pp. 20–28 (2017)

11. Ko, B.C., Kim, H.H., Nam, J.Y.: Classification of potential water bodies using Landsat 8 OLI and a combination of two boosted random forest classifiers. Sensors **15**(6), 13763–13777 (2015)
12. Wang, H., Zhao, Y., Pu, R.L., Zhang, Z.Z.: Mapping RobiniaPseudoacacia forest health conditions by using combined spectral, spatial and textural information extracted from IKONOS imagery and random forest classifier. Remote Sens. **7**(7), 9020–9044 (2015)
13. Pu, R.L., Landry, S., Yu, Q.Y.: Assessing the potential of multi-seasonal high resolution Pleiades satellite imagery for mapping urban tree species. Int. J. Appl. Earth Obs. Geoinf. **71**, 144–158 (2018)
14. Berbstein, L.S., Jin, X.M., Gregor, B., Adler-Golden, S.M.: Quick atmospheric correction code: algorithm description and recent upgrades. Opt. Eng. **51**, 111719 (2012)
15. Liu, Q.S., Li, X.Y., Liu, G.H., Huang, C., Li, H., Guan, X.D.: Sharpening of the VNIR and SWIR bands of the wide band spectral imager onboard Tiangong-II imagery using the selected bands. In: the International Archives of the Photogrammetry, Remote Sensing and Spatial Information Sciences, Beijing, China, vol. XLII-3, pp. 1085–1092 (2018)
16. Liu, Q.S.: Sharpening the WBSI imagery of Tiangong-II: Gram-Schmidt and principal components transform in comparison. In: 14th International Conference on Natural Computation, Fuzzy Systems and Knowledge Discovery (ICNC-FSKD 2018), Huangshan, China, pp. 524–531 (2018)
17. Ismail, R., Mutanga, O.A.: A comparison of regression tree ensembles: predicting *Sirexnoctilio* induced water stress in Pinuspatula forest of KwaZulu-Natal, South Africa. Int. J. Appl. Earth Obs. Geoinf. **12**, 45–51 (2010)

Object-Based Classification from Tiangong-2 Using Support Vector Machine Optimized with Evolutionary Algorithm

Huijiao Qiao[1] and Xue Wan[2,3](✉)

[1] School of Remote Sensing and Information Engineering, Wuhan University, Wuhan 430079, China

[2] Technology and Engineering Center for Space Utilization, Chinese Academy of Science, Beijing, China
wanxue@csu.ac.cn

[3] Key Laboratory of Space Utilization, Chinese Academy of Sciences, Beijing 100094, China

Abstract. Land cover information extraction using object-based image classification has become a widely accepted trend in remote sensing. Segmented objects have abundant features including spectrum, shape and texture, and feature selection affects the precision and efficiency of classification. However, how to select less features while achieving high classification accuracy remains unsolved. In this paper, an object-based image classification method based on Support Vector Machine (SVM) combined with Particle Swarm Optimization (PSO)-based feature selection evolutionary algorithm is proposed to evaluate its potential on land cover classification. The experiments results using Tiangong-2 Wide-band images demonstrate that the proposed SVM classification combined with PSO for low-middle resolution images can yield a preferable classification outcome especially for grassland, water and forest with the user's accuracy higher than 90%.

Keywords: Tiangong-2 · Object · PSO · SVM · Feature selection Classification

1 Introduction

Plentiful Earth's surface information can be acquired from remote sensing data using different feature extraction methods. Tiangong-2 Space Laboratory was successfully launched on 15th September 2016 and carried out a number of new sensors for earth observation. One of them is Wide-band Imaging Spectrometer (WIS) which integrates visible, near infrared, short wavelength infrared and thermal infrared spectrums [1]. It is beneficial to multi-spectral image classification. The main purpose of the WIS is for water monitoring, whether such abundant band information will be beneficial to other land cover types, such as forest, grass, farm, etc., will be investigated in this paper.

One of the most widely used land cover classification method is Support vector machine (SVM) [2], which is a classical machine learning algorithm based on VC dimension (Vapnik-Chervonenkis Dimension) theory and structural risk minimization

Y. Gu et al. (Eds.): *Proceedings of the Tiangong-2 Remote Sensing Application Conference*, LNEE 541, pp. 222–231, 2019.
https://doi.org/10.1007/978-981-13-3501-3_21

principle [3, 4]. The theory of SVM is solid, which can seriously balance model's complexity and leaning capability, and overcome the problems of "dimension disaster" and "over-learning" even when the number of samples is small [5, 6]. However, when using SVM to do classification, how to select the optimal feature subset to enable a fast and accurate classification is a problem [7, 8]. As feature selection can select smaller number of features in the mining process, computation time can be significantly saved and a better model can be trained for the test data [9].

Common methods for feature selection include Principle Component Analysis (PCA), Sequential Forward Selection (SFS), Sequential Float Forward Selection and Backtracking Search Optimization Algorithm (BSO). These methods, however, suffer from local convergence in high-dimensional feature space [10, 11]. Consequently, evolutionary algorithms with the advantages of high efficiency and low cost [12], such as Genetic Algorithm (GA), Simulated Annealing (SA), Ant Colony Optimization (ACO) and Particle Swarm Optimization (PSO), are introduced in feature selection. As a typical method in evolutionary algorithms, PSO has a higher global search ability approaching to the optimal solution and computes less costly than others [13, 14]. Fitness function plays a vital role in PSO-based feature selection algorithms [10] and the corresponding fitness value is used to evaluate the selected feature subset. Several PSO-based feature selection methods have been applied in the field of high-resolution remote sensing image classification [15, 16] and hyperspectral band selection [17]. However, as far as we know, it has not been used for Tiangong-2 WIS data, middle resolution remote sensing images, before. This paper aims to optimize the feature combination without reducing the object-based classification accuracy by combining PSO algorithm with SVM, and analyzes the optimal features to be selected, and its effects on land cover classification accuracy.

2 Methodology

2.1 Particle Swarm Optimization

Particle Swarm Optimization (PSO) is originally proposed by Dr. Kennedy and Dr. Eberhart and was first intended for simulating social behaviour, for example, foraging in a bird flock or fish school. PSO is initialized with a population of candidate solutions, called particles, and moves these particles around in the continuous search-space according to a few simple mathematical formulas over the particle's position and velocity, which is dynamically adjusted through searching the local best position of particles ($pbest_i$) in this space [17–20]. When improved position are being discovered, these will then come to guide the global best known position of the swarm ($gbest_i$) in the search space [21, 22]. The particle with high classification accuracy and few features can produce a high fitness value [23]. Thus, PSO algorithm has fast search speed and high efficiency.

PSO is an effective method for the optimization problem of continuous space, but feature selection of multispectral images is a combinatorial optimization problem and the fitness function value associated final global optimized position will be as optimized feature selection. Thus, this paper takes two different forms to encode nest positions: real number continuous space coding X and discrete binary coding BX, respectively. X is used to search the optimized position as usual and the values of BX indicates whether the feature is selected by assigning all elements of each dimensional to 0 or 1. The position variable Xi real number can be calculated into probabilities of BX_i being a value of 0 or 1, as Eqs. (1), (2). Equation (1) is a sigmoid function with the return value increasing from 0 to 1. Then, the final BX_i in Eq. (2) is obtained by comparing the return value and randomly generated value.

$$\mathrm{Sigmoid}(X_i) = 1/(1 + \exp(-X_i)) \tag{1}$$

$$BX_i = \begin{cases} 1, rand() < Sigmoid(X_i) \\ 0, rand() \geq Sigmoid(X_i) \end{cases} \tag{2}$$

2.2 Fitness Function

Classification accuracy, the number of features and the feature cost are the three criteria used to evaluate a fitness function. Thus, a high fitness value stands for high classification accuracy, small number of features and low feature cost. In this paper, a simple form of fitness function is used by combining two criteria [24].

$$\mathrm{F}(BX_i) = \mathrm{Accuracy}(BX_i)/(1 + \theta * \mathrm{n}(BX_i)) \tag{3}$$

$$\mathrm{Accuracy}(BX_i) = F_p/(F_p + F_N) \tag{4}$$

As defined in formula (3), the fitness function has a predefined weight: θ for the number of selected features and generally it is 0.01. $n(BX_i)$ is the number of selected features. Formula 4 is for the calculation of classification accuracy, where F_p is the number of objects with correct classification and conversely, F_N indicates the number of objects classified incorrectly. These numbers are counted by comparing the class labels from artificial visual interpretation and the algorithm mentioned in this paper.

2.3 PSO-Based Feature Selection for Object-Based Classification

When the PSO-based feature selection algorithm is used in this study, the features are extracted from the segmented objects and each feature are given an index. These feature indexes would be randomly selected by PSO-based feature selection algorithm. After several updating and iterations, the optimized results would be output. Figure 1 describes how it works:

Firstly, the parameters of the particle swarm, such as size *N*, position *X*, velocity *V*, and the maximum number of iterations *MaxDT*, are normalized and finalized. Then randomly select features and calculate the fitness of each particle with the selected features and determine the local best position of particle $pbest_i$ and global best known position of the swarm $gbest_i$. This is the vital step of feature selection for the purpose of retaining the better feature combination with higher fitness values. Consequently, update the position X_i and velocity of each particle V_i. After frequent iterations, when

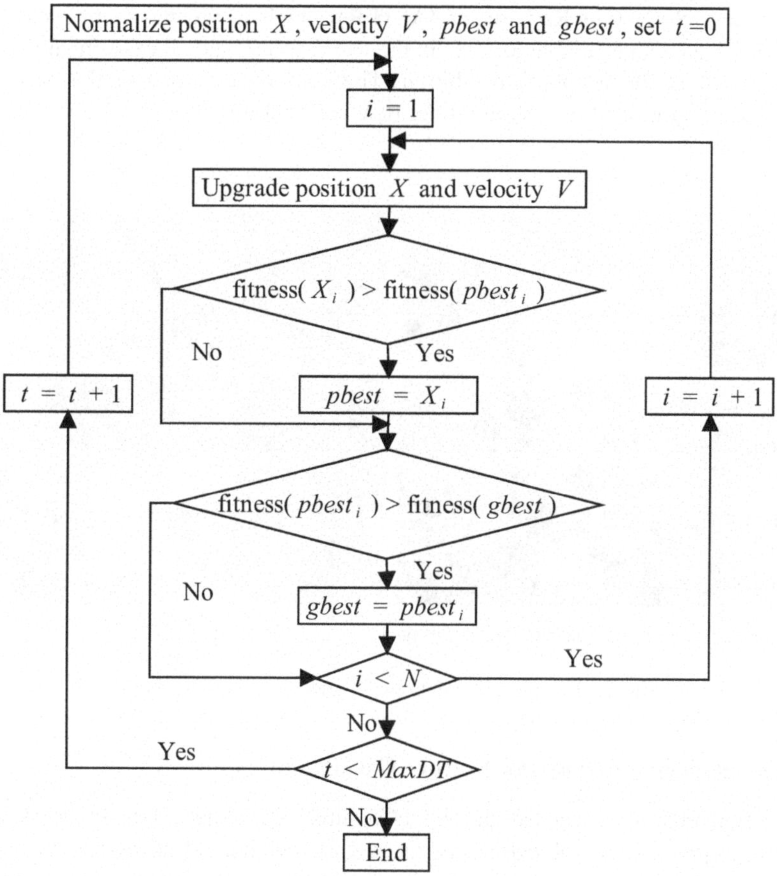

Fig. 1. The flowchart of PSO-based feature selection for object-based classification

the algorithm is converged, the iteration will stop, otherwise, the iteration will return to step 2. The ultimate *gbest* is the final solution.

3 Experiment and Analysis

3.1 Experimental Data

The experimental images were acquired in the region of Wuhan, China by Tiangong-2 Wide-band Imaging Spectrometer on March 22, 2018, with geometric resolution of 100 m, 200 m, and 300 m corresponding to visible, short wavelength infrared and thermal infrared spectrums respectively. This dataset is made up by 470*472 pixels and show a downtown area of Wuhan (see Fig. 2a). To acquire a higher classification accuracy, the matching Global Digital Elevation Model (GDEM) with spatial resolution of 30 m (see Fig. 2b) was downloaded from Advanced Spaceborne Thermal

Emission and Reflection Radiometer (ASTER) which is the most complete global digital elevation data available to users in the world so far, and all experimental images are resampled to 30 m spatial resolution. The land cover types of this study area include forest, grassland, water, building, bare soil and others.

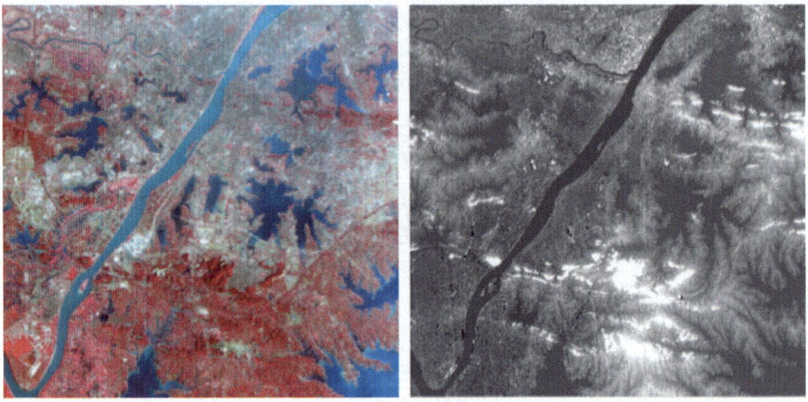

(a).Wide-band image data from Tiangong-2 (b).Aster GDEM

Fig. 2. Experimental data for object-based classification

3.2 Image Segmentation and Feature Calculation

Image segmentation is carried out by eCogniton Developer 9.0, which is a pre-processing operation for object-based image classification. Multi-resolution segmentation, which minimizes the average heterogeneity of objects and maximizes the inter-objects, was utilized for the segmentation in this study. The scale parameter determines the maximum heterogeneity for the segmented image objects [25]. Once the scale is determined, other four criteria define the heterogeneity of a single object: color, shape, smoothness and compactness. The combination of color and shape describes the homogeneity of the object. The compactness is a function of objects' perimeter and the number of pixels within an object, whereas the smoothness describes objects' perimeter and the perimeter of each object's bounding box [26]. Based on a trial-and-error analysis a set of parameter value was generated, which can ensure the final segmentation matched the visual interpretation (Table 1).

Table 1. A set of parameter values for segmentation

Scale	Shape	Compactness	Num. of objects
95	0.2	0.7	669

The typical objects obtained from segmentation are shown in the Fig. 3. After segmentation, the objects, such as grassland, water and some new high-rise buildings can be segmented in one single region. However, bare soil and old building are mixed with forest in the old town area and cannot by separately segmented.

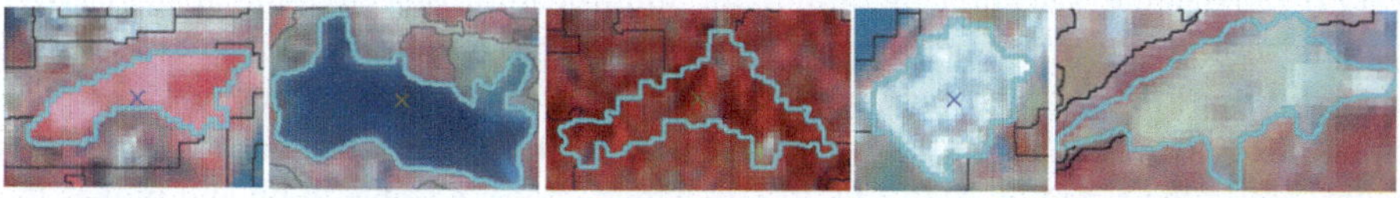

(a).Grassland (b).Water (c).Forest (d).Building (e).Bare soil

Fig. 3. Typical objects obtained from segmentation

The object features extracted from the segmented object are of great benefit to the final classification. In this paper, 110 features including spectral, texture and geometry feature are generated. It should be pointed out that some customized object features including NDVI, NDWI, NDSI and NDISI are beneficial to distinguish vegetation, water, bare soil and building, respectively.

3.3 Feature Selection and Classification Accuracy Assessment

After the segmentation, the samples size is 669. Among them, 466 training samples and 203 test samples are selected from six types of objects according to the ratio of 7:3, and different types with diverse numbers were randomly selected in a certain proportion, which can ensure all types of objects were uniform distribution. The number of iteration is 25, and the optimized settings of the learning factor and the inertia weight in PSO are set to 1.49 and 0.72 respectively, which go through experiments for many times. Then the SVM classifier is constructed, so the 203 test data can be re-labeled. The fitness value (72.5%) remains stable with the change of iterations (Table 2). As it is shown, grassland and water are classified with user and producer accuracy higher than 90%, however, the object types such as building, bare soil and others suggest unsatisfactory results with user accuracy about 60%. This is because these bare soils, old and squat buildings distribute sporadically in forest area, which is a common phenomenon in fringe area. Thus, they are classified as 'forest' due to the low-middle spatial resolution.

Table 2. Classification accuracy of SVM with PSO algorithm

Object's Types	Forest	Grassland	Water	Building	Bare soil	Others	Total	User accuracy
Forest	61						61	100%
Grassland		10					10	100%
Water			62				62	100%
Building	14			26			40	65%
Bare soil	8				16		24	66%
Others	2	1				3	6	50%
total	85	11	62	26	16	3	203	
Producer accuracy	71%	90%	100%	100%	100%	100%		

To validate the effectiveness of the PSO-based algorithm, the comparison with only using SVM classifier without any feature selections algorithm is made. The classification accuracy is 70.30%, which is less accurate than using PSO-based feature selection algorithm. Because some redundant features can be regarded as noise.

Moreover, this paper also analyzes the effects of sample size on the accuracy of PSO-based feature selection algorithm for multi-feature objective-based land cover classification. Four experiments using different sample numbers from 269 to 669 are carried out while the ratio between training data and testing data remains 7:3. The classification accuracies are shown in Fig. 4, and it can be seen that all the fitness values are in the range of 72.0% to 76.1%, which shows a stable trend and proves the reliability in different sample sizes.

Then, we analyze features which are selected most frequently by PSO for classification. In the experiments, 45 iterations, whose fitness values are larger than 70%, using PSO with different size numbers are carried out. The 9 most selected features are shown in Fig. 5. It shows NDVI and NDWI are the most popular features during this classification, because these two indicators are useful to distinguish forest and water, which are two main object types in the testing area. Meanwhile, these spectrum and geometry features, such as Hue, and GLCM_Ang.2nd, are also frequently used for classification, because in the case of buildings, high reflectivity (spectrum feature) and regular shape (geometry feature) are useful for interpreting. This result shows that the above features play a significant role in classification for Tiangong-2 data.

Meanwhile, the algorithm record the selected features, so we list these features and relevant classification information when fitness value is greater than 70% in the experiment with 469 sample numbers. As shown in Table 3, spectral, shape and texture, these three different kinds of features can be selected to classification and different

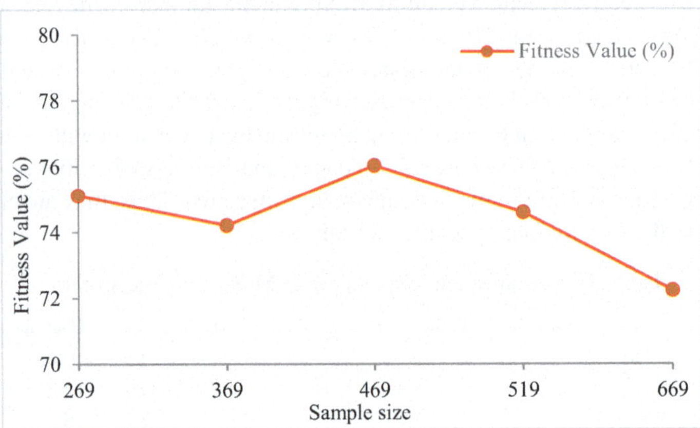

Fig. 4. Comparison analysis of accuracy with PSO-based feature selection algorithm in different sample sizes

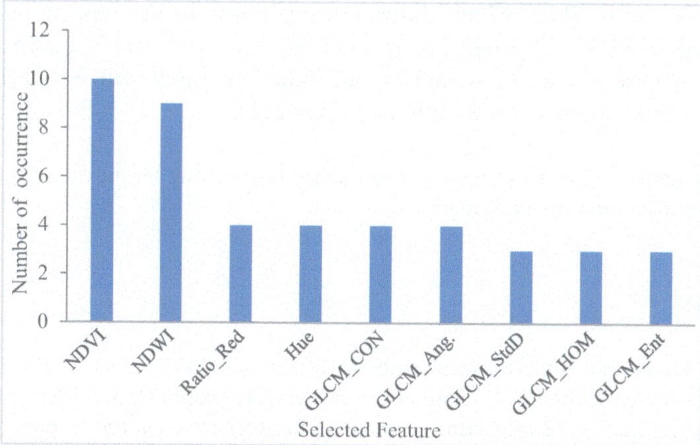

Fig. 5. The correlation between the selected features and the number of occurrence

Table 3. Experimental results summary of PSO approach on the test set

Iteration	Fitness value	Classification accuracy	Number of features	Selected features
1	71.27%	74.13%	4	NDVI, Standard, GLCM_Ang.2nd moment, GLCM_Contrast
2	71.93%	75.52%	5	Ratio_Red, GLDV_Mean, Hue, NDSI, Standard
3	71.98%	73.43%	2	GLCM_Contrast, NDVI
4	74.68%	76.92%	3	NDVI, GLCM_Entropy, Mean(Red)
5	75.41%	76.92%	2	GLCM_Standard, Hue
6	75.36%	77.62%	3	NDVI, NDWI, GLCM_Homogeneity
7	76.04%	78.32%	3	NDVI, NDWI, GLDV_Entropy

feature number and combination gain various precision. Just as what Fig. 5 shows, some features, such as NDVI, NDWI, GLCM_Entropy, GLCM_Ang.2nd moment and so on, are frequently-used, which shows that 6 types of objects are easily distinguished using these features.

4 Conclusion

In this paper, an improved object-based remote sensing imagery classification algorithm is used to combine PSO evolution algorithm for feature selection in order to optimize SVM classifier. The experiment results demonstrate that the improved method is able to achieve good classification accuracy in grassland, water and forest land cover types using Tiangong-2 WIS imagery. Two features, NDVI and NDWI are most frequently selected by PSO, which shows the superiority to the application of classification based on SVM. Although this method has good producer accuracy, it largely depends on precise results of segmentation. Future research will be carried out to segment the object types automatically and accurately.

Acknowledgements. Thanks to China Manned Space Engineering for providing space science and application data products of Tiangong-2.

References

1. MSADC Homepage. http://www.msadc.cn/en/sjfw/xzsm/mbxz/. Accessed 10 July 2018
2. Soliman, O.S., Mahmoud, A.S.: Remote sensing satellite images classification using support vector machine and particle swarm optimization. In: Innovations in Bio-Inspired Computing and Applications (IBICA), Third International Conference on IEEE, pp. 280–285 (2012)
3. Vapnik, V.N.: The nature of statistical learning theory. IEEE Trans. Neural Netw. **8**(6), 1564 (1997)
4. Vapnik, V.N.: Statistical Learning Theory. Wiley, New York (1998)
5. Cristianini, N., Shawe-Taylor, J, T.: An Introduction to Support Vector Machines. Cambridge University Press, Cambridge
6. Hsu, C.-W., Chang, C.-C.: A practical guide to support vector classification, 1–16 (2003)
7. Frohlich, H, Chapelle, O.: Feature selection for support vector machines by means of genetic algorithm. In: 15th IEEE International Conference, pp. 142–148. IEEE (2003)
8. Huang, C.L., Wang, C.J.: A GA-based feature selection and parameters optimization for support vector machines. Expert Syst. Appl. **31**(2), 231–240 (2006)
9. Yang, J., Honavar, V.: Feature Subset Selection Using a Genetic Algorithm. Feature Extraction, Construction and Selection, pp. 117–136. Springer, Boston, MA (1998)
10. Chen, Q., Chen, Y.: Genetic particle swarm optimization–based feature selection for very-high-resolution remotely sensed imagery object change detection. Sensors **16**(8), 1204 (2016)
11. Hall, M.A., Smith, L.A.: Feature selection for machine learning: comparing a correlation-based filter approach to the wrapper. In: FLAIRS Conference, pp. 235–239 (1999)
12. Huang, Y., Zhao, C.: Feature selection solution with high dimensionality and low-sample size for land cover classification in object-based image analysis. Remote Sens. **9**(9), 939 (2017)
13. Moradi, P., Gholampour, M.: A hybrid particle swarm optimization for feature subset selection by integrating a novel local search strategy. Appl. Soft Comput. **43**, 117–130 (2016)
14. Eberhart, R.C.: Particle swarm optimization: developments, applications and researches. In: Proceedings of the 2001 Congress on Evolutionary Computation, pp. 81–86 (2001)
15. Agrawal, R.K., Bawane, N.G.: Multiobjective PSO based adaption of neural network topology for pixel classification in satellite imagery. Appl. Soft Comput. **28**, 217–225 (2015)
16. Xue, B., Zhang, M.: Particle swarm optimization for feature selection in classification: Novel initialization and updating mechanisms. Appl. Soft Comput. **18**, 261–276 (2014)
17. Wang, X., Yang, J.: Feature selection based on rough sets and particle swarm optimization. Pattern Recogn. Lett. **28**(4), 459–471 (2007)
18. Bratton, D., Kennedy, J.: Defining a standard for particle swarm optimization. In: Swarm Intelligence Symposium, pp. 120–127. IEEE (2007)

19. Shi, Y., Eberhart, R.: A modified particle swarm optimizer. In: Evolutionary Computation Proceedings IEEE World Congress on Computational Intelligence, pp. 69–73. IEEE (1998)
20. Kennedy, J.: The particle swarm: social adaptation of knowledge. In: Evolutionary Computation IEEE International Conference, pp. 303–308. IEEE (1997)
21. Wikipedia Homepage. https://en.wikipedia.org/wiki/Particle_swarm_optimization. Accessed 25 Aug 2018
22. Nieto, P.J.G., Garcia-Gonzalo, E.: Hybrid PSO–SVM-based method for forecasting of the remaining useful life for aircraft engines and evaluation of its reliability. Reliab. Eng. Syst. Saf. **138**, 219–231 (2015)
23. Huang, C.L., Dun, J.F.: A distributed PSO–SVM hybrid system with feature selection and parameter optimization. Appl. Soft Comput. **8**(4), 1381–1391 (2008)
24. Li, P., Hong, Z.: Genetic feature selection for texture classification. Geo-spatial Inf. Sci. **7**(3), 162–166 (2004)
25. Imaging, D.: eCognition Developer Software:9.01. Reference Book. Trimble Germany GmbH, Raunheim (2012)
26. Chaofan, W., Jinsong, D.: Object-based classification approach for greenhouse mapping using Landsat-8 imagery. Int. J. Agric. Biol. Eng. **9**(1), 79–88 (2016)

Deep Learning Network Integrated Multi-spectral Data and Interferometric Imaging Radar Altimeter Data of Tiangong-2 for Land Use Classification

Kang Liu[1(✉)], Zhuang Zhou[1], Xue Wan[1], Baoqin Hei[1], Shaopan Xiong[1,2], and Chunying Wu[1]

[1] Key Laboratory of Space Utilization, Technology and Engineering Center for Space Utilization, Chinese Academy of Sciences, Beijing 100094, China
Liukang@csu.ac.cn

[2] University of Chinese Academy of Sciences, Beijing 100049, China

Abstract. Multispectral data and radar data contain abundant spectral and texture features, which are widely used in land use classification. High precision registration and information mining are the two major difficulties faced by the classification of these two data. In this paper, visible and near infrared spectrum range of Wide-band Imaging Spectrometer (VNI) and Interferometric Imaging Radar Altimeterdata (InIRA) are used as raw data, and data registration and combination is carried out (the combination data is called for VNI_InIRA). Then the information of each pixel before and after combination is extracted and transformed into gray-scale image which is easy to be input by convolution neural network (CNN). Finally, the classification of land use is carried out by trained CNN models, and the accuracy is verified. The results show that the classification results of VNI_InIRA data are better than those of VNI data. The overall classification accuracy is 94.29% and 91.83% respectively. Therefore, the CNN which synthesizes the spectral characteristics of VNI data and the texture topographic features of InIRA data is a feasible method to obtain accurate land use classification information, and to provide a reference for the study of land use classification extraction in coastal areas.

Keywords: Deep learning · Convolutional Neural Network · Tiangong-2 Multi-spectral data · Interferometric Imaging Radar Altimeter data Land use classification

1 Introduction

Multispectral remote sensing images contain abundant spectral information and are widely used in land use classification and change research. At present, most of the remote sensing data used in land classification research are multispectral remote sensing images with medium-high resolution, such as MODIS, Landsat8, TM/ETM+, SPOT data, etc. [1]. However, there are some limitations and shortcomings in classification based on multispectral remote sensing images. The main reason is that the probability of

Y. Gu et al. (Eds.): *Proceedings of the Tiangong-2 Remote Sensing Application Conference*, LNEE 541, pp. 232–240, 2019.
https://doi.org/10.1007/978-981-13-3501-3_22

cloud cover and bad weather in some areas is high, which makes it difficult to obtain multi-temporal and high-quality multispectral remote sensing images [2]. In addition, multispectral remote sensing images lack the penetration of vegetation canopy, which makes it difficult to use the information below the vegetation canopy [3, 4]. Radar is an active earth observation system, which can penetrate clouds, rain and fog, not affected by weather and time, and has the characteristics of all-weather and all-day. Radar echoes are sensitive to dielectric properties of objects, especially to changes in leaf water content and soil moisture, and can penetrate the canopy of vegetation to make up for the shortcomings of multispectral remote sensing. In addition, because of its special imaging method, it contains the shape, roughness and orientation information of the terrain, which is helpful to land use classification [5, 6]. However, because of the great difference between multispectral data and radar data, and none of these two kinds of data are acquired on the same platform, the registration between images is difficult. Tiangong-2 Space Laboratory is equipped with Wide-band Imaging Spectrometer and Interferometric Imaging Radar Altimeter to obtain multi-spectral data and radar data respectively, which makes it possible to achieve high-precision registration [7].

How can spectral and intensity information contained in multispectral and radar combined data be fully excavated? Deep learning network provides a new solution. Depth learning network can extract and analyze the rich information contained in image data with the advantage of multi-hidden layer structure. Convolutional Neural Network (CNN), as a feed forward neural network, has the characteristics of local perception domain, weight sharing domain and pooling, and has great advantages in image pattern classification. The advantages of multi-level and multi-capacity can meet the needs of processing large amounts of data in image classification. In recent years, it has achieved good results in the field of remote sensing image classification and processing [8, 9]. There are few researches on feature extraction and classification of multispectral and radar combination data using CNN. In this paper, CNN model is used to classify single multispectral data and combined data of multispectral and radar data, and the results of classification are evaluated and analyzed.

2 Methods Based on CNN Model

In order to classify with high accuracy, it is necessary to fully mine the information of multispectral and radar data, and the CNN model is adopted. The experimental procedure is shown as follows:

Firstly, Tiangong-2 multispectral data (VNI) and Interferometric Imaging Radar Altimeter data (InIRA) are pre-processed by atmospheric correction and filtering respectively. Then the two data are registered and combined, and the combination data is called VNI_InIRA. Then the information of each pixel of the VNI and VNI_InIRA is extracted, and the training data set is made to train the CNN model. Finally, the classification of land use is carried out by trained CNN models, and the accuracy is verified (Fig. 1).

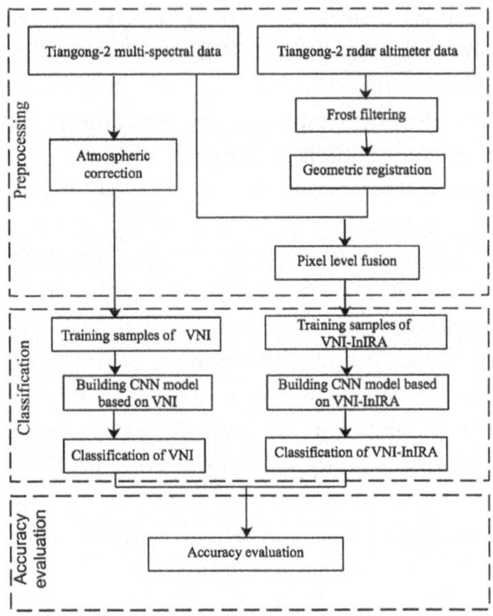

Fig. 1. Flow chart of the experimental method.

2.1 Image Preprocessing

The level 2 of VNI data has been calibrated by radiation and geometric correction. The FLAASH model is used for atmospheric correction to obtain the surface reflectance. The level 2 of InIRA data has been radiometric and geometric corrected. The value is the amplitude value (similar to SAR data), which can reflect the difference of surface roughness and dielectric constant of different objects. In order to eliminate speckle noise and coherent speckle noise, preserve the edge of image data and enhance its details, frost filter is used to filter the InIRA data [10]. Then, the InIRA data and VNI data are used for image registration, in which the VNI data is the reference image, the registration accuracy is within 0.5 pixels, and the resampling is 100 meters resolution (consistent with the VNI data). Bands of well-matched data are combined, and the combination data (VNI-InIRA) have 15 bands.

2.2 Image Classification

Training Samples. Combined with high-resolution images, the ENVI ROI tool is used to mark the classification samples in 8 categories.

The band information of each pixel of VNI and VNI-InIRA needs to be preprocessed to get the standard input of CNN. In order to reduce the classification accuracy of multi-spectral remote sensing images in different data ranges, normalization of data in different bands is necessary before training the CNN model. The calculation formula is as follows:

$$\mathrm{normalization}\left(x_{i,j}^{k}\right) = 255 \times \frac{x_{i,j}^{k} - \min(x^{k})}{\max(x^{k}) - \min(x^{k})} \tag{1}$$

In the formula, W, H and B represent the width, height and the total number of bands respectively, $1 \leq i \leq H$, $1 \leq j \leq W$, $1 \leq k \leq B$, x^k represents all pixels in k band, $x_{i,j}^{k}$ represents the pixels of row i and column j in band k.

The band information of each pixel is extracted and transformed into a 2 dimensional square matrix. Among them, the VNI data has 14 bands, repeating the first and second bands and converting them into 4 square arrays; the VNI-InIRA data have 15 bands, the InIRA data is repeated once, and converted into 4 square arrays. In order to extract pixel features from a larger scale, the nearest neighbor method is used to resample 4 of the square matrix into 32 square matrices. Finally, the square is saved as a grayscale image as the input of the pixel in the CNN model (Fig. 2).

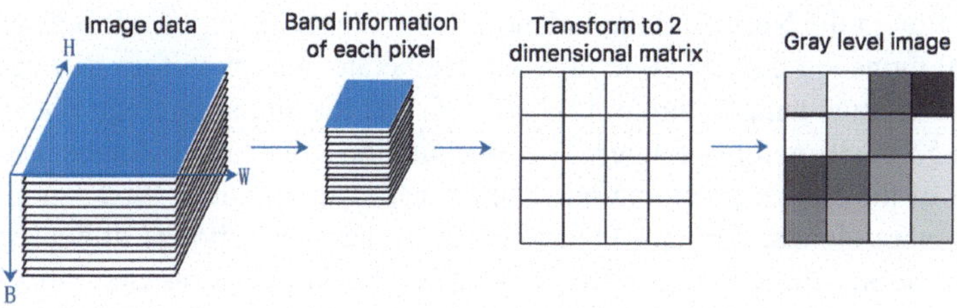

Fig. 2. Preprocessing of training samples.

CNN Model. In order to construct a CNN model, the depth of the model (number of layers), the size of the filter and the number of filters should be considered. The depth of the model is the main factor affecting the classification accuracy. Some studies have shown that when CNN models with a certain number of layers are constructed, the classification accuracy increases first and then decreases as the number of layers increases. The classification accuracy decreases with too many layers, and the time-consuming increases [11]. A seven-layer CNN model consisting of two convolution layers, two maximal pooling layers, and two fully connected layers and a softmax layer is constructed in this study (Fig. 3). The first convolution layer is filtered by six

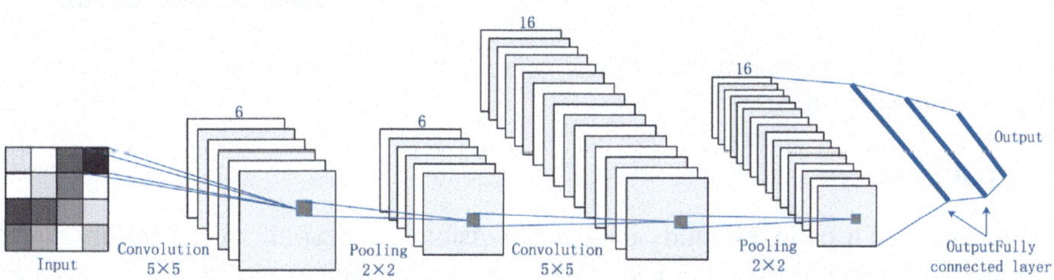

Fig. 3. Framework diagram of CNN model for classification.

convolution kernels (size 5) with step 1 for the input image. The second convolution layer is filtered by 16 convolution kernels (size 5) with step 1 for the feature image. Pooling can not only reduce the dimension of feature, but also preserve the relative position of feature points. The maximum value sampling (Max-Pooling) is performed in the down sampling area of size 2. Finally, the CNN model uses the ReLu (Rectified Linear Units) function as the activation function to solve the gradient dispersion problem of the traditional Sigmoid, Tanh and other activation functions. Its output is the input of the second fully connected layer. The second fully connected layer also passes through the ReLu activation function and outputs it as input to the softmax layer (classifier).

3 Experiment Results and Analysis

3.1 Research Area

The Yellow River Delta (118°33′ ~ 119°21′E and 37°35′ ~ 38°12′N) is bounded by the Bohai Bay in the North and the Laizhou Bay in the East (Fig. 4). It is a semi-humid continental monsoon climate zone in warm temperate zone with four distinct seasons. The annual average temperature is 11.7–12.6°C and the annual average precipitation is more than 530–630 mm in summer. The precipitation from July to August accounts for about half of the annual precipitation. The annual evaporation is 1962.1 mm [12, 13]. There are more clouds, rain and fog in coastal areas, which brings difficulties to land use classification. So the Yellow River Delta is chosen as the research area.

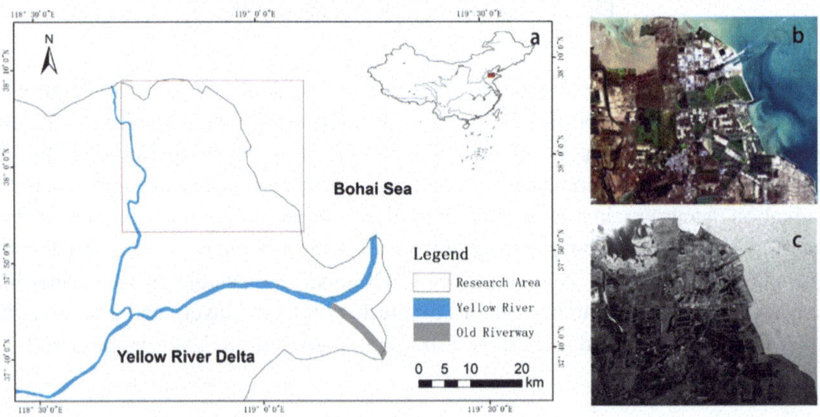

Fig. 4. (a) Location of research area; (b) True color image of VNI with band 8, 10, 12; (c) Two-dimensional images InIRA.

3.2 Data Sources

The data sources used in this study include the visible and near infrared spectrum range of Wide-band Imaging Spectrometer (VNI) and the two-dimensional image of radar altimeter (InIRA),whose product levels are both 2 levels. The VNI data was captured by Tiangong-2 Wide-band Imaging Spectrometer on October 14, 2016. It has 14 bands with spectral range 0.4–1.0 μm and the spatial resolution is 100 meters. The InIRA data was captured by Tiangong-2 interferometric imaging radar altimeter on October 17, 2017. It is Ku band and polarization mode is VV. The resolution is 40 m and backscattering coefficient measurement accuracy is better than 2.0 dB.

3.3 Model Training

According to the position information of the sample points, the CNN model is trained by using training data of VNI and VNI-InIRA. The change curve of loss function and classification accuracy of the test data set is shown in Fig. 5. Among them, the loss function change curve of the training set can reflect the error between the prediction result and the actual result of the model. The smaller the loss function, the better the model training is. The overall accuracy change curve of the test set can reflect the generalization ability of the model. The higher the test precision, the better the generalization ability of the model. Figure 5 shows that after 30 iterations, the loss function of the training set and the change curve of the overall accuracy of the test set tend to be stable and better. The training losses are less than 0.01 and the test accuracies are greater than 0.98.

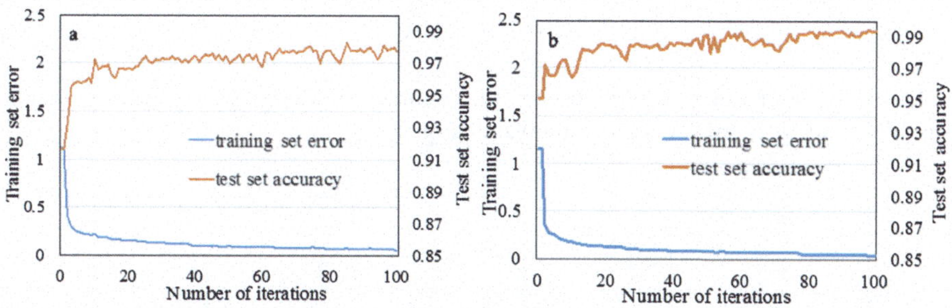

Fig. 5. The curves of loss function and classification accuracy versus iteration epochs of VNI (Fig. 2a) and VNI-InIRA (Fig. 2b)

The trained CNN model is applied to the classification of research area images. The classification results of VNI and VNI-InIRA are shown in Fig. 6.

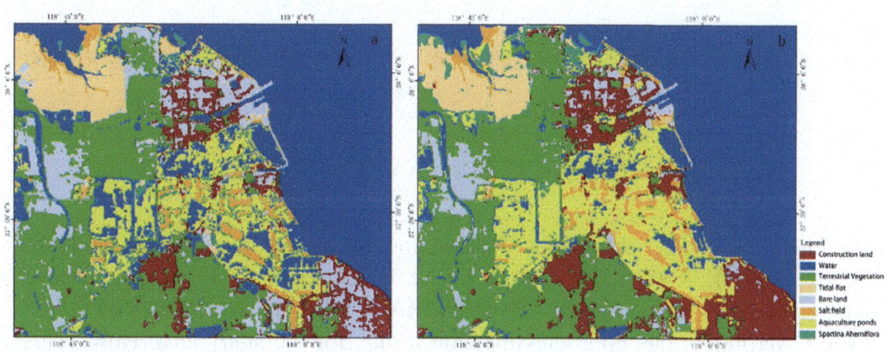

Fig. 6. Classification results based on CNN model; (a) is the result of VNI data; (b) is the result of VNI-InIRA data.

Figure 6 shows the classification results of VNI data and VNI-InIRA data using the trained CNN model. As shown in Fig. 6, the classification results of the two images are basically the same. The classification results based on VNI are more fragmented than those based on VNI-InIRA, and the phenomenon of salt and pepper is obvious. This is related to the data characteristics of InIRA. The value of InIRA reflects the difference between the roughness and dielectric constant of the terrain. These characteristics of the same terrain are more consistent. Therefore, the image classification results fused with InIRA data are more complete.

Producer's Accuracy (PA) indicates the proportion of those correctly classified samples of type i among all the measured type i samples. Producer's Accuracy corresponds to the Omission Error (OE), and the smaller the leakage error is, the higher the MC is.

$$OE = 1 - PA \quad (2)$$

User's Accuracy (UA) indicates the proportion of those correctly classified samples of type i among all the classified type i samples. User accuracy corresponds to the Commission Error (CE), and the smaller the mis-classified error is, the higher the UA is.

$$CE = 1 - UA \quad (3)$$

Based on independent sample points, the accuracy of land use classification results from VNI and VNI-InIRA data is analyzed (Fig. 7).

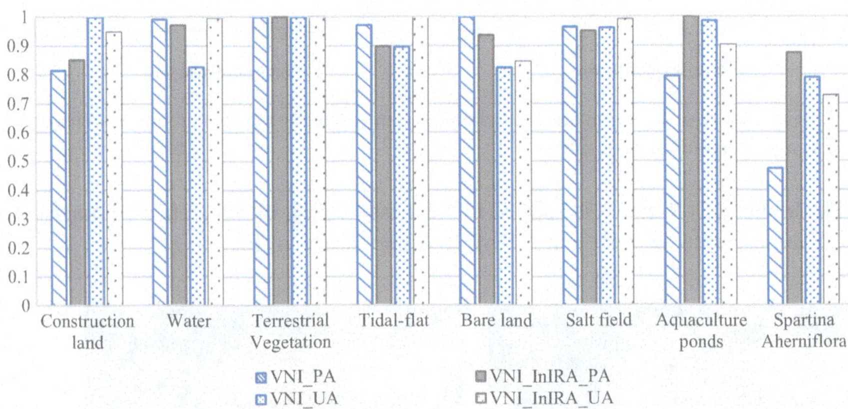

Fig. 7. The accuracy of CNN based on VNI data and VNI_InIRA data

In Fig. 7, VNI_PA and VNI_UA stand for the producer's accuracy and user's accuracy of VNI, VNI_InIRA_PA and VNI_InIRA_UA stand for the producer's accuracy and the user's accuracy of VNI_InIRA.

The CNN model has good classification results for the study area. However, the classification results of VNI-InIRA are better than those of VNI data. The overall classification accuracy is 0.9429 and 0.9183, and Kappa coefficients are 0.9340 and 0.9054 (Table 1). The results of consistency test are completely consistent and the classification reliability is high (Kappa coefficients are greater than 0.8).

Table 1. Accuracy of CNN based on VNI data and VNI_InIRA data

	VNI	VNI_InIRA
Overall Accuracy	0.9183	0.9054
Kappa	0.9429	0.934

Comparing the classification accuracy, we can get the extraction effect of each land type in the experiment (Fig. 7). For construction land, aquaculture pond and Spartina Aherniflora, the omission error of VNI is larger than that of VNI_InIRA. For water, tidal-flat, bare land and salt field, the commission error of VNI is larger than that of VNI_InIRA. Aquaculture ponds and water have similar spectral characteristics, which are easily confused based on spectral information. As the InIRA data contains more texture features, the fusion image based on VNI_InIRA outperforms the result of VNI. Spartina Alterniflora, as an invasive species, grows in the intertidal zone and is easily confused with tidal-flat wetlands and terrestrial vegetation. The surface roughness of Spartina Alterniflora is obviously different from that of tide-flat, and the moisture content of Spartina Alterniflora canopy is different from terrestrial vegetation. InIRA data is sensitive to surface roughness and moisture content. Therefore, for Spartina Alterniflora, the extraction accuracy of VNI_InIRA is higher than that of VNI data.

4 Conclusion

Based on VNI data and InIRA data, we used CNN method to classify land use in northern Yellow River Delta. The classification results of using VNI data and VNI-InIRA data are compared. The main conclusions are as follows:

(1) Through training, the CNN model has high accuracy and generalization ability. the loss function of the training set and the change curve of the overall accuracy of the test set tend to be stable and better after 30 iterations.
(2) The degree of fragmentation of the classification results based on VNI data is higher than the results based on VNI-InIRA data, and the phenomenon of salt and pepper is obvious.
(3) The classification results of VNI-InIRA data are better than those of VNI data. The overall classification accuracy is 0.9429 and 0.9183, and the Kappa coefficients are 0.9340 and 0.9054, respectively.
(4) The classification of coastal areas based on VNI and InIRA has achieved high accuracy. Further research on the spatial-spectral information of the images will be carried out to improve the classification accuracy and give full play to the application value of the data.

Acknowledgement. Thanks to China Manned Space Engineering for providing space science and application data products of Tiangong-2. Thanks to the National R&D Infrastructure and Facility Development Program of China, "Fundamental Science Data Sharing Platform" (DKA2018-12-02-23) who funded the project.

References

1. Yang, H.X., Zhao, J.C., Gao, L., et al.: China's space laboratory and manned space station. J. Nat. **69**(4), 1–5 (2017)
2. Mao, K.B., Shi, J.C., Li, Z.L., et al.: A physics-based statistical algorithm for retrieving land surface temperature from AMSR-E passive microwave data. Sci. China, Ser. D Earth Sci. **50** (7), 1115–1120 (2007)
3. Hadria, R., Duchemin, B., Baup, F., et al.: Combined used of optical and radar satellite data for the detection of tillage and irrigation operations: case study in central morocco. Agric. Water Manag. **7**(96), 1120–1127 (2009)
4. Haldar, D., Patnaik, C.: Synergistic use of multi-temporal Radarsat SAR and AWiFS data for Rabi rice identification. J. Indian Soc. Remote Sens. **38**(1), 153–160 (2010)
5. Ji, Y.J., Yue, G.R., Zhang, W.F.: Use fusion of SAR and optical images for land cover classification. J. Southwest For. Univ. **36**(3), 158–162 (2016)
6. Xu, W.: Classification of the intertidal saltmarsh using Sentinel-1 and Landsat-8 data. East China Normal University (2017)
7. Feng, L.Y.: Research on Construction Land Information Extraction from High Resolution Images with Deep Learning Technology. Zhejiang University (2017)
8. Qin, G.F.: Study of Multi-spectral Remote Sensing Imagery Classification and Urban Expansion Based on Machine Learning. Chongqing University (2012)
9. Zhang, H.L., Jiang, J.J., Wu, H.A., et al.: The BP neural network classification based on the fusion of SAR and TM images. Acta Geodaetica Cartogr. Sin. **35**(3), 229–233 (2006)
10. Ye, C.: Study of Shallow Covering Geological Information based on the Polarized Wave Penetrability. Jilin University (2015)
11. He, K.M., Sun, J.: Convolutional neural networks at constrained time cost. In: Proceedings of the 2015 IEEE Conference on Computer Vision and Pattern Recognition (CVPR), pp. 5353–5360. IEEE, Boston (2015)
12. Wu, J.J., Li, Y.Z., Yu, L.J., et al.: Dynamic changes and driving factors of landscape connectivity for natural wetland in Yellow River Delta. Ecol. Environ. Sci. **27**(1), 71–78 (2018)
13. Wang, X.H., Li, Y.Z., Meng, H., et al.: Distribution pattern of plant community in new-born coastal wetland in the yellow river delta. Scientia Geogr. Sin. **35**(8), 1021–1026 (2015)

Comparison of Land Cover Types Classification Methods Using Tiangong-2 Multispectral Image

Lei Yu, Jinhui Lan(✉), Yiliang Zeng, and Jinlin Zou

School of Automation and Electrical Engineering,
University of Science and Technology, Beijing 100083, China
lanjh@ustb.edu.cn

Abstract. In this paper, Qinghai Lake and Taihu Lake are used as experimental areas, and the visible and near infrared spectrum range of Tiangong-2 Wide-band Imaging Spectrometer are selected for classification research. On the basis of preprocessing, the images are classified by several common classification methods such as Minimum Distance Classification (MDC), Maximum Likelihood Classification (MLC), Spectral Angle Mapping (SAM) and Support Vector Machine (SVM). The classification results are verified using confusion matrices. In the land cover types classification of Qinghai Lake area, the overall classification accuracy of SVM is the highest, which is 99.04%, followed by SAM of 98.78%, MDC of 97.84%, and MLC of 86.89%. In the land cover types classification of Taihu Lake area, the overall classification accuracy of SVM is the highest, which is 92.44%, followed by MDC of 88.90%, SAM of 84.01%, and MLC of 71.01%. After comparative analysis, the practicality and superiority of the SVM method in the image classification of visible and near infrared spectrum range of Wide-band Imaging Spectrometer are proved, which provides a technical reference and theoretical basis for the classification research of Tiangong-2 data.

Keywords: Tiangong-2 · Multispectral · Classification · SVM

1 Introduction

Multispectral imaging technology, collectively referred to as multi-channel spectral imaging technology, utilizes a spectroscopic element for multi-channel imaging of a target sample. The image acquired by each channel has a higher spectral resolution, which represents the spectral reflection of the target sample on this channel. The images acquired by all channels are combined into an X-Y-λ data cube, which visually displays the two-dimensional shape information of the target sample. At the same time, the reflectance spectrum information of the sample is also recorded [1]. Since multispectral remote sensing images contain a large amount of ground information, they have unique advantages in the classification of land cover types. The use of multispectral remote sensing images for land cover classification research has become an important research direction.

Y. Gu et al. (Eds.): *Proceedings of the Tiangong-2 Remote Sensing Application Conference*, LNEE 541, pp. 241–253, 2019.
https://doi.org/10.1007/978-981-13-3501-3_23

Many scholars have carried out extensive research on multispectral image classification [2], including traditional minimum distance method and maximum likelihood method based on statistical characteristics of image data, spectral angle mapping method based on characteristics of ground objects, neural network classification method and decision tree classification method based on target decomposition. Compared with the above methods, SVM has its unique advantages. SVM is an improved learning method proposed by Vapnik et al. according to statistical theory. It shows many unique advantages in solving small sample, nonlinear and high dimensional pattern recognition problems [3]. Zhu et al. used SVM-based algorithms to classify ASTER data. The classification results show that the SVM-based method has higher performance in terms of convergence, training speed and classification accuracy [4]. Zhang Jinshui et al. used SVM to classify multi-source information of single-source high spatial resolution images. The research results show that the multi-source data composite SVM high spatial resolution remote sensing image classification method can effectively solve the problem of single source data information image classification effect fragmentation [5]. Chandani Joshi et al. used the SVM classification method in a content-based satellite image retrieval system. The retrieval efficiency is further increased by using the SVM classifier by classifying the satellite images based on Urban area, Water body and Vegetation. The experimental results show that the fusion technique gives better result and more accuracy can be obtained by classifying the dataset using SVM [6].

The ecological environment changes in Qinghai Lake can reflect the climate change in the entire Qinghai-Tibet Plateau. In recent years, due to global warming, the area of Qinghai Lake is increasing year by year. In order to strengthen the ecological environment protection around Qinghai Lake, the government has taken many measures. For example, returning farmland to forests, returning grazing to grassland, and reducing human activities have played a significant role in improving the ecological environment around Qinghai Lake. The classification of land cover types in Qinghai Lake area is of great significance for understanding the ecological environment changes around Qinghai Lake [7].

The Taihu Basin is one of the most developed regions in China, with an urbanization rate of over 50%. In such a highly developed and densely populated plain river network area, the quality of environmental protection and ecological construction is the key to sustainable economic and social development. The classification research of land cover types in the Taihu area can effectively help the government to grasp the land use situation in the Taihu area, and has important reference value for the government to formulate environmental protection and land use policies [8].

In this paper, the visible and near infrared multispectral images of Qinghai Lake area and Taihu Lake area collected by the Tiangong-2 Wide-band Imaging Spectrometer are used as data sources. Multispectral image data is processed in the ENVI software environment. By using several classification methods (such as MDC, MLC, SAM and SVM) to classify the multispectral images and verifying the classification results, the applicability and superiority of SVM method are proved. The work provides a technical reference and theoretical basis for the classification research of Tiangong-2 data.

2 Remote Sensing Data and Research Area

2.1 Remote Sensing Data

The Tiangong-2 Space Laboratory was launched by China on September 15, 2016 and was equipped with a Wide-band Imaging Spectrometer. In this paper, the visible and near infrared spectrum range multispectral images of Wide-band Imaging Spectrometer are used for the land cover types classification. The channels from 1 to 5 are the near infrared spectrum, the channels from 6 to 8 are the red spectrum, the channels from 10 to 11 are the green spectrum, and the channel 12 is the blue spectrum. The spectrum range of each channel of visible and near infrared spectrum range of Wide-band Imaging Spectrometer is shown in Table 1.

Table 1. The spectrum range of each channel of visible and near infrared spectrum range of wide-band Imaging spectrometer.

Band number	Spectral range (μm)	Band number	Spectral range (μm)
V14	0.403–0.423	V7	0.6775–0.6875
V13	0.433–0.453	V6	0.740–0.760
V12	0.480–0.500	V5	0.810–0.830
V11	0.510–0.530	V4	0.845–0.885
V10	0.555–0.575	V3	0.895–0.915
V9	0.610–0.630	V2	0.930–0.950
V8	0.655–0.675	V1	0.970–0.990

2.2 Qinghai Lake Area

Qinghai Lake Area is located in the northeast of Qinghai-Tibet Plateau and in Qinghai Province. Its geographical extent is 99°28′23″ ~ 101°3′48″E, 36°19′42″ ~ 37°27′42″ N, with a total area of about 17656 sq km. Qinghai Lake is the largest inland lake and saltwater lake in China. The circumference of Qinghai Lake is more than 360 km, the lake area is 4435.69 sq km, and the lake is 3196 meters above sea level. Figure 1 shows the visible and near infrared multispectral image of the Qinghai Lake area collected by the Tiangong-2 Wide-band Imaging Spectrometer on August 13, 2017. We use the secondary product data after radiometric correction and geometric correction. The land cover types in the image are mainly divided into vegetation, bare soil, water and cloud.

2.3 Taihu Lake Area

Taihu Lake Area is located in the provinces of Zhejiang and Jiangsu. Its geographical extent is 119°37′48″ ~ 120°58′13″E, 30°39′25″ ~ 31°46′4″N, with a total area of about 12612 sq km. Taihu Lake is one of the five largest freshwater lakes in China. The lake area is 2427.8 sq km and the lakeshore line is 393.2 km long. The west and

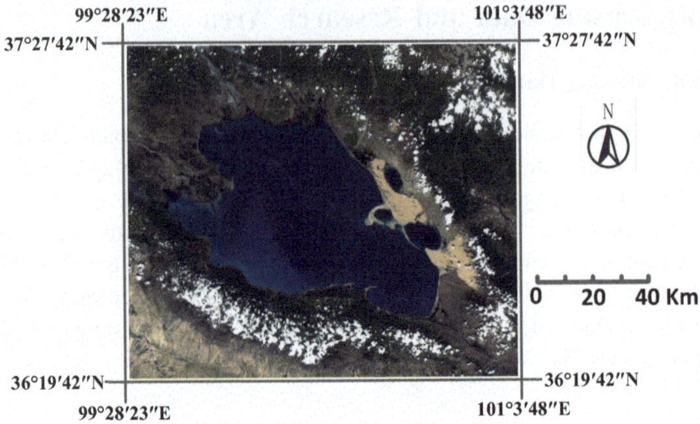

Fig. 1. Visible and near infrared spectrum range multispectral image of wide-band Imaging spectrometer of Qinghai Lake Area.

southwest sides are hilly mountains, and the east side is dominated by plains and water networks. Figure 2 shows the visible and near infrared multispectral image of the Taihu Lake area collected by the Tiangong-2 Wide-band Imaging Spectrometer on March 14, 2017. We use the secondary product data after radiometric correction and geometric correction. The land cover types in the image can be roughly divided into four categories: urban area, water, farmland and forests.

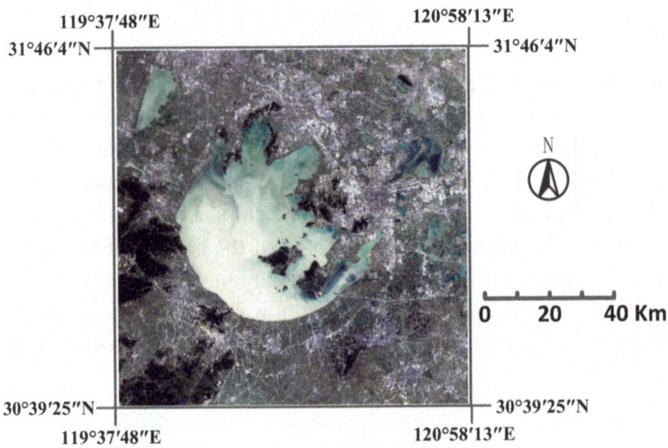

Fig. 2. Visible and near infrared spectrum range multispectral image of wide-band imaging spectrometer of Taihu Lake Area.

3 Classification Methods

In this paper, the MLC, the MDC, the SAM and the SVM classification methods are used to classify the visible and near infrared spectrum range multispectral images of Wide-band Imaging Spectrometer. The various methods are as follows.

Minimum Distance Classification (MDC) method extracts the mean vector of each type in the training sample, and then calculates the distance of the classified pixel spectral vector to each mean vector, and finally divides the classified pixels into the least distance class by comparison [9]. The MDC method has obvious defects due to the direct classification using Euclidean distance and absolute distance. Because the range of variation of the gray values of different categories, that is, the magnitude of the variance is different, the distance between the pixels and the center of the class cannot be simply used to divide the ownership of the pixels.

Maximum Likelihood Classification (MLC) is a classical classification method. A statistical method is often used to establish a discriminant function set. Then, according to this discriminant function set, the attribution probability of each sample to be classified is calculated. Finally, the category of the sample is determined according to the magnitude of the attribution probability. The biggest advantage of MLC is that it can quickly specify the class of pixels to be classified [10]. However, when the amount of data is large and the number of categories is large, the calculation speed of the MLC method will be significantly slowed down, and more training samples are needed. The MLC classification, also known as Bayesian discriminant classification, is based on Bayesian criteria for remote sensing image classification [11].

Spectral Angle Mapping (SAM) is to treat the spectrum as a multidimensional vector and calculate the generalized angle of the two spectra (pixel spectrum and reference spectrum). The smaller the angle is, the more similar the spectra are. A given similarity threshold is used to classify unknown spectra [10]. Moreover, the measure of similarity between the two spectra is not affected by the gain factor. Because the angle between the two vectors is not affected by the length of the vector itself. The laboratory spectra can be directly used to compare the reflectance spectra of remote sensing images to achieve spectral classification and recognition. However, this method is sensitive to spectral characteristics. When mixed pixels are ubiquitous, the classification accuracy will be low.

Support Vector Machine (SVM) is a machine learning method based on statistical learning theory [12]. The SVM can automatically find support vectors that have a greater ability to distinguish between classifications, thereby constructing a classifier that maximizes the interval between class and class, thus having better generalization and higher classification accuracy. The basic idea of SVM is to find a classification hyperplane so that the two types of samples in the training samples can be separated and as far as possible from the plane [13]. Compared with the traditional method, SVM not only overcomes the large sample requirements of the traditional method, but also effectively overcomes the dimensionality disaster and the local minimum problem, and shows its superiority when dealing with nonlinear problems [14, 15].

4 Classification Results and Analysis Evaluation

4.1 Qinghai Lake Area Classification Results

The land cover types in Qinghai Lake area can be roughly divided into vegetation, bare soil, water and cloud. The digital number (DN) value curves are shown in the Fig. 3. In this paper, the data we use is secondary product data that has been corrected for radiation and geometry. In this paper, combined with the high-resolution Google Earth image, the interest areas of various land cover types are selected as the training samples from the image in the ENVI software environment using the visual interpretation method. The numbers of selected training samples are: vegetation 984, bare soil 876, water 943, and cloud 785. Then, the visible and near infrared multispectral images of the Qinghai Lake Area are respectively classified by the MDC method, the MLC method, the SAM method and the SVM method. The classification results are shown in Fig. 4.

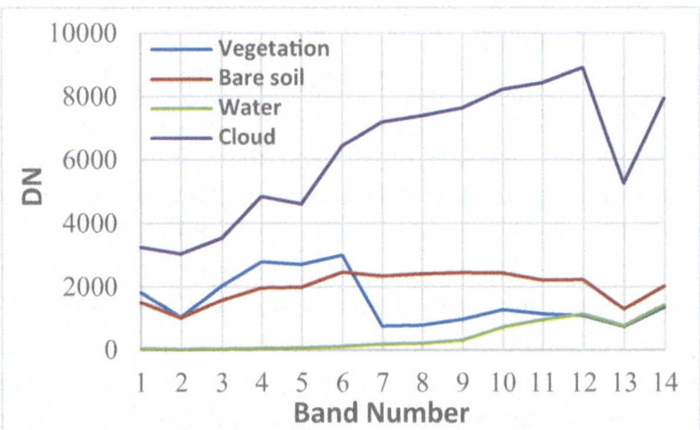

Fig. 3. DN value curves of four land cover types in Qinghai Lake Area.

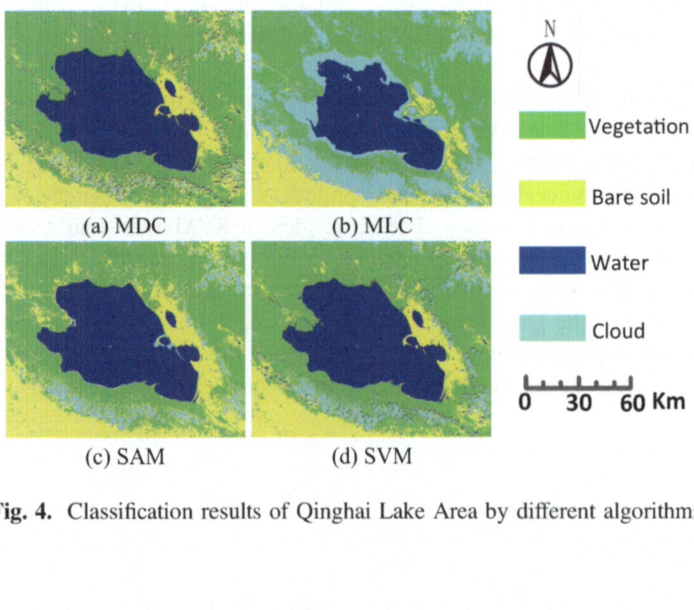

Fig. 4. Classification results of Qinghai Lake Area by different algorithms.

In order to verify the classification accuracy, we visually interpret the land cover types in the research area based on the high-resolution Google Earth data and select 724,478 verification samples from the remote sensing image in the ENVI software environment. The numbers of selected verification samples are: vegetation 196276, bare soil 108471, water 393816, and cloud 25915. The accuracy of the classification results of the four algorithms is evaluated by the confusion matrix, as shown in Tables 2, 3, 4 and 5.

Table 2. MDC method classification confusion matrix for Qinghai Lake Area.

Class	Vegetation	Bare soil	Water	Cloud	Total	User Acc./%
Vegetation	195821	623	43	1140	197627	99.09
Bare soil	385	107754	14	12996	121149	88.94
Water	70	94	393759	256	394179	99.89
Cloud	0	0	0	11523	11523	100
Total	196276	108471	393816	25915	724478	
Prod.Acc./%	99.77	99.34	99.99	44.46		
OA	97.84%					
Kappa	0.9643					

Table 3. MLC method classification confusion matrix for Qinghai Lake Area.

Class	Vegetation	Bare soil	Water	Cloud	Total	User Acc./%
Vegetation	191274	8024	79	249	199626	95.82
Vare soil	88	100344	0	255	100687	99.66
Water	0	0	312493	0	312493	100
Cloud	4914	103	81244	25411	111672	22.76
Total	196276	108471	393816	25915	724478	
Prod.Acc./%	97.45	92.51	79.35	98.06		
OA	86.89%					
Kappa	0.8028					

Table 4. SAM method classification confusion matrix for Qinghai Lake Area.

Class	Vegetation	Bare soil	Water	Cloud	Total	User Acc./%
Vegetation	194285	98	0	798	195181	99.54
Bare soil	1987	108099	99	5404	115589	93.52
Water	0	8	393542	0	393550	100
Cloud	4	266	175	19711	20156	97.79
Total	196276	108471	393816	25915	724478	
Prod.Acc./%	98.99	99.66	99.93	76.06		
OA	98.78%					
Kappa	0.9799					

Table 5. SVM method classification confusion matrix for Qinghai Lake Area.

Class	Vegetation	Bare soil	Water	Cloud	Total	User Acc./%
Vegetation	195814	373	27	1151	197365	99.21
Bare soil	395	108001	61	4619	113076	95.51
Water	40	92	393728	158	394018	99.93
Cloud	27	5	0	19987	20019	99.84
Total	196276	108471	393816	25915	724478	
Prod.Acc./%	99.76	99.57	99.98	77.13		
OA	99.04%					
Kappa	0.9842					

Combining the classification result images of various algorithms and the corresponding confusion matrices, it can be found that the classification errors of the MDC, the SAM and the SVM are mainly to misjudge the cloud layer as bare soil. And the producer accuracy of cloud recognition using MDC, SAM and SVM is 44.46%, 76.06% and 77.13% respectively. For the MLC method, the case where the cloud layer is misjudged as bare soil is not serious, but the case where the water body is misjudged as the cloud layer is very serious. By comparing the producer accuracy and user accuracy of the classification results of several algorithms, it can be found that the classification effects of several algorithms from good to bad satisfy SVM > SAM > MDC > MLC.

4.2 Taihu Lake Area Classification Results

The land cover types of Taihu Lake Area can be roughly divided into forest, farmland, water and urban area. The DN value curves are shown in the Fig. 5. The data we use is secondary product data that has been corrected for radiation and geometry. The same method is used to select the training samples. The numbers of selected training samples are: forest 843, farmland 826, water 784, and urban area 796. Four classification methods are used to classify the visible and near infrared multispectral image of the Taihu Lake Area. The classification results are shown in Fig. 6.

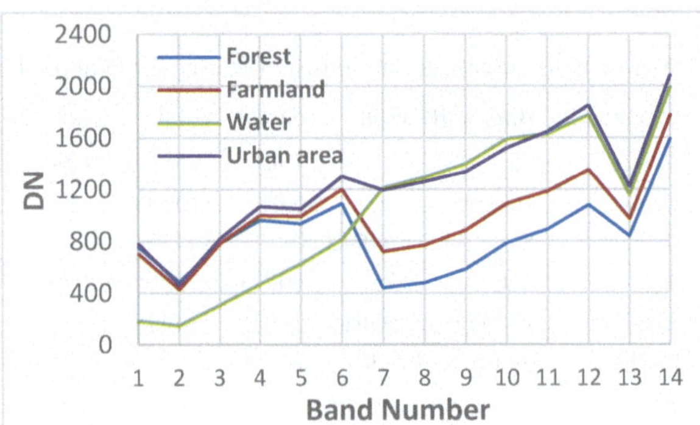

Fig. 5. DN value curves of four land cover types in Taihu Lake Area.

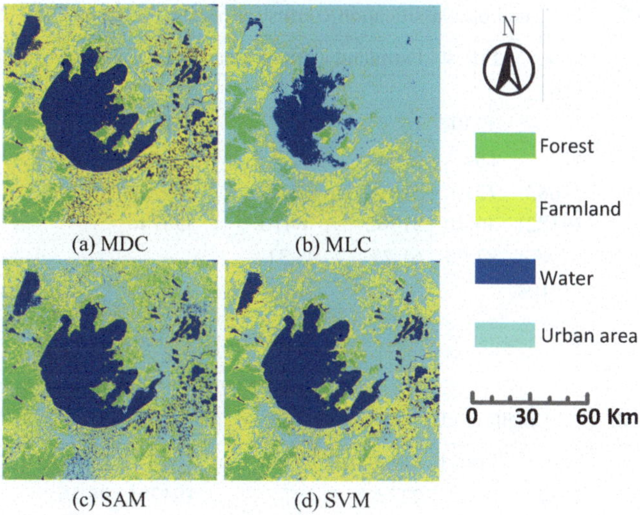

Fig. 6. Classification results of Taihu Lake Area by different algorithms.

In order to verify the classification accuracy, visual interpretation is still used to select 398740 verification samples from the remote sensing image in the ENVI software environment. The numbers of selected verification samples are: water 191821, forest 46801, farmland 81702, and urban area 25915. The accuracy of the classification results by the four algorithms is evaluated by the confusion matrix, as shown in Tables 6, 7, 8, 9.

Table 6. MDC method classification confusion matrix for Taihu Lake Area.

Class	Water	Forest	Farmland	Urban area	Total	User Acc./%
Water	191472	24	3578	817	195891	97.74
Forest	12	43406	87	33	43538	99.7
Farmland	207	3347	71268	29214	104036	68.5
Urban area	130	24	6769	48352	55275	87.48
Total	191821	46801	81702	78416	398740	
Prod.Acc./%	99.82	92.75	87.23	61.66		
OA	88.90%					
Kappa	0.8344					

Combining the classification results images of various algorithms and the corresponding confusion matrices, it can be found that the classification errors of the MDC method, the SAM method and the SVM method are mainly to misjudge farmland and urban area. The SAM method is the most serious case of misjudging farmland and urban area, followed by the MDC method. For the support vector machine method, the phenomenon of misjudging farmland as urban area is still serious, and the phenomenon of misjudging urban area as farmland is much better than the SAM method and the MDC method. The SAM method is more serious than the other three methods in

Table 7. MLC method classification confusion matrix for Taihu Lake Area.

Class	Water	Forest	Farmland	Urban area	Total	User Acc./%
Water	116727	0	0	0	116727	100
Forest	0	39307	16	0	39323	99.96
Farmland	412	184	50042	1329	51967	96.3
Urban area	74682	7310	31644	77087	190723	40.42
Total	191821	46801	81702	78416	398740	
Prod.Acc./%	60.85	83.99	61.25	98.31		
OA	71.01%					
Kappa	0.6012					

Table 8. SAM method classification confusion matrix for Taihu Lake Area.

Class	Water	Forest	Farmland	Urban area	Total	User Acc./%
Water	191491	36	2277	487	194291	98.56
Forest	46	40281	7317	1421	49065	82.1
Farmland	77	6340	44202	17489	68108	64.9
Urban area	207	144	27906	59019	87276	67.62
Total	191821	46801	81702	78416	398740	
Prod.Acc./%	99.83	86.07	54.1	75.26		
OA	84.01%					
Kappa	0.7625					

Table 9. SVM method classification confusion matrix for Taihu Lake Area.

Class	Water	Forest	Farmland	Urban area	Total	User Acc./%
Water	191280	6	190	107	191583	99.84
Forest	8	43554	71	31	43664	99.75
Farmland	190	3155	62185	6701	72231	86.09
Urban area	343	86	19256	71577	91262	78.43
Total	191821	46801	81702	78416	398740	
Prod.Acc./%	99.72	93.06	76.11	91.28		
OA	92.44%					
Kappa	0.8878					

misjudgements between farmland and forest. The classification error of the MLC is mainly to misidentify water and farmland as urban area, and some forests are also misclassified into urban area. By comparing the producer accuracy and user accuracy of the classification results of several algorithms, it can be found that the classification effects of several algorithms from good to bad satisfy SVM > MDC > SAM > MLC.

4.3 Overall Evaluation of Experimental Data

Further comparing the classification result data in the table, the classification results of various algorithms can be evaluated by the overall classification accuracy and the Kappa coefficient. As shown in Tables 10 and 11, we compared the overall classification accuracies and Kappa Coefficients of Qinghai Lake Area and Taihu Lake Area. For the Qinghai Lake Area, the overall classification accuracies of the four classification algorithms from large to small satisfy SVM > SAM > MDC > MLC, and the Kappa coefficients from large to small satisfy SVM > SAM > MDC > MLC. For the Taihu Lake Area, the overall classification accuracies of the four classification algorithms from large to small satisfy SVM > MDC > SAM > MLC, and the Kappa coefficients from large to small satisfy SVM > MDC > SAM > MLC.

Table 10. Comparison of overall accuracies and kappa coefficients of land cover types classification in Qinghai Lake Area.

Method	Overall accuracy	Kappa coefficient
SVM	99.04%	0.9842
SAM	98.78%	0.9799
MDC	97.84%	0.9643
MLC	86.89%	0.8028

Table 11. Comparison of overall accuracies and kappa coefficients of land cover types classification in Taihu Lake Area.

Method	Overall accuracy	Kappa coefficient
SVM	92.44%	0.8878
MDC	88.90%	0.8344
SAM	84.01%	0.7625
MLC	71.01%	0.6012

In summary, SVM is superior to MLC, MDC and SAM in terms of overall classification accuracy, which shows the practicability and superiority of the SVM classification method in the image classification of visible and near infrared spectrum range of Wide-band Imaging Spectrometer.

4.4 Analysis and Discussion

In this paper, the visible and near infrared spectrum range multispectral images of Tiangong-2 Wide-band Imaging Spectrometer are used as the data source. According to the characteristics of the data, several common classification methods are used to classify and compare the images. Through repeated classification training, satisfactory classification results are obtained. Finally, various classification results were verified and analyzed.

Comparing the results of several classification methods, the overall classification accuracy of SVM classification method is the best, followed by MDC and SAM, and the classification result of MLC is the worst. SVM, MLC and MDC are based on the statistical properties of remote sensing image data, while the SAM method is based on land cover type characteristics. In the SVM classification, the setting of the kernel function type and its parameters has a great influence on the classification performance of the SVM. In this study, the kernel function and its parameters are repeatedly set, and finally the classification result with the highest classification accuracy is selected. Because the distribution of land cover types in the Taihu Lake area is more complicated than that in the Qinghai Lake area, mixed pixels are common. Compared with the MDC method, the SAM method is more sensitive to spectral features, which is why the SAM is better than the MDC in the Taihu Lake Area classification, and the SAM is worse than the MDC in the Qinghai Lake Area classification. MLC method obtains the mean and variance parameters of each category by statistics and calculation of the interest region, and then determines a classification function. Therefore, the MLC method has high requirements for training samples. The spatial resolution of the image data used in this paper is large, and mixed pixels are ubiquitous. The method of obtaining training samples by visual interpretation has a great influence on the estimation of spectral mean vector and covariance matrix parameters by MLC method, so the classification effect of MLC method is poor.

5 Conclusion

In this paper, the visible and near infrared spectrum range multispectral images of Tiangong-2 Wide-band Imaging Spectrometer are used as the data source. The commonly used classification methods, such as the MDC, the MLC, the SAM and the SVM, are used to classify the multispectral remote sensing images of the Qinghai Lake Area and the Taihu Lake Area. Through the comparative analysis of the classification results, it can be concluded that the SVM method has the highest classification accuracy among the four classification methods, followed by the MDC method or the SAM method, and the MLC method has the worst classification accuracy. The classification research of land cover types in Qinghai Lake area through Tiangong-2 data has important research significance for mastering the ecological environment changes in Qinghai Lake area. The land cover types classification research of the Taihu Lake area is conducive to understanding the overall land use, vegetation cover and water area changes in the Taihu Lake Basin, which has a strong guiding significance for the formulation of the government's land policy and environmental protection policy. In summary, through the research in this paper, the practicability and superiority of the SVM method in the classification of visible and near infrared spectrum range multispectral images of Wide-band Imaging Spectrometer are proved. The work provides a technical reference and theoretical basis for the classification research of Tiangong-2 data.

Acknowledgments. The work is supported by the Major Special Project of the China High-Resolution Earth Observation System. Thanks to China Manned Space Engineering for providing Wide-band Imaging Spectrometer data products of Tiangong-2. We would like to thank the editor and reviewers for their reviews that improved the content of this paper.

References

1. Wang, X.P., Zhao, H.X., Li, Q.H.: Relevant fundamental research of colored artworks by multispectral imaging technology. J. Opt. **35**(10), 304–311 (2015)
2. Yu, J., Zhang, Z., Guo, P., Qin, H., Zhang, J.: Multispectral remote sensing image classification based on PSO-BP considering texture. In: Intelligent Control and Automation, 2008, vol. 2008, pp. 6807–6810. IEE, Chongqing (2008)
3. Li, H.T., Gu, H.Y., Zhang, B.: Based on SVM and MNF for classification of hyperspectral remote sensing images. Remote Sens. Inf. **2007**(5), 12–15 (2007)
4. Zhu, G.B., Dan, G.B.: Classification Using ASTER Data and SVM Algorithms: The Case Study of Beer Sheva, Israel. Remote Sens. Environ. **80**(2), 233–240 (2002)
5. Zhang, J.S., He, C.Y., Pan, Y.Z.: Research on classification of multi-source and high spatial resolution of remote sensing data based on SVM. J. Remote Sens. **10**(1), 49–57 (2008)
6. Chandani, J., Saurabh, M.: Empirical analysis of SIFT, Gabor and fused feature classification using SVM for multispectral satellite image retrieval. In: International Conference on Image Information Processing, 2017, vol. 2017, pp. 542–547. IEEE, India (2017)
7. Li, X.D., Li, L., Wang, G.X.: Water level and agriculture and husbandry characteristics in Qinghai Lake Basin: the response to climate warming and wetting, vol. 34, no. 26, pp. 119–127 (2018)
8. Wu, F., Wu, J.F., Ling, H.: Analysis on land use change of Taihu Basin. China Popul. Resour. Environ. **28**(7), 143–145 (2018)
9. Tao, Q.X.: Research on classification method of vegetation hyperspectral remote sensing. J. Shandong Univ. Sci. Technol. **26**(5), 61–65 (2007)
10. Tong, Q.X., Zhang, B., Zheng, L.F.: Principle, Technology and Application of Hyperspectral Remote Sensing. Higher Education Press, Beijing (2006)
11. Fan, L.H., Lv, J.W., Yu, Z.T.: A multi-spectral remote sensing image classification technique based on improved ML algorithm. Electron. Opt. Control **2014**(10), 52–56 (2014)
12. Tang, J.T., Hu, D., Gong, Z.M.: Study of classification by support vector machine on synthetic aperture radar image. Remote Sens. Technol. Appl. **23**(3), 341–345 (2008)
13. Jia, P., Li, H.T., Gu, H.Y.: Research on classification of multi-source remote sensing image based on SVM. J. Surv. Mapp. **33**(4), 21–23 (2008)
14. Zhao, G.S., Zhang, X.R.: Research on Support Vector Machine (SVM) method of classification based on the principal component analysis. Comput. Eng. Appl. **40**(3), 37–38 (2004)
15. Xu, H.M., Yang, T.X.: Research on quality evaluation of lakes based on support vector machine classification algorithm. Jilin Univ. **36**(4), 570–573 (2006)

Land Use Change Monitoring in Angkor Wat Based on Tiangong-2 Wide Band Imaging Data

Shuwen Peng[1,2], Xiaohuan Xi[1(✉)], and Cheng Wang[1]

[1] Key Laboratory of Digital Earth Science, Institute of Remote Sensing and Digital Earth, Chinese Academy of Sciences, Beijing 100094, China
xixh@radi.ac.cn

[2] University of Chinese Academy of Sciences, Beijing 100049, China

Abstract. Angkor Wat is one of the most important archaeological sites in the world. However, due to its changes in natural environment and human activities, the Angkor Wat and its surrounding environment are facing great challenges. We select the data of Tiangong-2 Wide-band Imaging Spectrometer in 2017 and 2018 and to obtain the land use in Angkor area with the use of the maximum likelihood method. By analyzing the change information within one year, the monitoring application of the Angkor Wat and its surroundings is carried out. The results show that the overall accuracy is about 82.7% and the urbanization and tourism blowout have caused the over-exploitation of the resources of Angkor Wat and its ecological imbalance. The area of forest greatly reduced and was converted into urban and agricultural land; the agricultural land increased greatly, and it has continued to expand to the Angkor heritage area. The urban area around the heritage has increased. Tiangong-2 Wide-band Imaging Spectrometer is capable for land cover classification, and can provide important information to support the protection and sustainable development of various heritages.

Keywords: Angkor wat · Tiangong-2 wide-band imaging data
Land use · Change detection

1 Introduction

Angkor Wat is the symbol of Cambodia. It is well-known as the Great Wall of China, the Taj Mahal of India, and the Borobudur of Indonesia. The Angkor Ruins were included in the World Cultural Heritage Lists by the United Nations Educational, Scientific and Cultural Organization (UNESCO) in 1992 and were also designated as "endangered heritage". Under the initiative of the Cambodian government and UNESCO, a number of countries and international organizations launched a conservation campaign to save Angkor Wat [1–6].

In 1999, French scholar Pottier produced complete and detailed archaeological drawings of the Angkor area based on multi-source data, including predecessor archaeological data, aerial photographs and field surveys, etc. The contents contain

Y. Gu et al. (Eds.): *Proceedings of the Tiangong-2 Remote Sensing Application Conference*, LNEE 541, pp. 254–263, 2019.
https://doi.org/10.1007/978-981-13-3501-3_24

temples, pools, rivers networks, settlements, roads and Ruins [7]. The University of Florida conducted a study on land use change in Angkor using the Landsat image from 1989 to 2005 [8]. The University of Helsinki used remote sensing and GIS technology to analyze the impact of the Angkor water system and its environment [9]. Keio University used the Landsat and SPOT fusion images to study the geomorphology of Angkor area [10].

In recent years, the rise of tourism in Angkor has greatly promoted the urbanization process in Siem Reap and enhanced the development of infrastructure construction. The rapid construction of various service facilities at the Angkor site, and the massive deforestation of the surroundings have significantly changed the original human landscape and destroyed the local ecological balance. To promote the sustainable development of Angkor, it is particularly important to scientifically protect and develop the Angkor heritage and carry out research on environmental changes in the Angkor area and surrounding areas.

Remote sensing technology has been widely used in land use classification and its change detection because of its real-time, fast information acquisition, wide coverage and periodicity. The Tiangong-2 wide band imaging spectrometer contains a wealth of spectral information that provides better data support for classification. In this paper, we can use the visible and near-infrared data of Tiangong-2 wide band imaging spectrometer to accurately obtain the land use situation in Angkor area, and carry out the monitoring and application work of the Angkor site. It can verify the application potential of Tiangong-2 data in change monitoring.

2 Materials and Methods

2.1 Study Area

The Angkor site is located in Siem Reap, Kampuchea province. It covers an area of more than 400 km^2 and 5,000 km^2 including the buffer zone. The type of land in the site is complex. Besides the precious site itself, there are a lot of forests, rivers, sandstone geology, and human activities. These are not only the foundation of Angkor's existence, but also accelerate the extent and process of its destruction. The sprawl of urban development in Siem Reap and the overexploitation of tourism have threatened Angkor Site.

The study area of this paper is 103.3721°E ~ 104.2505°E, and 13.0944° N ~ 13.7469°N, mainly including the Coulomb Mountain and the Tonle Sap Lake in Siem Reap and its surrounding areas [11, 12]. This area is a typically tropical monsoon climate zone, with high temperature and rainy. The research area is relatively flat. The main water systems include Tonle Sap and three rivers (Sluen, Siem Reap and Zollens), all located in the northern part of Tonle Sap. Tonle Sap Lake is a reservoir of the Mekong River during the rainy season [13] (Fig. 1).

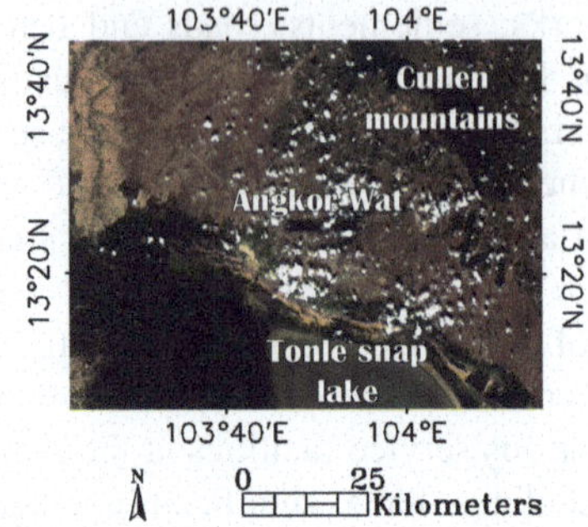

Fig. 1. Multi-band combined image (synthesized by visible near-infrared spectrum 8/10/12 channel)

2.2 Remote Sensing Data Sources

This paper selected visible near-infrared images acquired by Tiangong-2 wide-band imaging spectrometer in January 2017 and June 2018 (data from the manned space application data promotion service platform) [14], monitoring the land use change in Angkor site.

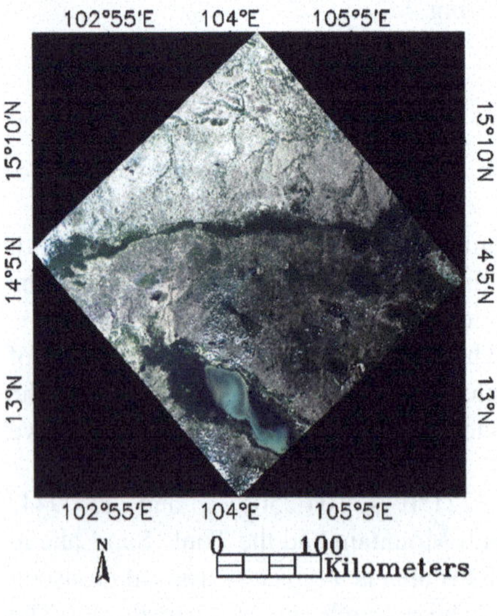

Fig. 2. Tiangong-2 image data in January, 2017

Fig. 3. Tiangong-2 image data in June, 2018

The Wide-band Imaging Spectrometer of the Tiangong-2 is a new generation of wide-band, wide-field and "map-in-one" optical remote sensors. This is the first time at home and abroad to realize the combined integration function of visible near-infrared,

Table 1. Wide-band imaging spectrometer application technical indicators (400 km track height)

Data indicator	Visible near infrared spectrum	Shortwave infrared spectrum	Thermal infrared spectral
Spectral range (μm)	0.4 ~ 1.0	1 ~ 1.7	8 ~ 10
Number of channels	14	2	2
Band width (μm)	V1:0.970 ~ 0.990 V2:0.930 ~ 0.950 V3:0.895 ~ 0.915 V4:0.845 ~ 0.885 V5:0.810 ~ 0.830 V6:0.740 ~ 0.760 V7:0.6775 ~ 0.6875 V8:0.655 ~ 0.675 V9:0.610 ~ 0.630 V10:0.555 ~ 0.575 V11:0.510 ~ 0.530 V12:0.480 ~ 0.500 V13:0.433 ~ 0.453 V14:0.403 ~ 0.423	S1:1.23 ~ 1.25 S2:1.63 ~ 1.65	T1:8.125 ~ 8.825 T2:8.925 ~ 9.275
Resolution (μm)	100	200	400
Coordinate system	WGS84		
Projection method	UTM		
Revisit period	About 3 days		
Swath (μm)	300		

short-wave infrared and thermal infrared multi-spectral large field of view full-sweep imaging on a single instrument [15].

Landsat-5 is one of the world's most widely used and most effective source of remote sensing information for Earth resources satellites. Compared with its Thematic Mapper, the Tiangong-2 wide-band imaging spectrometer has the following advantages: (1) It can obtain fine-grained spectral information. The wide-band imaging spectrometer has a total of 18 bands, far exceeding the 7 bands of the TM. Each band has a narrower range, which allows for better segmentation of the type of similar classes. (2) The revisit period is short and the time resolution is high. It is much smaller than the 16 - days period of TM, which can obtain multiple images in a short time. It can achieve global coverage faster and keep abreast of dynamic changes. (3) A wide range of bands. Covering different wavelength ranges from thermal infrared to visible light extends the range of short-wave infrared in TM.

The visible near-infrared spectrum of the imager can be used in many fields such as agricultural and forestry ecology, land resources, glacial desert and other land use classification information. In addition, the short-wavelength infrared spectrum of the imager can be used in soil moisture, crop growth monitoring, and cloud monitoring, etc. Besides, thermal infrared spectrum can be used in many fields such as surface temperature inversion and fire monitoring. The specific data indicators of the instrument are shown in Table 1.

2.3 Method

The Tiangong-2 Wide-band Imaging Spectrometer obtained in this paper is a L2 product processed by geometrical and radiation correction. In this study, the images are first atmospherically corrected and registered. Then, based on the study area, they are cut to obtain the same geographic location.

Different land use patterns can reflect the economic characteristics of the land. Therefore, this paper classify the land use types into five categories, including forest, water, grassland, crop and urban. Through multi-band combination, different characteristics of each band are integrated, and different feature information is enhanced to facilitate visual interpretation. For example, 4/12/10 channels are false color compound, which can be used to highlight vegetation features, while 8/10/12 is true color compound, the color of the image is basically the same as the actual color of the original area or scene. Due to the extremely narrow band range, the V7 channel can subdivide the vegetation type, and the difference between grassland, crop and forest is significant. The water has a distinct absorption peak in the V13 channel which can be used for shoreline monitoring. When selecting samples, we pay attention to the uniform distribution of the samples and ensure that the full image is included as much as possible.

Combining Google Earth's high spatial resolution image and multi-band combined image, sample selection of five types of features are carried out. The Maximum Likelihood Method (MLC) is one of the most commonly used methods in remote sensing image classification [16]. Compared with other methods (such as K-means, neural network), it has the advantages of clear parameters, simple algorithm and easy implementation. The maximum likelihood classification method in the supervised classification of low- and medium-space resolution remote sensing data is considered to be the best method. Therefore, this paper chooses the maximum likelihood classifier for classification. Finally, the accuracy is evaluated according to the calculation of the overall accuracy and Kappa coefficient. We ignore the transfer of clouds and shadows to other classes to reduce the cloud noise and then compare the differences between the two thematic images.

3 Results and Discussions

Figures 2 and 3 show the classified results of land use in Angkor wat in January 2017 and June 2018 respectively. The accuracy evaluation results show that the overall classification accuracy in 2017 is 88.98%, and the kappa coefficient is 0.8575. The overall classification accuracy in 2018 is 76.43%, and the kappa coefficient is 0.6818.

Table 2 reflects the confusion matrix of the validation set for 2017. It can be found that the classification of forests in the five species is the best, with a misclassification

Table 2. Confusion matrix in 2017

Ground truth (Pixels)									
Prediction (Pixels)	Class	Cloud	Shadow	Urban	Forest	Water	crop	Grassland	Total
	Cloud	373	0	0	0	0	0	0	373
	Shadow	0	921	8	2	30	5	0	966
	Urban	0	0	687	0	0	0	0	687
	Forest	0	9	0	3653	144	0	747	4553
	Water	0	0	0	0	2048	0	0	2048
	crop	0	0	175	20	6	1786	8	1995
	Grassland	0	0	0	51	0	0	260	311
	Total	373	930	870	3726	2228	1791	1015	10933

rate of about 2%. The classification effect of grassland is the worst, and the confusion with forest land and farm land is more serious. Water is often confused with farm land (Figs. 4 and 5).

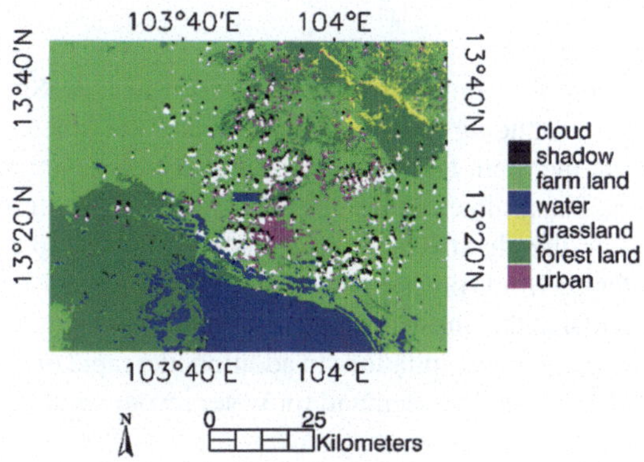

Fig. 4. Land use classification of Angkor Wat, January, 2017

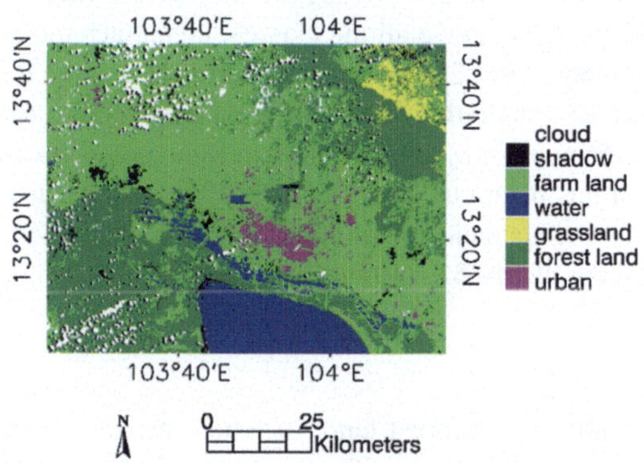

Fig. 5. Land use classification of Angkor Wat, June, 2018

Table 3. Area transition matrix of land use in Angkor Site from 2017 to 2018

Land use area in 2017/km^2							
		Water	Crop	Grassland	Forest	Urban	Sum
Land use area in 2018/km^2	Water	516.13	18.21	0	24.1	0.72	559.16
	crop	91.08	2000.75	4.16	450.22	40.33	2586.54
	Grassland	0	12.57	38.15	69.09	0.28	120.09
	Forest	71.38	209.13	67.39	1662.44	11.35	2021.69
	Urban	1.95	75.68	0	4.72	36.25	118.6
	Sum	680.54	2316.34	109.7	2210.57	88.93	
	Difference	−121.38	270.2	10.39	−188.88	29.67	

Table 3 shows the transfer and change of land use types from 2017 to 2018. The amount of forest and water has been greatly reduced, and the area of farm land has increased significantly. The urban area continues to expand to the Angkor heritage site, threatening the protection of the remains of Angkor.

3.1 Water

The water mainly contains the Tonle Sap Lake, the Barai Lake, the Sluen River, the Siem Reap River and the surrounding water systems. Since the spatial resolution of the spectrometer is 100 m, the Siem Reap River is difficult to be extracted in images. The most important natural factor affecting the change of water is the climate. Especially the precipitation in summer directly affects the water level of Tonle Sap Lake. From 2017 to 2018, water is transferred from forest land and forest land to 24.1 km^2 and 18.21 km^2 respectively. It is speculated that the drop in water level is mainly due to the arrival of the dry season and the decrease in precipitation. In addition, the rapid expansion of farm land and urban areas has also boosted the demand for water resources at the Angkor site [17].

3.2 Forest

Forest is the largest land use type in the Angkor site, accounting for about 50% of the total area. From 2017 to 2018, above all the classes, it changes most, with a decrease of 188.88 km^2. Among them, 450.22 km^2 was transferred out to the farm land. The dense forests of Angkor and Mount Coulomb made a great contribution to the preservation of the site. However, as the number of tourists in Siem Reap increases continuously, a large number of forest land has been cut down, converted into crop land and urban land in order to meet the growing demand for tourism. Forest experienced the largest decline during the year, reducing nearly 5% in ratio of total area.

3.3 Farmland

Farmland area is second only to forest land, of which the main crop is rice [18]. This agricultural area has grown by nearly 5%, the largest increase among all types. The growth of area mainly comes from forest and water body, which are distributed in the Tonle Sap Lake, on both sides of the Sloan River and near Angkor Wat. Among them, it was transferred from the forest land to 450.22 km^2 and it was transferred from the water to 91.08 km^2. The rapid development of urban areas and tourism increased the demand for food, resulting in a large number of forests and grasslands being cultivated for agricultural use [19, 20].

3.4 Urban

Urban increased by nearly 29.67 km^2, and is mainly derived from forest. The spatial distribution of increased area is mainly concentrated in the urban area of the Siem Reap and the foothills of the Coulomb Mountain, and is concentrated around the road. It's worth noting that, due to the small amount of cloud noise in the two used images of the Tiangong-2, the classification accuracy of the urban area was affected to some extent. The tourism industry has driven the economic development of Angkor, but the urban development and over-development tourism in Siem Reap has threatened the site of Angkor Wat. The northern urban part of Siem Reap is adjacent to the Angkor site, and the eastern part is near to the Roluos site. Urban development should be limited in the northward and eastward.

3.5 Grassland

The distribution of grassland and the edge of forest land is the transition zone between forest land and crop land. The grassland area has increased by 10.39 km^2, mainly from forest land.

4 Conclusions

As a world cultural heritage, Angkor Wat has a wide area and complex and diverse land types. Forests, water systems and geological features are the basis for the existence of the Angkor wat. However, human activities and natural environment changes make these environmental elements constantly changing, which has brought tremendous pressure on the surrounding environment of Angkor site and severe challenges to the protection of the Angkor site [21].

Based on the data of Tiangong-2 wide-band imaging spectrometer, we carried out the change monitoring of the Angkor site. The land use of the study area is divided into five categories: forest, water, grassland, crop and urban. In the classification process, the image is first pre-processed such as band selection and registration. Then the image is cropped according to the scope of the study area. In order to reduce to cloud noise, we classified them separately and ignored. Combined with high-resolution images for visual interpretation, the labeled examples of four types are collected. The Maximum Likelihood Method (MLC) are used to perform supervised classification. Finally, according to the classification results, the characteristics of land use changes in Angkor area are analyzed.

The accuracy evaluation results show that the overall classification accuracy in 2017 is 88.98%, and the kappa coefficient is 0.8575. The overall classification accuracy in 2018 is 76.43%, and the kappa coefficient is 0.6818.The results show that the forest land of the Angkor site has been reduced by nearly 5%, mainly into crop land and urban land. The development of urbanization has led to a surge in crop land and continues to expand to the Angkor heritage site. Water levels declining and deforestation have threatened the protection of the site of Angkor. Therefore, we should reasonably control the urbanization development of Angkor, scientifically develop site, and promote the coordinated development of local economy and site protection.

Tiangong-2 wide band imaging data provides effective data for land use change monitoring in Angkor site, and has broad application prospects in the identification of feature elements, information extraction and change monitoring of heritage sites.

Acknowledgement. The authors are grateful to the reviewers for constructive comments and advice on the manuscript, and also, we greatly thank China Manned Space Engineering for providing the Tiangong-2 space science and application data products.

References

1. Uchida, E., Ogawa, Y., Maeda, N., et al.: Deterioration of stone materials in the Angkor monuments Cambodia. Eng. Geol. **55**(1), 101–112 (1999)
2. Banno, A., Masuda, T., Oishi, T., et al.: Flying laser range sensor for large-scale site-modelling and its applications in Bayon digital archival project. Int. J. Comput. Vis. **78**(2–3), 207–222 (2008)
3. Maeda, T., Sugiura, Y., Hirai, T.: FEM modeling of the towers in Bayon temple in Cambodia based on micro-tremor measurements. Adv. Eng. Softw. **39**(4), 346–355 (2008)
4. Kummu, M.: Water management in Angkor: human impacts on hydrology and sediment transportation. J. Environ. Manag. **90**(3), 1413–1421 (2009)
5. Fletcher, R., Johnson'S, I., Bruce'S, E., et al.: Living with heritage: site monitoring and heritage values in Greater Angkor and the Angkor world heritage site Cambodia. World Archaeol. **39**(3), 385–405 (2007)
6. Kvamme, K., Ernenwein, E.G., Hargrave, M., et al.: New Approaches to the use and integration of multi- sensor remote sensing for historic resource identification and evaluation (2006)
7. Evans, D., Pottier, C., Fletcher, R., et al.: A comprehensive archaeological map of the world's largest preindustrial settlement complex at Angkor, Cambodia. Proc. Natl. Acad. Sci. **104**(36), 14277–14282 (2007)
8. Watanabe, N., Usuda, Y., Kubo, S.: Extracting ruins in angkor region using satel-lite image data. In: CIPA International Symposium, pp. Recife/ Olinda, Brazil (1999)
9. Gaughan, A.E.: Spatial and Temporal Land-cover Transformation in the Angkor Basin: A Changing Landscape in Cambodia. University of Florida (2006)
10. Fletcher, R., Penny, D., Evans, D., et al.: The water management network of Angkor Cambodia. Antiquity **82**(317), 658–670 (2008)
11. Evans, D., Pottier, C., Fletcher, R., et al.: A comprehensive archaeological map of the world's largest preindustrial settlement complex at Angkor, Cambodia. Proc. Ntl Acad. Sci. U. S. A. **104**(36), 14277–14282 (2007)
12. Wager, J.: Environmental planning for a world heritage site: case study of Angkor, Cambodia. J. Environ. Plann. Manag. **38**(3), 419–434 (1995)
13. Thongmanivong, S., Fujita, Y., Fox, J.: Resource use dynamics and land-cover change in Ang Nhai Village and Phou Phanang National Reserve Forest Lao PDR. Environ. Manag. **36** (3), 382–393 (2005)

14. The manned space application data promotion service platform homepage. http://www.msadc.cn/. Accessed 25 Aug 2018
15. Liu, K., Qin, B.Y., Ling-Li, M.U., et al.: Monitoring and analysing of reclamation change based on multi-data remote sensing images: a case study of partial area of Liaoning province. Mar. Environ. Sci. (2017)
16. Jiancheng, L.U.O., Qinmin, W.A.N.G., Jianghong, M.A., et al.: The EM- based maximum likelihood classifier for remotely sensed data. Acta Geodaetica Cartogr. Sin. **3**(3), 234–239 (2002)
17. Kaitao, L., Cheng, W., Xiaohuan, X., Shuhua, Q.: KHUON Khun-neay. Remote sensing analysis of land use/land cover change in Angkor heritage site. Remote Sens. Inf. **30**(1), 119–124 (2015)
18. Gaughan, A.E., Binford, M.W., Southworth, J.: Tourism, forest conversion, and land transformations in the Angkor basin Cambodia. Appl. Geogr. **29**(2), 212–223 (2009)
19. Scientific Steering Committee and International Project Office of LUCC, Nunes C, Augé J I. Land-Use and Land-Cover Change (LUCC): Implementation Strategy. Environ. Policy Collection (1999)
20. Traviglia, A.: A View of Greater Angkor: a Multi-scalar Approach for Investigating the Khmer Forests. Advances in Remote Sensing for Archaeology and Cultural Heritage Management, pp. 23–26. Aracne Press, Rome (2008)
21. Wager, J.: Developing a strategy for the Angkor World Heritage site. Tour. Manag. **16**(7), 515–523 (2005)

Identification of Cotton Using Random Forest Based on Wide-Band Imaging Spectrometer Data of Tiangong-2

Xiaojun She[1,2(✉)], Kangyu Fu[1,2], Jie Wang[1,2], Wenchao Qi[3,4], Xiaolu Li[1,2], and Shuangling Fu[5]

[1] Chongqing Engineering Research Center for Remote Sensing Big Data Application, School of Geographical Sciences, Southwest University, Chongqing 400715, China
shexiaojun@swu.edu.cn
[2] Research Base of Karst Ecoenvironments at Nanchuan in Chongqing, Ministry of Nature Resources, School of Geographical Sciences, Southwest University, Chongqing 400715, China
[3] State Key Laboratory of Remote Sensing, Institute of Remote Sensing and Digital Earth, Chinese Academy of Sciences, Beijing 100101, China
[4] University of Chinese Academy of Sciences, Beijing 100049, China
[5] China Coal Technology and Engineering Group Chongqing Research Institute, Chongqing 400039, China

Abstract. Crop identification needs the relatively high spatial, temporal and spectral resolution due to the similarity spectral or spatial features of crops. Tiangong-2 (TG-2) Space Laboratory which payloads the Wide-band Imaging Spectrometer (WIS) can provide 14 bands covering from visible to near-infrared (VIS-NIR) wavelength with spatial resolution of 100 m and a repeat circle of 3 days. The high resolution of spatial-spectral-temporal is of promising advantage in precision crops identification. Random forest (RF) is a powerful machine learning classifier and has a good potential in the remote sensing classification. In this paper, TG-2 WIS image was used to extract the cotton in Xinjiang area with field experimental validation. RF along with support vector machine (SVM) and maximum likelihood classification (MLC) were utilized and compared in the cotton classification study. Three types of features of principal component, texture feature and spectral index are extracted for the three classifiers. A total of 29 features are used as input bands for MLC, SVM and RF. The results showed that RF was superior to the other methods. The overall accuracy and Kappa coefficients of RF are 91.2% and 0.871, respectively. This paper proves that the RF classification method has a good effect in the vegetation extraction of TG-2 data and positive significance for expanding the application of TG-2 data.

Keywords: Tiangong-2 · Cotton extraction · Random forest

Y. Gu et al. (Eds.): *Proceedings of the Tiangong-2 Remote Sensing Application Conference*, LNEE 541, pp. 264–276, 2019.
https://doi.org/10.1007/978-981-13-3501-3_25

1 Introduction

Accurately and timely identifying the distribution and growth of cotton, which is an important economic plant in China, is of great significance for crops management and to provide support for formulating corresponding agricultural plans and economic policies [1]. Xinjiang has a long history of cotton planting, accounting for about a quarter of the country's total cotton planting area, and its output shares more than one-third of the country's total cotton production [2]. As a pillar industry in Xinjiang, the cotton industry is closely related to Xinjiang's economic development and stability, and also plays an important role in the national economy [3]. With the advantages of rapid and economic data acquisition over large scale, remote sensing imagery has been widely applied in mapping crop's distribution and growth monitoring [4–7].

The use of remote sensing technology for regional crop planting monitoring can be generally concluded into two aspects. On the one hand, single image acquired from the crop's key growing period was used to generate the classification algorithms, such as support vector machine (SVM), or to derive greening coefficients, for instance, vegetation index, and to identify the crop categories or to assess the crop growth status [4, 8–11]. On the other hand, multi-temporal images were used to catch the plants' phenology and then to classify different crops [1, 6, 7, 12]. However, optical remote sensing data cannot guarantee full coverage during crop growth season due to the cloudy observations. As a result, multi-temporal solutions will be affected in some areas. For example, phenology of Xinjiang cotton starts at June and ends at September, the acquisition of remote sensing images will be influence by the summer rains in the northern China. In the limited area, the way of key image was usually adopted. Vegetation index derived by hyperspectral remote sensing images is recommended when the area is small and the hyperspectral data is available. This is because the hyperspectral and spectral width are mutually restricted, data coverage and spatial consistency are highly limited [13].

Tiangong-2 (TG-2) Space Laboratory is newly launched which loaded a Wide-band Imaging Spectrometer (WIS) including 14 bands from visible to near infrared, short-wave infrared and thermal infrared bands. Related to the vegetation research, visible and near-infrared bands are widely used because of the reflectance spectrum of vegetation [7, 9]. During the growth of cotton, it has strong absorption in the red band and strong reflection in the near-infrared band. Images with abundant spectral bands typically achieve better separation for crops with similar spectral characteristics [3].

Random forest (RF) is a promising machine learning classifier and becoming popular in the remote sensing classification [14–16]. In this study, TG-2 data combined with RF methods and field sample site was used to evaluate the cotton extraction accuracy and then as a way to effectively extract crop planting information.

2 Study Area and Data Sets

2.1 Study Area

The study area (92°45′E—94°15′E, 42°30′N—43°1′N, Fig. 1) is located in the alluvial oasis plain on the south side of the Tianshan Mountains, on the northern edge of the Hami Basin, belonging to Hami City and covering 7145.6 km^2. Its terrain is high in the north and low in the south. The climate is typical of the temperate continental arid climate. The annual average temperature and precipitation are 9.8 °C and 35 mm respectively, the annual sunshine hours are about 3350 h, and the frost-free period is 182 days. The main crops in the study area are cotton, red dates, grapes and other fruits. The phenology of cotton is as follows: Mid-to-late April is the sowing season, May is the emergence period, June is the bud period, July to August is the flowering and boll period, September is the boll opening period, and October is the harvest period.

Fig. 1. The location of study area

2.2 WIS of TG-2 Data

WIS of TG-2 has stable state and excellent performance, which has obtained a large amount of high-quality observation data. It has 14 bands in the visible and near infrared (VIS-NIR) spectrum with spatial resolution of 100 m. The bands in VIS-NIR are used in this study and the details of the bands' information are presented in Table 1. The study area is located in sparsely northwest area where farmland concentrated and uniformly distributed in large area, which is conducive to vegetation extraction. The growth period of cotton in research area is April to October. Cotton thrive during the blooming period and we can use the image of this period to accurately extract cotton of this region because of its apparent texture feature. WIS of TG-2 data with less cloud and high quality is chosen in this paper and its imaging time is July 29, 2017 (Fig. 2).

Table 1. The VIS-NIR spectrum data in WIS of TG-2 used in this paper.

Bands	Spectral range (μm)
B1	0.970–0.990
B2	0.930–0.950
B3	0.895–0.915
B4	0.845–0.885
B5	0.810–0.830
B6	0.740–0.760
B7	0.6775–0.6875
B8	0.655–0.675
B9	0.610–0.630
B10	0.555–0.575
B11	0.510–0.530

Fig. 2. Study area obtained from TG-2 on July 29, 2017 on false color display.

TG-2 data is radiometrically and geometrically corrected and the atmospheric correction is needed. We used the FLAASH module in ENVI 5.3 software to conduct atmospheric correction for the WIS of TG-2 data involved in the study and to obtain the image of the study area.

Both field survey and visual interpretation from Google Earth are adopted to collect samples to qualify the accuracy of classified data and verified data. Firstly, we got a total of 86 GPS cotton samples from the field investigations. Secondly, based on these field investigations, this study collected more cotton and other land types samples on Google

Earth by visual interpretation. In total, we collected 19874 pixels from the Google Earth, including 5335 cotton pixels, 3777 other crops pixels, 2289 construction land pixels, 196 waters pixels and 8277 bare land pixels. All these samples were distributed evenly. There are more bare land pixels and fewer waters pixels because gobi is widely distributed and lakes are scarce in the study area. About half of the total samples are used as the training area samples, and the remaining samples are used to verify.

3 Method

3.1 Random Forest

RF is an ensemble learning method for classification [17], which adopts the bootstrap resampling strategy to extract multiple samples randomly from the original samples. It conducts decision tree model for each bootstrap sample, and combines the predictions of multiple decision trees to obtain the final prediction result through voting. A large number of theoretical and empirical studies have proved that RF has superiority of high prediction accuracy, good tolerance to outliers and noise, and it is not prone to over fitting. The simply principle of RF is as follows: Firstly, k sample sets were selected from the original training set by bootstrap sampling, and the size of each sample set was the same as that of the original training set. Secondly, k decision tree models are established for k samples to obtain k classification results. Finally, each record is voted based on k classification results to determine its final classification map. The flow chart of RF was displayed in Fig. 3. Among the chart, there are three feature libraries used for randomization which demonstrated in the following Sect. 3.3.

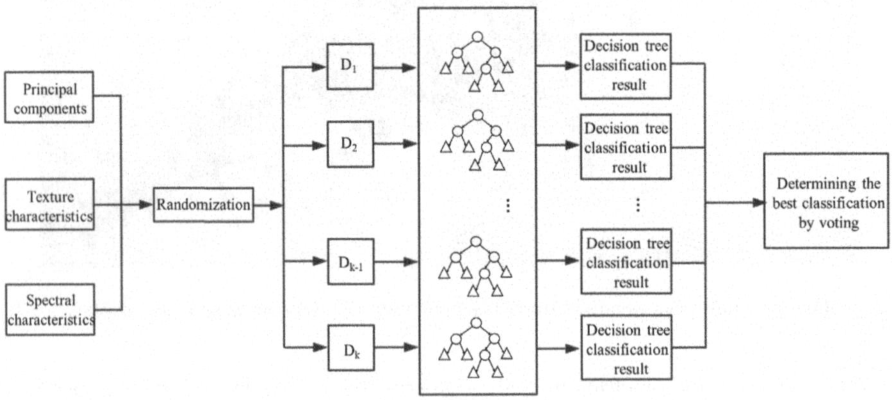

Fig. 3. The flow chart of RF model.

The learning process known as the decision tree for random forest is as follows: ①n random samples from training set including N samples are put back as next input samples of decision tree. ②m features are randomly selected from M sample features as input sample features of each node, where m is far less than M. ③ The optimal splitting of m features is taken as the splitting rule of this node. ④Each decision tree is cultivated to the maximum extent, and the algorithm is not pruned during the growth process and the development progress.

To apply the random forest classifier, two important parameters need to be set: ① the number of trees, which determines the number of decision trees integrated in the model. The larger the value is, the better the convergence of the model at the expense of running time is. In addition, when the number of trees is too large, the model will be supersaturated. ② Randomly extracting the maximum number of features (max features), which represents the maximum number of features randomly selected from the feature space when generating each decision tree. The larger the value is, the greater the strength of each decision tree in the model is, but the correlation among decision trees is also increased. Therefore, max features need to be tuned according to the OOB error rate to achieve a certain precision.

Research shows that random forest classifier can deal with high-dimensional data and is suitable for classification operations of large data volumes. Especially in high-dimensional data classification, it can better reflect the advantages of high speed and accuracy as well as good stability [18].

The advantages of random forest classifier include: ① High precision algorithm; ② Processing big data sets without deleting a large number of input feature variables; ③ The importance estimation of variables can be given. ④ In the process of modeling, an unbiased estimate of the general error can be produced without over-estimation. ⑤ The algorithm can effectively handle missing data. ⑥ Forest models generated by training data are easy to be preserved and reused in the future. ⑦ Model can be extended to unclassified data for unsupervised classification [17].

3.2 Maximum Likelihood Classification and SVM

Maximum Likelihood Classification (MLC), also known as Bayesian classification, is one of the most frequently used supervised classification methods and one of the most stable classifiers [19]. It is a method of finding the attribution probability of each pixel for each class and then classifying the pixel into the category of the maximum attribution probability. MLC uses a Gaussian probability density function. Simply put, MLC obtain the probability density function value by calculating the statistical value of the sample. In practical applications, MLC is one of the classifiers with the highest classification accuracy.

SVM is a supervised parametric statistical learning method based on VC dimension theory and structural risk minimization (SRM) [11, 20]. SVM can automatically find support vectors that have a greater ability to distinguish between classifications, thereby constructing a classifier that maximizes the interval between classes and classes, thus having better generalization and higher classification accuracy. After a series of research and improvement, the method has many unique advantages in solving small sample, nonlinear problems and high-dimensional pattern recognition applications and can be applied to other machine learning problems such as function fitting.

MLC requires that the classified data be as close as possible to the statistical center in each sample space, which is fundamentally different from the idea of maximizing RF edges. SVM is similar to RF in finding the best hyperplane and it is the pursuit of splitting or edge maximization. Using RF to classify land cover and comparing it with iterative algorithm, integrated learning algorithm and SVM classification algorithm in classification accuracy and classification speed and the experiment shows that RF has more advantages and is a good classification algorithm for remote sensing image [14].

In this paper, the integrated spectral features and texture features were used as the initial classification features and the RF, MLC, SVM were used to extract cotton.

3.3 Feature Library Construction

The phenomenon of "foreign matter with the same spectrum" of the ground objects makes the cotton extraction by the single-phase remote sensing image obviously insufficient, and it is difficult to meet the accuracy requirement. The multi-feature synthesis method greatly enriches the information volume of remote sensing data and is beneficial to improve the discrimination of features.

Considering the features of TG-2 remote sensing image, the spectral information is fully utilized. Vegetation is generally sensitive to near-infrared and infrared bands so that vegetation index can reflect different vegetation types and it is more stable than monitoring vegetation with single-band information. Two representative vegetation indexes are selected: Normalized Difference Vegetation Index (NDVI) and Normalized Difference Water Index (NDWI).

Texture is the intrinsic property of the surface of an object, and describes the spatial distribution of the gray of the neighborhood of the image. Compared with other land types, the spatial texture characteristics of vegetation are more distinct. Cotton planting in Xinjiang is planted in large areas and has certain texture characteristics. Adding texture features is conducive to the accurate extraction of cotton information.

The gray level co-occurrence matrix is a common method for describing textures by studying the spatial correlation properties of gray scales. Principal component transformation and K-T transformation can enhance the spectral information of vegetation, in which principal component transformation can separate the main information and noise and reduce the data redundancy and the correlation between bands. The gray level co-occurrence matrix is used to calculate the eight texture features of the first three components: Mean, Variance, Homogeneity, Contrast, Dissimilarity, Entropy, Second Moment, Correlation. Therefore, this paper constructed feature variables and extracted 8 texture features respectively on the basis of original multi-spectral band data to expand the static band of the original image based on WIS of TG-2data. In summary, this paper extracts three types of features: spectral characteristics of vegetation index (NDVI, NDWI), t three principal components and texture features, a total of 29 features, the specific information of each feature is shown in Table 2 Then the MLC, SVM and RF are used for classification.

Table 2. Characteristic variables for classification.

Characteristic variable	Number of bands	Band explanation
The principal components	3	The first three components after principal component transformation (PCA1, PCA2, PCA3)
Texture characteristics	24	Texture features of the first three principal components calculated by gray level co-occurrence matrix (Mean, Variance, Homogeneity, Contrast, Dissimilarity, Entropy, Second Moment, Correlation)
Spectral characteristics	2	NDVI, NDWI

Study area is divided into five land types, including cotton, other crops, construction land, waters and bare land and the training sample is 2652 cotton pixels, 2363 other crops pixels, 777 construction land pixels, 119 waters pixels and 3403 bare land pixels. The separability (Table 3) between the different land types are calculated in order to ensure the accuracy of the classification results. When the separability is closer to 2, which indicates the classification result is more accurate.

Table 3. Separability between different land types

Classification	Cotton	Other Crops	Construction land	Waters	Bare land
Cotton	0				
Other crops	1.9271	0			
Construction land	1.9914	1.9598	0		
Waters	1.9999	1.9995	1.9771	0	
Bare land	2.0000	1.9999	1.9961	1.9997	0

4 Results and Discussion

RF, MLC and SVM are used in classification and Fig. 4 is the classification result map of the study area. Remaining samples are used to verify the classification results in order to compare the classification accuracy of these three classification methods. Four coefficients including the overall classification accuracy, the overall Kappa coefficient, the producer's accuracy, and the user's accuracy of the five land types are calculated and the accuracy of the classification results of the three classification methods is given in Table 4.

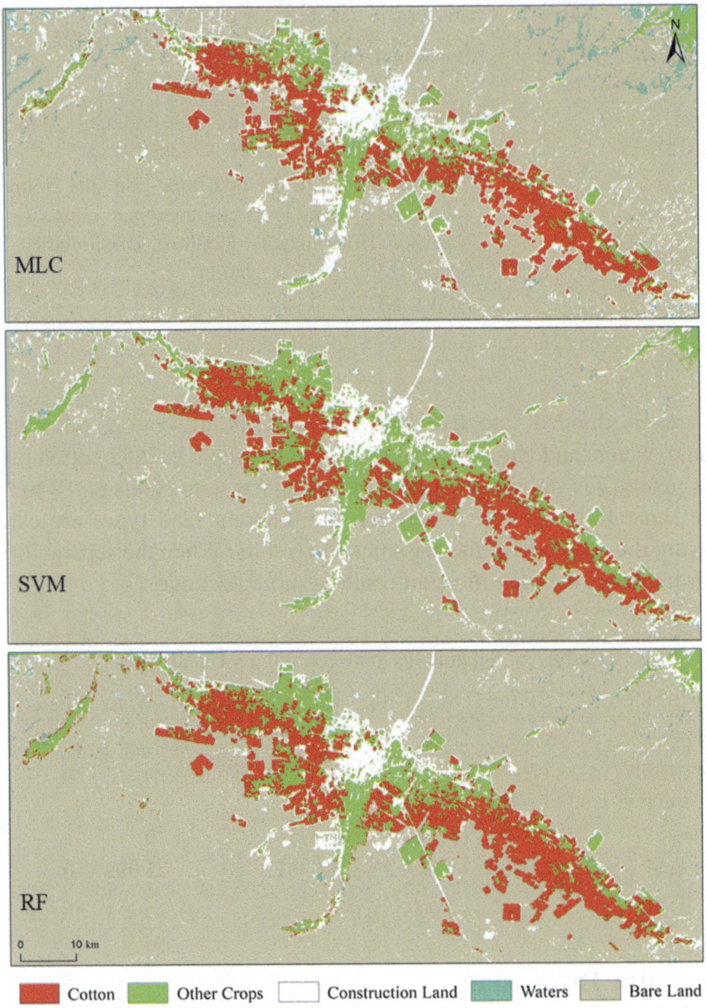

Fig. 4. Classification results mapping based on MLC, SVM and RF respectively.

It can be seen from Table 4 that the MLC, SVM and RF have high classification accuracy. The overall classification accuracy is 88.7% for MLC, 89.1% for SVM and 91.2% for RF and the overall Kappa coefficient is 0.837 for MLC, 0.842 for SVM and 0.871 for RF. In terms of the effectiveness of the classification methods, the overall accuracy and Kappa coefficient of RF are the highest, followed by SVM and MLC, which proves RF has excellent performance in classification of different land types. In terms of cotton extraction precision, RF, SVM and MLC all have high producer's accuracy and user's accuracy. Especially in RF, both the producer's accuracy and the user's accuracy of cotton are above 90%, reaching 91.6% and 90.4% respectively. In other land types, the producer's accuracy of waters in SVM is as low as 57.1%, and the user's accuracy of waters in MLC is as low as 52.4%. Rivers and lakes are scarce in the

Table 4. Accuracy evaluation of classification results of each method

Methods	Overall accuracy (%)	Overall Kappa	Producer's Accuracy (PA) and User's Accuracy (UA) (%)									
			Cotton		Other Crops		Construction		Waters		Bare land	
			PA	UA	PA	UA	PA	UA	PA	UA	PA	UA
MLC	88.7	0.837	89.2	90.6	76.6	72.9	82.3	74.2	84.4	52.4	94.0	98.9
SVM	89.1	0.842	89.7	91.7	83.0	69.4	74.2	78.3	57.1	88.0	95.7	97.9
RF	91.2	0.871	91.6	90.4	82.6	73.1	78.7	89.4	72.7	87.5	97.6	98.1

study area and both SVM and MLC are not sensitive to waters, which results in poor precision of waters identification.

RF have the highest accuracy in the three classification methods and the confusion matrix based on RF is given in Table 5. Among the 2683 pixels of cotton, the number of correctly classified pixels is 2458 and the number of pixels misclassified into other crops is 212. Among the 1414 pixels of other crops, the number of correctly classified pixels is 1168 and the number of pixels that are misclassified into cotton is 186. The pixels that misclassified into cotton are in other crops samples and the pixels that misclassified into other crops are in cotton samples. Thus, the confusion between cotton and other crops is serious. The phenology of cotton and other crops is similar in the study area and Table 3 shows the Separability between cotton and other crops is the lowest, which affects the identification precision of cotton. Bare land is closely connected to construction land in the study area, which also lead to the misclassification between bare land and construction land.

Table 5. RF classification confusion matrix

Classes	Cotton	Other crops	Construction	Waters	Bare land	Total
Cotton	2458	186	35	0	41	2720
Other crops	212	1168	211	1	6	1598
Construction	11	59	1189	10	61	1330
Waters	0	0	0	56	8	64
Bare land	2	1	77	10	4758	4848
Total	2683	1414	1512	77	4874	10560

The multi-feature synthesis method is beneficial to improve the discrimination of different land type and RF based on multi-feature method shows fast speed, high precision and good stability in the classification operation of high-dimensional data. RF has high precision and achieves better performance in cotton extraction and it is an effective classification method. Still, there are some shortcomings in this paper [14]. Firstly, the resolution of the VIS-NIR band in WIS of TG-2 is 100 m and the research area is located in the sparsely northwest region of China where the cotton planting is a concentrated contiguous area, which is conducive to vegetation extraction. While the classification effect needs further study in other small and scattered regions. Secondly, limitations exist in the visual interpretation, which will affect the selection accuracy of the training samples, and thus might affect the possible final classification accuracy.

5 Conclusions

In this paper, the cotton in the Hami, Xinjiang was identified based on WIS of TG-2 and GPS cotton samples by three machine learning classifiers, RF, SVM and MLC. In order to improve the classification accuracy, three characteristic variables are extracted:

principal component, texture feature and spectral index, a total of 29 features are used as input bands. From the research results we have obtained the following conclusions:

(1) RF has better effect and higher precision than MLC and SVM in cotton extraction and also has better stability and applicability. In addition, RF has a good effect in the identification of different land types.
(2) The multi-feature synthesis method greatly enriches the information of WIS of TG-2, which conduces to improving the discrimination and classification accuracy of different land types and has positive significance for expanding the application of TG- 2 data.

Acknowledgments. Thanks to China Manned Space Engineering for providing space science and application data products of Tiangong-2. This research is jointly sponsored by the Doctoral Fund of Southwest University (Grant No. SWU116083) and the Fundamental Research Funds for the Central Universities of China (Grant No. XDJK2018C016).

References

1. Wang, W., et al.: Cotton extration method of integrated multi-features based on multi-temporal Landsat 8 images. J. Remote Sens. **21**(1), 115–124 (2017)
2. Wang, C., Isoda, A., Wang, P.: Growth and yield performance of some cotton cultivars in Xinjiang, China, an arid area with short growing period. J. Agron. Crop Sci. **190**(3), 177–183 (2004)
3. Dong, H., et al.: Effects of plastic film residue on cotton yield and soil physical and chemical properties in Xinjiang. Trans. Chinese Soc. Agric. Eng. **29**(8), 91–99 (2013)
4. Wardlow, B., Egbert, S., Kastens, J.: Analysis of time-series MODIS 250 m vegetation index data for crop classification in the U.S. Central Great Plains. Remote Sens. Environ. **108**(3), 290–310 (2007)
5. Pan, Y., et al.: Winter wheat area estimation from MODIS-EVI time series data using the Crop Proportion Phenology Index. Remote Sens. Environ. **119**, 232–242 (2012)
6. Sakamoto, T., Gitelson, A.A., Arkebauer, T.J.: MODIS-based corn grain yield estimation model incorporating crop phenology information. Remote Sens. Environ. **131**, 215–231 (2013)
7. Wang, M., Tao, F.-L., Shi, W.-J.: Corn Yield Forecasting in Northeast China Using Remotely Sensed Spectral Indices and Crop Phenology Metrics. J. Integr. Agric. **13**(7), 1538–1545 (2014)
8. Moran, M.S., et al.: Estimating Crop Water Deficit Using the Relation between Surface-Air Temperature and Spectral Vegetation Index. Remote Sens. Environ. **49**(3), 246–263 (1994)
9. Thenkabail, P.S., Smith, R.B., Pauw, E.D.: hyperspectral vegetation indices and their relationships with agricultural crop characteristics. Remote Sens. Environ. **71**(2), 158–182 (2000)
10. Kern, A., et al.: Statistical modelling of crop yield in Central Europe using climate data and remote sensing vegetation indices. Agric. For. Meteorol. **260**, 300–320 (2018)
11. Cortes, C., Vapnik, V.: Support-vector networks. Mach. Learn. **20**(3), 273–297 (1995)
12. Ding, Y., et al.: Temporal dynamics of spatial heterogeneity over cropland quantified by time-series NDVI, near infrared and red reflectance of Landsat 8 OLI imagery. Int. J. Appl. Earth Obs. Geoinf. **30**, 139–145 (2014)

13. Haboudane, D., et al.: Integrated narrow-band vegetation indices for prediction of crop chlorophyll content for application to precision agriculture. Remote Sens. Environ. **81**(2–3), 416–426 (2002)
14. Pal, M.: Random forest classifier for remote sensing classification. Int. J. Remote Sens. **26** (1), 217–222 (2005)
15. Liaw, A., Wiener, M.: Classification and regression by randomForest. R News **2**(3), 18–22 (2002)
16. Rodriguez-Galiano, V.F., et al.: An assessment of the effectiveness of a random forest classifier for land-cover classification. ISPRS J. Photogramm. Remote Sens. **67**, 93–104 (2012)
17. Breiman, L.: Random forests. Mach. Learn. **45**(1), 5–32 (2001)
18. Verikas, A., Gelzinis, A., Bacauskiene, M.: Mining data with random forests: a survey and results of new tests. Pattern Recogn. **44**(2), 330–349 (2011)
19. Mingguo, Z., Qianguo, C., Mingzhou, Q.: The Effect of Prior Probabilities in the Maximum Likelihood Classification on Individual Classes. Photogramm. Eng. Remote Sens. **75**(9), 1109–1117 (2009)
20. Mountrakis, G., Im, J., Ogole, C.: Support vector machines in remote sensing: A review. ISPRS J. Photogramm. Remote Sens. **66**(3), 247–259 (2011)

Drought Monitoring Using Tiangong-2 Wide-Band Spectrometer Data

Lingli Mu(✉), Shengyang Li, Bangyong Qin, and Kang Liu

Key Laboratory of Space Utilization, Technology and Engineering Center for Space Utilization, Chinese Academy of Sciences, Beijing 100094, People's Republic of China
mulingli@csu.ac.cn

Abstract. Drought monitoring is an important research direction for the remote sensing application. In this paper, the red and near infrared data acquired by the Tiangong-2 Space Laboratory on 28th June in 2017 were used to monitor drought in the south of America. To calibrate the drought monitoring inversion model, the in situ soil moisture data about 15 observation sites from 21th June to 28th June collected by the Soil Climate Analysis Network (SCAN) were analyzed. In addition, the hysteresis effect between PDI and the soil moisture were discussed. The result shows: the band8 and band11 of the Wide-band Imaging Spectrometer (WIS) aboard Tiangong-2 is more sensitive to the soil moisture at 50 cm depth, and the PDI has a hysteresis effect on the drought response about 1–2 days. The red and near infrared data from the Tiangong-2 is a valid data source for the drought monitoring and will play an important role in drought monitoring.

Keywords: Drought monitoring · Tiangong-2
Wide-band imaging spectrometer · PDI · ISMN · SCAN

1 Introduction

Drought is one of the world's largest natural disasters with the longest duration and the greatest loss. Drought not only has a major impact on crop growth and food security, but also influences on urban production and life. So, more and more governments, scientists and public attach great importance to drought monitoring. And a number organizations and institutions have carried out some research. In traditional approach, meteorological, hydrological or agricultural ground observation data have being used and played an important role in the drought monitoring for a long time. But it can't effectively monitor the extent and severity of water shortage at regional scale.

Remote sensing can observe wide range and get huge data with high resolution. Compared with ground observation, remote sensing method has advantages for large-scale drought monitoring. Since the 1970s, researchers have used visible-near infrared [1, 2], thermal infrared [3] and microwave remote sensing data to monitor large-area drought [4]. And now, a number of organizations and institutions have established many effective methods and drought monitoring systems, such as the VegDRI [5] in USA and Canada, Drought Watch in China and Mongolia, South Asia Drought

Y. Gu et al. (Eds.): *Proceedings of the Tiangong-2 Remote Sensing Application Conference*, LNEE 541, pp. 277–285, 2019.
https://doi.org/10.1007/978-981-13-3501-3_26

Monitoring System (SADMS) at the International Water Management Institute (IWMI) and so on. But the current remote sensing monitoring system requires a long term data accumulation to create monitoring index and models. For example, the NOAA-AVHRR, MODIS remote sensing data are used to generate VCI, TCI, VTCI, VHI, and monitor the large-scale droght. But this approach needs data accumulation with low resolution and effectiveness. With the technique development, the high-resolution remote sensing data also has been adopted in the small-scale area. So, it becomes necessary to use moderate satellite data to carry out regional drought monitoring.

Wide-band Imaging Spectrometer (WIS) aboard Tiangong-2 Space Laboratory has a wide view range with visible and near infrared bands (VNIR, resolution 100 m), shortwave infrared bands (SWIR, resolution 200 m) and thermal infrared bands (TIR, resolution 400 m) [6]. It provides the possibility to monitoring drought with moderate spatial resolution. In this paper, VNIR data and soil moisture released by International Soil Moisture Network (ISMN, website: http://ismn.geo.tuwien.ac.at/) was used to monitoring the drought in the south of USA. Also, the sensitivity and hysteresis effect of the drought monitoring index PDI was analyzed.

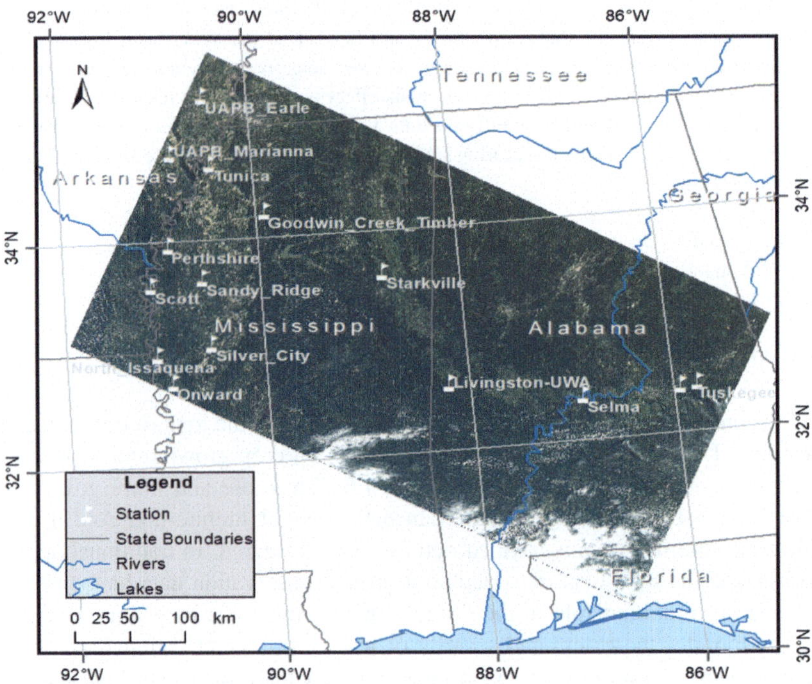

Fig. 1. Location of the study area, ground observation sites and image acquired by Tiangong-2 on 28th June, 2017. In the south of the image, some clouds can be seen.

2 Study Area and Data

2.1 Study Area

The study area covering most areas of Mississippi and Alabama is located in the southern America, on the north of the Mexico Gulf with subtropical climate, shown in Fig. 1. The annual precipitation is 1300–1600 mm. The main crops are peanuts, soybean, cotton, maize and rice. The crop growth period is about 200 days in the north and 300 days in the south. And there are sometimes tornado disasters every year from March to April, and hurricanes in the late summer. In recent years, an increase in average drought severity has been observed over the southwest of America and parts of the Southeast, including Alabama and Mississippi [7]. The annual average temperature is 17 °C, and the highest temperature in summer is 38 °C. Therefore, this paper selected summer as the time for drought monitoring.

2.2 Data and Research Area

2.2.1 Soil Moisture Data

The in situ soil moisture data were collected the Soil Climate Analysis Network (SCAN) [8]. The SCAN is a State-level climatic networks in United States and provides the soil moisture data daily since 2000. The soil moisture is measured in mm^3/mm^3, at the depths: 5 cm, 10 cm, 20 cm, 50 cm and 100 cm. In the static variables files for each site released by the International Soil Moisture Network (ISMN, website: http://ismn.geo.tuwien.ac.at/) records the soil saturation moisture, soil composition, land cover classification and climate information. The soil moisture (sm) files records the time, location (longitude and latitude), altitude, soil depth, soil moisture value, quality and so on. In this paper, about 15 ground observation sites were selected shown in Fig. 1. All of the data were downloaded from ISMN. For comparison, the soil moisture data are converted into relative soil moisture using the formula (1).

$$RSM = SM/SSM \quad (1)$$

Where RSM is the Relative Soil Moisture, SM is in situ Soil Moisture measured by SCAN and SSM is the Saturation Soil Moisture released by IIASA (v1.1) [9]. In order to analyze the delayed response of remote sensing data to soil moisture data, the data between 21th June and 29th June, 2017 were selected.

2.2.2 Remote Sensing Data

WIS was designed to acquire the wide spectrum data to monitor the inland lake, land, atmosphere, ocean with moderate resolution. The WIS consists of 14 visible and near infrared channels (spectrum: 0.40–1.04 um, resolution 100 m), 2 shortwave infrared channels (spectrum: 1.232–1.252 um, 1.630–1.654 um, resolution 200 m), and 2 thermal infrared channels (spectrum: 8.125–8.825 um, 8.925–9.275 um, resolution 400 m) with 300 km swath and 42° field of view [6].

In this paper, the visible and near infrared bands were selected, shown in Table 1. And the Level-2 imageries obtained on June 28th, 2017 were applied and downloaded from http://www.msadc.cn. The Level-2 products have get the radiometric calibration and navigation in Technology and Engineering Center for Space Utilization, Chinese Academy of Sciences. In order to reduce the atmospheric influence, atmospheric correction was performed by the FLAASH module in ENVI (parameter configure: Mid-Latitude Summer Atmospheric model, Rural Aerosol Model and 40 km Initial Visibility). The image in Fig. 1 is the reflectance of band 8/10/12.

Table 1. Parameters for the WIS spectrum

Band num	Spectral range (μm)	Resolution (m)	Band num	Spectral range (μm)	Resolution (m)
Band1	0.403–0.423	100	Band10	0.810–0.830	100
Band2	0.433–0.453	100	Band11	0.845–0.885	100
Band3	0.480–0.500	100	Band12	0.895–0.915	100
Band4	0.510–0.530	100	Band13	0.930–0.950	100
Band5	0.555–0.575	100	Band14	0.970–0.990	100
Band6	0.610–0.630	100	Band15	1.23–1.25	200
Band7	0.655–0.675	100	Band16	1.63–1.65	200
Band8	0.6775–0.6875	100	Band17	8.125–8.825	400
Band9	0.740–0.760	100	Band18	8.925–9.275	400

2.3 Methodology

The vegetation has strong absorption in the Red and reflection in near infrared (Nir), so the vegetation index constructed by Red and Nir can reflect the health of the vegetation. The growth of vegetation can indirectly reflect the moisture information of the soil. Therefore, in the 1970s, researchers began to study the relation between Red, Nir and the soil moisture. Some drought indices were derived from NOAA AVHRR or MODIS for global or regional drought monitoring with low resolution [2, 10].

According to the distribution of vegetation spectrum in the Red and Nir, the vertical coverage of Red and Nir spectral feature space to the soil line was used to describe the vegetation coverage, and the vertical vegetation index (PVI) was proposed by Richardson [11]. Because of the water body's strong absorption in Red and Nir, the soil water content is the main factor affecting the Red and Nir reflectivity of the soil. Ghulam and Qin [12] analyzed the scatter of Red and Nir reflectance using the ETM + data. They found that the scatter points was in a triangle. In the Fig. 2, the points at the bottom of the triangle usually stands for bare soil, and the soil line can be derived with those bare soil points [11]. Based on the scatter plot, Zhan [13] interpreted the drought severity distribution in the triangle, and Ghulam [12, 13, 14] developed the Perpendicular Drought Index (PDI). As shown in Fig. 2, BC is the soil line, and L is perpendicular to soil line. The PDI is the perpendicular distance from this point to L. So the PDI is presented by the following formula [12].

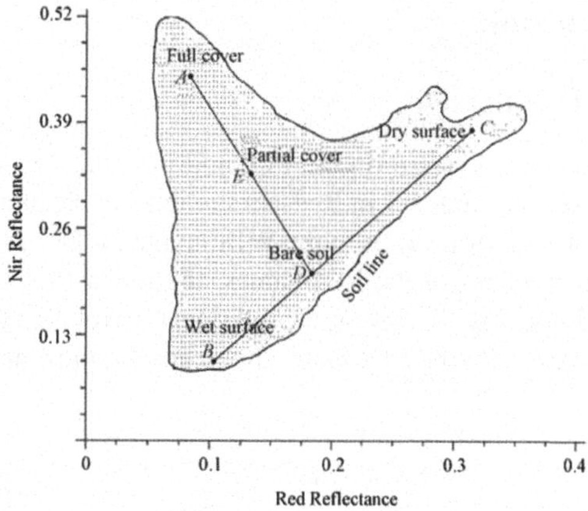

Fig. 2. Construction of Red-Nir space using ETM+ data [12]

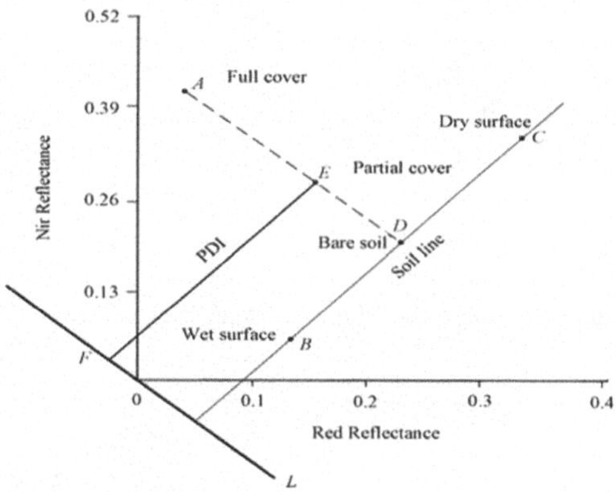

Fig. 3. Sketch map of PDI [12]

$$PDI = (Rred + M^{*}RNir)/sqrt(M^2 + l) \tag{2}$$

$$Rred = M^{*}RNir + b \tag{3}$$

Where Rred is reflectance in red, and RNir in near infrared. M is the slope of soil line which is derived from the Red-Nir distribution of bare soil points. PDI covers from 0.0 to 1.0. In the Fig. 3, higher PDI represents more water stress. Now the PDI is used widely in drought monitoring. In this paper, the most suitable bands and soil moisture depth will be selected and compared by analyzing the relation between the PDIs and soil moisture data.

3 Analysis and Result

3.1 Soil Moisture Calculating

The soil moisture inversion is based on the PDI and the relative soil moisture. Band7, Band8, Band10, Band11 and Band12 in Red-Nir are used to calculate the PDI. The PDI values of the 15 sites were extracted from the WIS image on 28th June 2017. In order to evaluate the representativeness of the PDI values, the mean PDI value were extracted within 3 * 3 window. In the Fig. 4, there is a good linear correlation between the PDI and Mean PDI with $R^2 = 0.98$. So, the PDI values have good representativeness (Fig. 5).

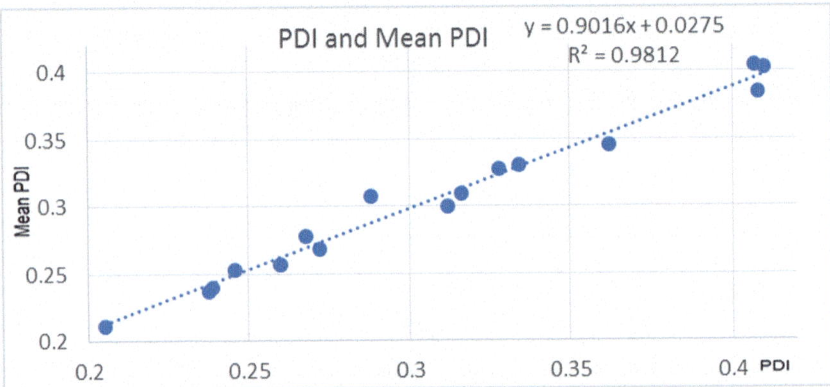

Fig. 4. Relation between the PDI and Mean PDI within 3 * 3 window

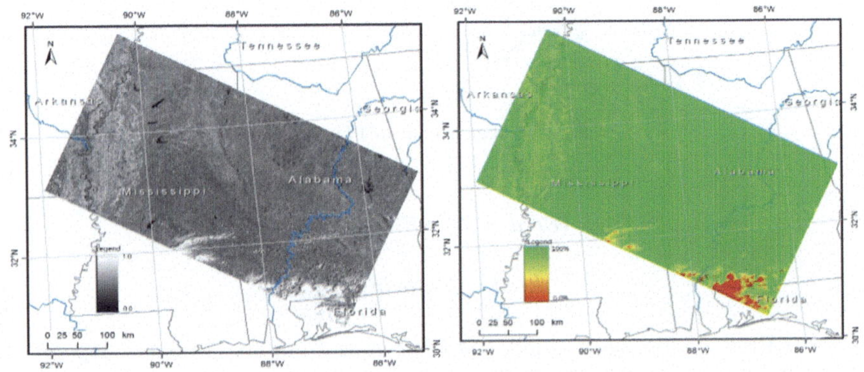

Fig. 5. PDI (left) and drought monitoring result (right) derived from the PDI

Because the response of vegetation to soil moisture maybe delayed, so the relative soil moisture data from June 21st to 28th were selected for sensitivity and hysteresis analysis. In this paper, the sensitivity of PDI to soil moisture was analyzed at 5 cm, 10 cm, 20 cm, 50 cm, and 100 cm depth.

The Table 2 shows the correlation coefficients between the PDIs and the relative soil moisture at different depth. In the table, it can be seen that the correlation

Table 2. Correlation coefficient R^2 between the PDIs and the relative soil moisture

Time	R-Nir	5 cm	10 cm	20 cm	50 cm	100 cm	R-Nir	5 cm	10 cm	20 cm	50 cm	100 cm
6/21	B7–B10	0.00	0.27	0.44	0.44	0.13	B7–B11	0.00	0.28	0.44	0.45	0.13
6/22		0.00	0.14	0.37	0.35	0.08		0.00	0.14	0.37	0.37	0.08
6/23		0.01	0.26	0.29	0.37	0.18		0.02	0.26	0.31	0.40	0.18
6/24		0.07	0.21	0.35	0.23	0.12		0.07	0.22	0.35	0.24	0.12
6/25		0.02	0.18	0.36	0.32	0.12		0.03	0.18	0.36	0.34	0.12
6/26		0.04	0.19	0.38	0.48	0.18		0.05	0.18	0.40	0.49	0.19
6/27		0.05	0.20	0.32	0.48	0.18		0.06	0.20	0.31	0.50	0.20
6/28		0.06	0.20	0.32	0.38	0.19		0.07	0.20	0.32	0.38	0.20
6/21	B7–B12	0.01	0.32	0.45	0.42	0.13	B8–B10	0.00	0.27	0.44	0.46	0.14
6/22		0.01	0.16	0.36	0.35	0.08		0.00	0.14	0.37	0.37	0.10
6/23		0.04	0.28	0.30	0.36	0.18		0.01	0.27	0.31	0.40	0.19
6/24		0.07	0.23	0.36	0.23	0.12		0.07	0.23	0.36	0.25	0.14
6/25		0.04	0.19	0.37	0.34	0.12		0.02	0.19	0.36	0.35	0.14
6/26		0.07	0.20	0.41	0.49	0.18		0.05	0.20	0.40	0.50	0.20
6/27		0.08	0.21	0.34	0.49	0.18		0.05	0.21	0.32	0.50	0.21
6/28		0.09	0.22	0.34	0.40	0.19		0.06	0.21	0.34	0.40	0.22
6/21	B8–B11	0.00	0.28	0.44	0.46	0.14	B8–B12	0.01	0.32	0.45	0.45	0.14
6/22		0.00	0.14	0.38	0.40	0.10		0.01	0.16	0.37	0.37	0.10
6/23		0.02	0.27	0.34	0.44	0.20		0.04	0.29	0.31	0.40	0.19
6/24		0.07	0.23	0.36	0.25	0.14		0.07	0.24	0.36	0.24	0.13
6/25		0.03	0.18	0.36	0.36	0.14		0.04	0.20	0.37	0.35	0.13
6/26		0.05	0.19	0.40	0.52	0.21		0.07	0.21	0.41	0.50	0.20
6/27		0.06	0.20	0.31	0.52	0.22		0.08	0.22	0.34	0.50	0.20
6/28		0.07	0.20	0.32	0.40	0.22		0.09	0.22	0.34	0.40	0.21

coefficients of the soil moisture at 50 cm depth and the PDI constructed by the band 8 and band11 are the highest. And the coefficients become less and less at 20 cm, 10 cm, 100 cm and5 cm.

So, the PDI derived from band8 and band11 and the relative soil moisture at 50 cm depth were selected to build the soil moisture inversion model using the following linear equation.

$$RSM = PDI^{*}a + b \tag{4}$$

Where *RSM* is the relative soil moisture, a is the slope and b is the offset.

At the same time, the correlation coefficients on 28th are almost less than those on the 26th and 27th. That is to say a hysteresis effect was observed. The delay time is about 1–2 days for the PDI to reflect the water stress.

3.2 Drought Monitoring

In the PDI image (in Fig. 1), the values in most area are low, but the water body is dark with very low value (less than 0.05). In west and north-west of the study area, the PDI values (above 0.25) are higher than those in other area, especially along the Mississippi River. But the values in the cloud polluted area are invalid.

In this paper, the relative soil moisture derived from the formula (4) were used to evaluate the drought. And the relative moisture less than 60% is defined as drought, above 60% is normal. The relative soil moisture in most area is above 60%. Especially in east and north-east, the color is green. And the water bodies including the rivers are in deep green. According to the relative soil moisture, there is light drought in the yellow area. The red area is cloud polluted and invalid.

The drought monitoring results shows that most area is normal, and there is light drought in the west and northwest of the study area.

4 Conclusion

In this paper, WIS data acquired by Tiangong-2 Space Laboratory and in situ soil moisture measured by SCAN were used for soil moisture inversion and monitor drought in Mississippi and Alabama. The results show that the PDI index derived from band8 (0.6775–0.6875 um) and Band11 (0.845–0.885 um) is more sensitive to soil moisture at 50 cm, and the highest correlation coefficient r^2 is 0.52. At the same time, the PDI has a certain hysteresis effect on the drought response about 1–2 days. The drought monitoring results using the relative soil moisture inversion demonstrated the distribution of water stress and light drought in west and northwest of the study area. Though the short-wave infrared and thermal infrared data were not used in this paper, the PDI had reflected the water stress. In future research, it is possible to develop other indices using more bands to monitor drought with high accuracy.

Acknowledgements. Thanks to China Manned Space Engineering for providing space science and application data products of Tiangong-2.

References

1. Kogan, F.N.: Droughts of the late 1980s in the United States as derived from NOAA polar orbiting satellite data. Bull. Am. Meteorol. Soc. **76**, 655–668 (1995)
2. Mu, L.: Suitability and uncertainty analysis of agricultural drought indicator with remote sensing (2006)
3. Price, J.C.: Estimating surface temperatures from satellite thermal infrared data—a simple formulation for the atmospheric effect. Remote Sens. Environ. **13**, 353–361 (1983)
4. Schmugge, T.J., O'Neill, P.E., Wang, J.R.: Passive microwave soil moisture research. IEEE Trans. Geosci. Remote Sens. **24**, 12–22 (1986)
5. https://lta.cr.usgs.gov/VegDRI
6. Manned Space Application Data Promotion Service Platform. http://www.msadc.cn/rw
7. Ganguli, P., Ganguly, A.R.: Space-time trends in U.S. meteorological droughts. J. Hydrol. Reg. Stud. **8**, 235–259 (2016)
8. Coopersmith, E.J., Cosh, M.H., Bell, J.E.: Extending the soil moisture data record of the climate reference network (CRN) and the soil climate analysis network (SCAN). Adv. Water Res. **79**, 80–90 (2015). https://doi.org/10.1016/j.advwatres.2015.02.006
9. http://webarchive.iiasa.ac.at/Research/LUC/External-World-soil-database/HTML/
10. She, W.M., Ye, C.H.: Research on drought monitoring model based on MODIS data. J. Yangtze River Sci. Res. Inst. **21**(3), 11–15
11. Richardson, A.J., Wiegand, C.L.: Distinguishing vegetation from soil background information. Photogramm. Eng. Remote Sens. **43**(12), 1541–1552 (1977)
12. Ghulam, A.: Remote monitoring of farmland drought based on n-Dimensional spectral feature space. Ph.D. thesis, Peking University (2006)
13. Ghulam, A., Qin, Q., Zhan, Z.: Designing of the perpendicular drought index. Environ. Geol. **52**(6), 1045–1052 (2007)
14. Ghulam, A., Qin, Q., Teyip, T., Li, Z.L.: Modified perpendicular drought index (MPDI): a real-time drought monitoring method. ISPRS J. Photogramm. Remote Sens. **62**, 150–164 (2007)

Temporal and Spatial Changes of the Yellow River Delta Wetland Based on Multi-Source Data During 30 Years

Pifu Cong[1(✉)], Kexin Chen[2], Limei Qu[1], and Jianbo Han[1]

[1] National Marine Environmental Monitoring Center, Dalian Liaoning 116023, China
congpifu@163.com
[2] Liaoning Normal University, Dalian Liaoning 116029, China

Abstract. Estuary wetlands are important and vulnerable. In order to demonstrate the performance of the imaging data of Wide-band Imaging Spectrometer of Tiangong-2 in wetland classification and detect the changes of the Yellow River Delta wetland, we used object-oriented classification, landscape center of gravity transfer model and landscape index to depict the temporal and spatial changes of the Yellow River Delta wetland in the past 30 years from 1986 to 2016. Landsat 5 TM data in 1986 and imaging data of Wide-band Imaging Spectrometer of Tiangong-2 in 2016, as well as Landsat 8 OLI in 2016, were utilized. The imaging data of Wide-band Imaging Spectrometer of Tiangong-2 was fused with Landsat image by Gram-Schmidt spectral sharpening algorithm for analysis. It shows that: (1) After fusion with the higher resolution image, Tiangong-2 multi-spectral data is helpful for wetland study. (2) The natural wetlands decreased from 2404.24 km^2 in 1986 to 1666.47 km^2 in the past 30 years. Human-made wetlands increased from 42.41 km^2 in 1986 to 836.58 km^2 in 2016. (3) During the past 30 years, the gravity centers of natural wetlands, human-made wetlands and non-wetlands have been shifting towards the sea respectively. The shifting distance of human-made wetlands was larger. (4) The wetland landscape index indicated that the landscape pattern dominated by natural wetlands tended to be dominated by non-wetlands.

Keywords: Tiangong-2 · Yellow River Delta wetland
The Object-oriented classification · Landscape center of gravity transfer model
Landscape index

1 Introduction

Wetlands, together with forests and oceans, are recognized as the three major global ecosystems. Wetlands contain lots of animal and plant resources and enjoy the reputation of the kidney of the earth, bird paradise, cradle of life and so on. They play an important role in protecting biodiversity [1], regulating climate [2] and conserving water [3]. So they have good social, economic and ecological benefits. With the rapid development of the economy, people's demand for resources has increased. The development and utilization of wetland resources have accelerated, causing serious

Y. Gu et al. (Eds.): *Proceedings of the Tiangong-2 Remote Sensing Application Conference*, LNEE 541, pp. 286–299, 2019.
https://doi.org/10.1007/978-981-13-3501-3_27

damage to the ecological environment, thus affecting the space for human survival. Usually, it is difficult to monitor the changes of wetland timely. But with the rapid development of computer technology and remote sensing technology, the monitoring of wetland changes has changed from traditional local field survey to large-scale monitoring by remote sensing satellites. The improvement of satellite operation cycle, resolution, width and bandwidth provides a powerful help for the identification and extraction of wetland landscape types.

Wetland has progressively become the focus of international academic researchers, governments and even the public [4]. High-resolution remote sensing image data have played an important role in many fields such as land use, agricultural monitoring, ecological environment monitoring, disaster monitoring and many other fields [5]. The use of remote sensing images for large-area wetland monitoring has been the mainstream of research on wetland ecology. Satellite sensors were launched earlier in European and American countries, and their scholars studied wetland monitoring earlier. In 2002, Ozesmi, S.L. and others discussed the characteristics of various remote sensing satellites used for wetland change detection [6]. These satellites include Landsat MSS, Landsat TM, SPOT, NOAA AVHRR, and IRS-1B liss- II. In 2007, Klemas, V. compared the cost and practicability of high, medium and low resolution images as well as aerial data respectively. They concluded that the long-term trend and short-term change of wetland vegetation and hydrology could be determined effectively by using satellite images and aerial images combined with ground observation and survey [7].

Chinese scholars have also carried out many kinds of research on the changes of wetland landscape by remote sensing, especially in the Yellow River Delta [8]. The data they used are mainly based on a series of Landsat satellites. There are many methods of visual interpretation and the traditional classification. The classification categories of wetlands are also different. In 2009, Zong, X.Y. used Landsat satellite data to monitor changes in the Yellow River Delta wetland in 1986, 1996 and 2006 using a visual interpretation method [4]. In 2010, Zhao, H. et al. introduced the future trend of wetland research, including multi-source data fusion [9]. In 2014, Liu, G. et al. utilized the Landsat satellite data to examine the wetland changes in the Yellow River Delta in 2000, 2005 and 2010 using an object-oriented classification method [10]. In 2017, Han, M. et al. studied the impact of human activity intensity on wetland change by visual interpretation utilizing remote sensing images of 1989, 1999, 2009 and 2014 [11]. However, there are few studies on the monitoring of landscape types in the Yellow River Delta based on multi-source remote sensing data.

Tiangong-2 is a space laboratory independently developed by China. The imaging data of Wide-band Imaging Spectrometer (WIS) of Tiangong-2 is a new generation of optical remote sensor with wideband, wide field of view and image-spectrum emerging technology. In this paper, Tiangong-2 data, Landsat satellite, Sentinel 2 data and field survey data were integrated to monitor the temporal-spatial dynamic changes of wetland landscape in the Yellow River Delta from 1986 to 2016 by using object-oriented classification method. The results can provide a reference for wetland protection and restoration.

2 Study Area and Data

2.1 Study Area

The Yellow River Delta is located in Dongying City, Shandong Province, bordering the Bohai Sea to the north and Laizhou Bay to the east [12]. It is the second largest estuary Delta in China, after the Yangtze River Delta. It is located at 118°07′ – 119°10′ E and 36°55′ – 38°10′ N. The Yellow River Delta is a new estuary wetland with the relatively gentle terrain. It has four distinct seasons. It is cold in winter and hot in summer. It belongs to semi-humid continental monsoon climate in the warm temperate zone. The region is rich in flora and fauna and it is a transit station for international migration of birds. It is also one of the regions with the fastest growing natural land area in the world [13]. A large amount of sediment carried by the upper reaches of the Yellow River was deposited at the estuary [14]. The study area is the modern Yellow River Delta. The modern Yellow River Delta takes the fishing rafts in Kenli County as the pivot point. It picks the estuary from the north, the Songchunrong Mouth in the south, and the negative 6 m isobath in the east, as shown in Fig. 1.

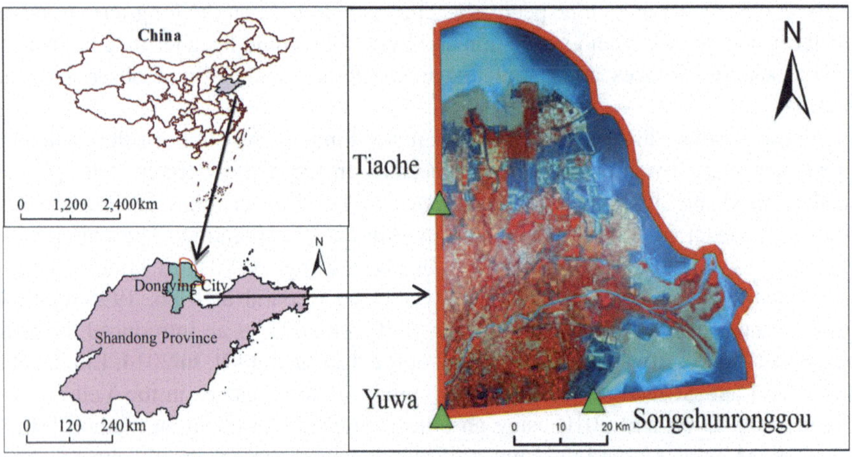

Fig. 1. The study area

2.2 Data

In this paper, imaging data of WIS of Tiangong-2 in 2016 (http://www.Msadc.cn/) (100 m, October 14th), Landsat 8 OLI in 2016 (30 m, October 13th), and Landsat 5 TM in 1986 (30 m, May 20th) were used to study the changes of the wetland landscape in the Yellow River Delta in the past 30 years. The imaging data of WIS of Tiangong-2 has 14 channels with a spectral range of 0.4 to 1.0 μm [15]. It is the first time for a single instrument integrating the function of visible-near infrared, short-wave infrared and thermal-infrared multi-spectral as well as full-push-scan imaging in the large field of view [16]. It has narrow bandwidth and rich information and very helpful for the accurate identification of wetland types. The spatial resolution is 100 m and the swath is 300 km. Landsat 5 has 7 bands with 30 m spatial resolution except for the sixth bands. Level 2 data of imaging data of WIS of Tiangong-2 is a product after radiation correction and geometric correction. So only atmospheric correction is needed. In this paper, the FLAASH model in ENVI is used for atmospheric correction. Radiometric calibration and atmospheric correction are carried out in the Landsat-5 image. In order to make the two kinds of data source images have higher consistency in geographic space, the geometric correction of Tiangong-2 data and Landsat 5 data was carried out by using 10 m resolution pseudo-color (NIR, R, G) synthetic band data of Sentinel 2 (October 14th, 2016) which was corrected. Imaging data of WIS of Tiangong-2 and Landsat8 data are cross-calibrated so the geographical location is better. The correction error was 0.0046. The images were subset according to the boundary of the study area. The corrected Tiangong-2 data has a good advantage of spectrum for wetland identification, while the Landsat 8 data has the relatively high spatial resolution. In order to integrate their respective advantage, corrected Tiangong-2 data of 100 m and Landsat 8 OLI data of 30 m are fused into 30 m data in GS (Gram-Schmidt spectral sharpening). The fused data not only have the resolution of Landsat data, but also have multi-band information of Tiangong-2 data, the fusion result is shown in Fig. 2.

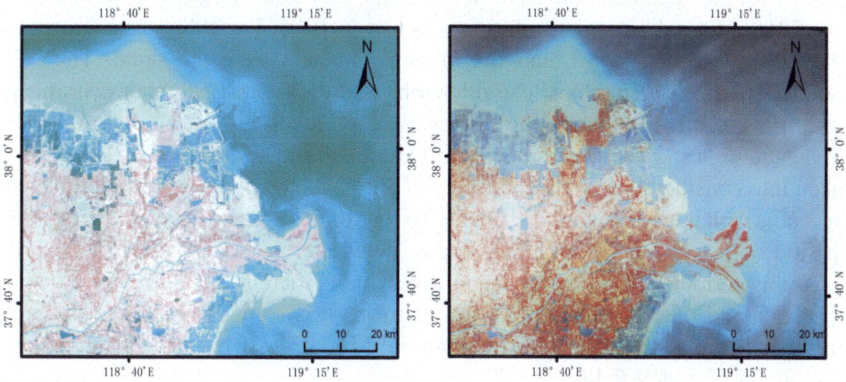

(a)Before fusion (b)After fusion

Fig. 2. Imaging data of WIS of Tiangong-2

3 Method

3.1 Classification Category

There are numerous studies on coastal wetland classification [17]. Depending on different research purposes, wetland classification categories are also different. In the study, on the basis of field survey data of wetland in 2016 and classification category of Ramsar Convention, wetland type was divided into three categories: natural wetlands, human-made wetlands, and non-wetlands. Natural wetlands include river wetlands, intertidal muds, shallow marine waters, and intertidal marshes. Human-made wetlands include reservoir and ponds, salt pans, aquaculture ponds and paddy fields. Non-wetlands include forest lands, dry farmlands, construction lands, and unused lands.

3.2 Object-Oriented Classification

Traditional wetland classification methods are based on pixels, which only consider the spectral information of pixels [18]. And they require a larger spectral difference between the types of objects. While object-oriented analysis can take the spectral information, texture information and spatial location information of objects into account in classification, thus avoiding some pepper and salt noise generated by pixel classification. The object-oriented method is to segment the image into spectral information, texture features, and other similar objects, and then classify the objects. The process involves various attributes of the object, such as spectral features, shape features, texture features, etc. [19]. The object-oriented method improves the accuracy and efficiency of the classification.

In this paper, object-oriented classification method is adopted in order to avoid pixel-based fragmentation. The most important part of object-oriented classification is image object segmentation. In this paper, the multi-scale segmentation algorithm is utilized to set the segmentation scale to 50, shape factor to 0.3, compactness factor to 0.7 for 2016. The radiometric resolution was low in 1986. For 1986, the segmentation scale of 1986 was set to 10, and the other parameters were the same. The NDVI threshold is used for hierarchical classification. And the nearest neighbor classification algorithm is performed by selecting objects on each level.

When configuring the nearest neighbor feature, all the information of 14 bands can be taken into consideration, so the accuracy of automatic recognition of the influence of ground objects has been improved. Owing to the different growth cycle of vegetation, the period of recognition is different. Paddy fields are usually planted from April to May. At this period of small seedling and watery field, there is a big difference between paddy fields and other vegetation. But in October, it is sometimes difficult to extract paddy fields from a single image. The image of May 22th, 2016, was utilized to assist the extraction of paddy fields.

3.3 Landscape Center of Gravity Transfer Model

The landscape types were merged into three categories: natural wetlands, human-made wetlands, and non-wetlands. The landscape center of gravity transfer model [20] was used to analyze the shifting trends of the center of gravity of the three landscape types over the past 30 years. The coordinate calculation formula is as follows:

$$X_T = \frac{\sum_{i=1}^{n} (C_{Ti} \times X_{Ti})}{\sum_{i=1}^{n} C_{Ti}}, \quad Y_T = \frac{\sum_{i=1}^{n} (C_{Ti} \times X_{Ti})}{\sum_{i=1}^{n} C_{Ti}} \tag{1}$$

Among them, X_T and Y_T respectively represent the latitude and longitude coordinates of the gravity center of a certain wetland type distribution in the T year. C_{Ti} represents the area of the *i-th* small patch of the wetland type. X_{Ti} and Y_{Ti} represent the geographic longitude and latitude coordinates of the geometric center of the *i-th* small patch of the wetland type, and n represents the number of patches contained in the wetland type.

3.4 Landscape Index

The spatial changes of wetland landscape types form a new landscape pattern. The wetland landscape pattern affects many ecological processes [21], so it has significant research value. The landscape index provides a powerful help for this study. Thirteen indices were selected at the class level and landscape level respectively. The landscape indices were calculated by Fragstats 4.2 software. At the class level, the proportion of patches in total landscape area (PLAND), number of patches (NP), landscape shape index (LSI), patch cohesion index (COHESION), splitting index (SPLIT) were selected. Number of patches (NP), landscape shape index (LSI), splitting index (SPLIT), contagion index (CONTAG), Shannon's diversity index (SHDI), Shannon's evenness index (SHEI), aggregation index (AI), patch cohesion index (COHESION) were chosen at the landscape level.

4 Results and Discussion

4.1 The Classification Results

The classification results in 1986 and 2016 are shown in Fig. 3.

The 1986 data is far away and has not been compared with high-resolution data, and no precision verification has been carried out. More than 200 precision verification points were randomly selected from 2017 field survey data and Sentinel 2 10 m resolution data. This paper selects Error Matrix based on Samples method in the Accuracy Assessment module of the eCognition software. Overall Accuracy gets 0.93. KIA gets 0.92. The confusion matrix is given in Table 1.

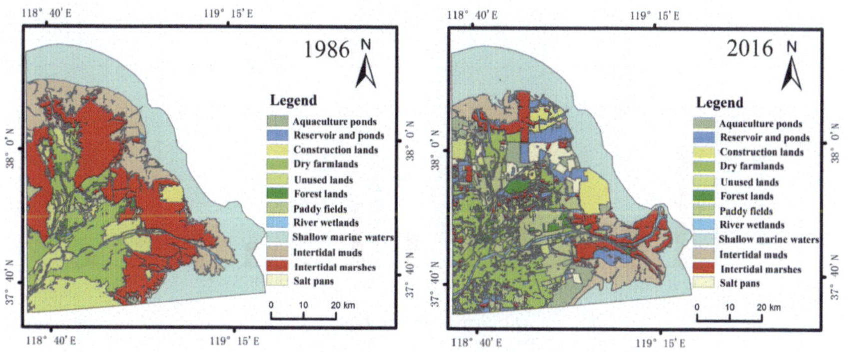

Fig. 3. Classification results of 1986 and 2016

Table 1. The confusion matrix

Class sample	SMW	CL	AP	IMD	IMH	UL	DF	SP	RAP	FL	PF	RW	KIA per class
SMW	13	0	0	0	0	0	0	0	0	0	0	0	1.00
CL	0	19	0	0	0	0	0	0	0	0	1	0	0.94
AP	0	0	17	0	0	1	0	1	0	0	0	0	0.84
IMD	0	0	0	19	2	0	0	0	0	0	0	0	0.94
IMH	0	1	0	1	25	0	0	0	0	0	0	0	0.81
UL	0	0	0	0	0	15	0	0	0	0	0	0	0.93
DF	0	0	0	0	1	0	16	0	0	2	0	0	1.00
SP	0	0	1	0	0	0	0	19	0	0	0	0	0.94
RAP	0	0	2	0	0	0	0	0	15	0	0	0	1.00
FL	0	0	0	0	0	0	0	0	0	17	0	0	0.84
PF	0	0	0	0	2	0	0	0	0	1	19	0	0.94
RW	0	0	0	0	0	0	0	0	0	0	0	7	1.00

Note: SMW: Shallow marine waters; IMD: Intertidal muds; AP: Aquaculture ponds; DF: Dry farmlands; PF: Paddy fields; RAP: Reservoir and ponds; IMH: Intertidal marshes; RW: River wetlands; CL: Construction lands; FL: Forest lands; UL: Unused lands; SP: Salt pans

The accuracy of classification results is the basis for further analysis. This result can be used for future analysis.

4.2 Analysis of Temporal and Spatial Changes of Landscape Types

(1) For the analysis of temporal change, the area statistics are shown in Figs. 4 and 5. From the Figs. 4 and 5, we can clearly see that natural wetlands showed a downward trend, with a total reduction of 737.77 km^2 in 30 years. The change of intertidal marsh was the largest, which reduced by 550.78 km^2, indicating that the vegetation in the intertidal muds of the Yellow River Delta was seriously damaged. The intertidal muds decreased by 123.47 km^2. The wetland continued to expand to the sea, so the reduction of intertidal muds indicated that the intertidal

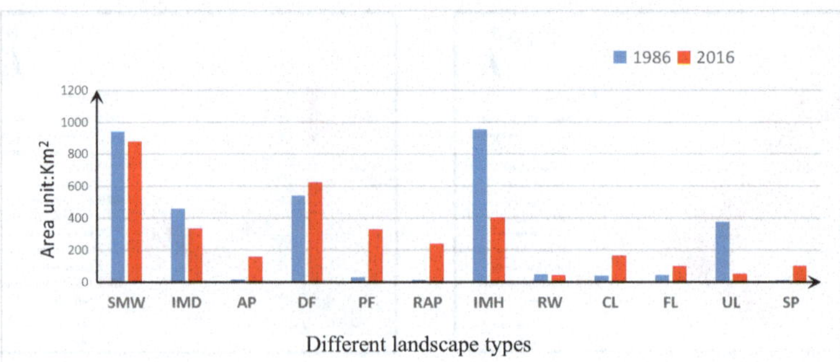

Fig. 4. Variation of different landscape types (Abbreviations are the same as in Table 1)

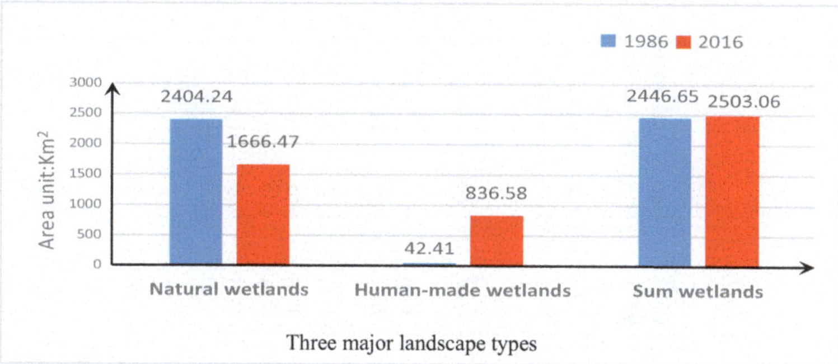

Fig. 5. Change of wetland area

muds were affected by both human activities and coastal erosion. Human-made wetlands increased by 794.18 km^2, reflecting that human activities have a greater impact on the changes of wetland landscape types in the past 30 years. Among them, paddy fields increased by 298.64 km^2, reservoir and ponds increased by 237.89 km^2, aquaculture ponds increased by 154.76 km^2, and salt pans increased by 102.88 km^2. Due to the huge increase of human-made wetlands, the total area of wetland increased by 56.41 km^2. As for non-wetlands, unused lands decreased by 323.63 km^2. And on basis of the large base, the dry farmlands increased by 83.57 km^2, reflecting the acceleration of human development of cultivated land.

(2) The transfer matrix of different landscape types is shown in Table 2. Natural wetlands have a large transfer part. 14.45% of the intertidal marshes have been transferred to dry farmlands, 11.16% of the intertidal marshes have been transferred to reservoir and ponds, and 10.99% of the intertidal marshes have been transferred to aquaculture ponds. 73.49% of the intertidal muds have been transferred. The transfer indicates that the intertidal muds are greatly influenced by man. 16.45% of the river wetland was transferred to the dry farmlands, indicating that the water flow in the Yellow River decreased compared with the original. In the transfer of human-made wetlands, fewer aquaculture ponds and reservoir and ponds were transferred out, and more paddy fields were transferred out. 49.54% of paddy fields were transferred to dry farmlands, and 6.05% of paddy fields to construction lands. Only 0.27% aquaculture ponds in 1986 were transferred. 14.66% of reservoir and ponds were transferred to dry farmlands and 1.26% of reservoir and ponds to forest lands. In the process of non-wetlands transfer, 98.46% of unused lands were transferred out. Construction lands and dry farmlands were relatively less transferred, 36.94% of forest lands were transferred to dry farmlands, 46.93% of unused lands were transferred to dry farmlands, 80.42% of construction lands were not transferred, and the internal transfer of non-wetlands changed greatly.

Table 2. 1986–2016 Landscape transfer matrix (Unit: %)

1986–2016	Intertidal marshes	Dry farmlands	River wetlands	Construction lands	Reservoir and ponds	Forest lands	Shallow marine waters	Paddy fields	Intertidal muds	Unused lands	Salt pans	Aquaculture ponds
Intertidal marshes	21.43	14.45	1.29	6.79	11.16	2.26	0.44	10.82	11.32	3.70	5.34	10.99
Dry farmlands	8.80	46.48	1.59	3.53	2.37	6.98	0.01	26.06	0.29	1.84	0.94	1.12
River wetlands	11.69	16.45	11.86	1.26	19.60	3.94	11.38	4.35	12.94	0.19	2.55	3.80
Construction lands	4.82	12.79	0.00	80.42	0.73	0.55	0.03	0.06	0.00	0.00	0.00	0.61
Reservoir and ponds	2.47	14.66	0.00	0.00	79.86	1.26	0.00	1.34	0.00	0.40	0.00	0.00
Forest lands	11.44	36.94	1.53	0.99	1.72	32.30	0.00	14.61	0.29	0.09	0.00	0.08
Shallow marine waters	4.74	0.00	0.63	0.60	0.13	0.00	83.22	0.00	10.21	0.01	0.22	0.24
Paddy fields	25.05	49.54	0.00	6.05	9.90	4.11	0.00	4.60	0.00	0.75	0.00	0.00
Intertidal muds	12.13	0.36	0.58	7.67	16.13	0.69	19.29	0.14	26.51	0.14	8.65	7.71
Unused lands	8.45	49.63	2.25	1.71	7.92	5.60	0.00	20.09	0.57	1.54	1.39	0.85
Salt pans	0.00	0.00	0.00	0.00	0.00	0.00	0.00	0.00	7.91	60.14	0.00	31.94
Aquaculture ponds	0.00	0.00	0.00	0.01	0.00	0.00	0.00	0.00	0.27	0.00	0.00	99.72

(3) For the analysis of spatial variation, the landscape center of gravity transfer model is shown in Fig. 6 was obtained on the same scale.
From the figure, we can see that natural wetlands, human-made wetlands, and non-wetlands have a tendency to shift to the sea from 1986 to 2016. It can be observed that the movement of human-made wetlands is larger, followed by non-wetlands and natural wetlands. The natural wetlands were distributed near the coast in 1986. In 2016, it retreated, indicating that the development and utilization of natural wetlands by human activities have caused a reduction in the area of natural wetlands. In 1986, the human-made wetlands were less concentrated in the middle of the Yellow River Delta. In 2016, artificial development and utilization of coastal wetlands increased, resulting in a considerable movement of human-made wetlands to the sea. It can show up from the shift of the center of gravity of the non-wetlands that human activities have intensified during the past 30 years.

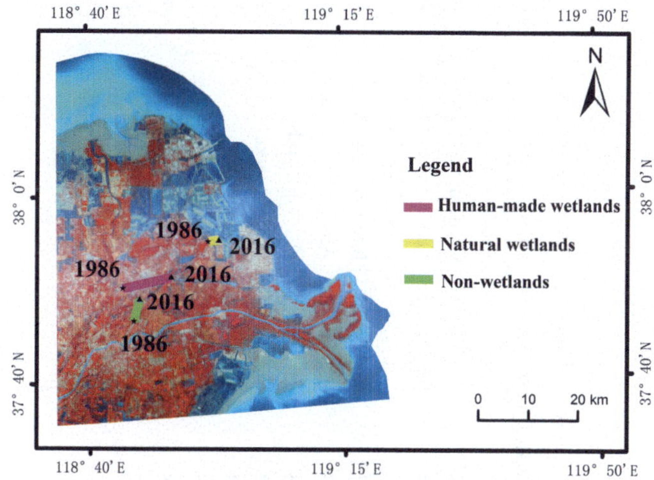

Fig. 6. Landscape center of gravity shift

(4) From the calculation of the class level index, the results are shown in Table 3. The indices at the landscape level are shown in Table 4.
Among them, PLAND represents the area percentage of a certain type of landscape in the total landscape area. Except for shallow marine waters, according to the PLAND index in 1986, natural wetlands dominated the landscape, and intertidal marshes accounted for 27.72% of the total landscape area. However, dry farmlands dominated the landscape in 2016, and dry farmlands accounted for 18.09% of the total landscape area. The intertidal marshes decreased to 11.77%, indicating that the natural wetlands have been damaged to a certain extent and the area is less. With the interference of human activities, the number of patches in human-made wetlands and non-wetlands has increased a lot. Landscape shape index is directly proportional to the complexity of patches. Most of the LSI index of landscape types increase, which indicates that the patches tend to be complex.

Table 3. Landscape indices at class level

1986	PLAND	NP	LSI	COHESION	SPLIT	2016	PLAND	NP	LSI	COHESION	SPLIT
Shallow marine waters	27.23	9	4.75	99.95	13.49	Shallow marine waters	25.52	28	4.70	99.95	15.40
Construction lands	1.19	27	8.01	98.73	15876.88	Construction lands	4.85	209	16.40	98.87	1478.74
Reservoir and ponds	0.11	20	6.58	94.29	6642626.53	Reservoir and ponds	7.00	399	22.66	98.51	2757.47
Intertidal muds	13.34	91	18.74	99.76	158.62	Intertidal muds	9.75	130	13.29	99.66	343.88
Intertidal marshes	27.72	175	20.14	99.84	33.75	Intertidal marshes	11.77	653	37.65	98.89	819.63
Aquaculture ponds	0.14	8	5.72	97.33	891972.45	Aquaculture ponds	4.63	71	8.21	99.28	1939.22
Dry farmlands	15.67	121	18.87	99.74	138.55	Dry farmlands	18.09	512	34.74	99.68	142.40
Unused lands	10.92	36	12.60	99.59	279.09	Unused lands	1.54	58	10.84	99.08	10090.60
Salt pans	0.05	3	1.80	96.53	6164599.25	Salt pans	3.03	27	6.80	99.29	3799.65
Forest lands	1.27	80	17.93	97.35	130548.37	Forest lands	2.94	206	24.52	97.96	14127.86
Paddy fields	0.93	21	9.05	98.78	49291.55	Paddy fields	9.59	167	17.02	99.15	1061.27
River wetlands	1.43	145	36.90	97.89	47243.62	River wetlands	1.29	19	10.50	99.44	9032.87

Table 4. Landscape indices at landscape level

	NP	LSI	SPLIT	CONTAG	SHDI	SHEI	AI	COHESION
1986	736	20.73	8.27	62.53	1.74	0.70	98.08	99.76
2016	2479	30.03	12.64	52.90	2.17	0.87	97.15	99.50

COHESION describes the degree of connectivity between patches. The overall trend of natural wetlands for connectivity decreased. The connectivity of human-made wetlands and non-wetlands increased, which indicated that human disturbance to wetland landscape was greater. The fragmentation of natural wetlands increases. The large-scale development and utilization of human-made wetlands and non-wetlands make the fragmentation of sporadic parts converge. And the fragmentation of human-made wetlands and non-wetlands decrease.

In Table 4, at the landscape level, the number of patches increased and COHESION index decreased. The connectivity between surface patches decreased. AI and CONTAG index decreased, indicating that the aggregation between patches became worse and the patches tended to fragment. The overall landscape shape index LSI increased, indicating that the landscape shape tended to be more complicated. On the whole, landscape tends to be fragmented and complicated.

5 Conclusions

(1) This study demonstrates the potential of imaging data of WIS of Tiangong-2 with abundant spectral information in wetland extraction and classification. It can be used extensively after fusion with the higher resolution image, which has both advantages of spectral and spatial resolution. It is an effective data source for coastal wetland extraction and change research.

(2) Over the past 30 years, the natural wetlands decreased from 2404.24 km^2 in 1986 to 1666.47 km^2 in 2016, with a total decrease of 737.77 km^2. Human-made wetlands increased from 42.41 km^2 in 1986 to 836.58 km^2 in 2016. The total wetland area increased by 56.41 km^2 for the fast and huge increase of human-made wetlands. In the process of wetland transfer, natural wetlands transfer changed greatly, 65.52% of intertidal marshes converted to human-made wetlands and non-wetlands, and 73.49% of intertidal muds changed. The paddy fields in the human-made wetlands also changed greatly. 95.40% of the paddy fields were transferred outward, and 49.54% of the paddy fields were transferred to the dry farmlands.

(3) Spatially, the centers of gravity of natural wetlands, human-made wetlands, and non-wetlands all have a tendency to shift towards the sea in the past 30 years, indicating that natural wetlands decreased. And human-made wetlands increased

and moved towards the intertidal muds. Both of the results show that human activities have a greater impact on the changes of wetland landscape types.

(4) From the perspective of landscape pattern changes, it can be seen from the index changes at the type level that the landscape pattern dominated by natural wetlands in 1986 turned to the non-wetlands landscape pattern in 2016, and the number of patches (NP) increased. The patch shapes of various wetland landscape tended to be complex, and the patches of natural wetlands tended to fragment. From the index changes of landscape level, we can see that the connectivity between patches was poor and the landscape also tended to fragment. If this trend is not limited, it will have a serious impact on the wetland ecological environment in the Yellow River Delta.

Acknowledgements. This work is under the auspices of National Key R&D Program of China (No.2017YFC0505901). Thanks to China Manned Space Engineering for providing space science and application data products of Tiangong-2.

References

1. Rao, B.R.M., Dwivedi, R.S., Kushwaha, S.P.S., et al.: Monitoring the spatial extent of coastal wetlands using ERS-1 SAR data. Int. J. Remote Sens. **20**(13), 2509–2517 (1999)
2. Zhang, Z., Craft, C.B., Xue, Z., et al.: Regulating effects of climate, net primary productivity, and nitrogen on carbon sequestration rates in temperate wetlands. NorthEast China. Ecol. Indic. **70**, 114–124 (2016)
3. Wang, W.F., Luo, J.M., Wang, Y.J., et al.: Water conservation function in Zhalong saline wetland at local scale. J. Qiqihar Univ. **26**(06), 71–75 (2010)
4. Zong, X.Y.: Analysis of wetland pattern and change in the Yellow River Delta. Taiyuan University of Technology (2009)
5. Li, S.Y., Zhang, W.F., Yang, S.: Intelligent fusion of multi-source high-resolution remote sensing images. J. Remote Sens. **21**(3), 415–424 (2017)
6. Ozesmi, S.L., Bauer, M.E.: Satellite remote sensing of wetlands. Wetl. Ecol. Manag. **10**(5), 381–402 (2002)
7. Klemas, V.: Using remote sensing to select and monitor wetland restoration sites: an overview. J. Coast. Res. **29**(4), 958–970 (2013)
8. Liu, J., Feng, Q., Gong, J., et al.: Land-cover classification of the Yellow River Delta wetland based on multiple end-member spectral mixture analysis and a Random Forest classifier. Int. J. Remote Sens. **37**(8), 1845–1867 (2016)
9. Zhao, H., Zhang, H.Y., Li, N.N., et al.: RS-based research of Chinese wetland: status and trend. Geogr. Geo-Inf. Sci. **26**(2), 62–66 (2010)
10. Liu, G., Zhang, L., Zhang, Q., et al.: Spatio - Temporal dynamics of wetland landscape patterns based on remote sensing in Yellow River Delta. Wetlands **34**(4), 787–801 (2014)
11. Han, M., Zhang, C., Lu, G., et al.: Gradient response of wetland landscape pattern to human activity intensity in the Yellow River Delta. J. Agric. Eng. **33**(6), 265–274 (2017)
12. Zhao, G., Ye, S., Gao, M., et al.: Analysis of land use and shoreline changes at the Dawenliu Nature Reserve of Yellow River Delta based on remote sensing. J. Geo-Inf. Sci. **15**(3), 408 (2013)
13. Zong, X.Y., Liu, G.H., Qiao, Y.L., et al.: Dynamic changes of wetland landscape pattern in the Yellow River Delta. J. Geo-Inf. Sci. **11**(1), 91–97 (2009)

14. Chen, J., Wang, S.Y., Mao, Z.P.: Remote sensing monitoring of wetland changes in the Yellow River Delta from 1976 to 2008. Prog. Geogr. Sci. **30**(5), 585–592 (2011)
15. Space application data promoting service platform for China Manned Space Engineering. http://www.msadc.cn. Last accessed 21 Sep 2018
16. Liu, K., Qin, B.Y., Mu, L.L., et al.: Monitoring and analysing of reclamation change based on multi-data remote sensing images: a case study of partial area of Liaoning province. Mar. Environ. Sci. **36**(6), 911–917 (2017)
17. Cao, Y., Mo, L.J., Li, Y., et al.: Wetland landscapes ecological classification: research progress. Chin. J. Appl. Ecol. **20**(12), 3084–3092 (2009)
18. Chang, X., Zhang, Q., Luo, M., et al.: Comparison of Qinzhou bay wetland landscape information extraction by three methods. In: ISPRS-International Archives of the Photogrammetry, Remote Sensing and Spatial Information Sciences xl-4, pp. 21–28 (2014)
19. Walter, V.: Object-based classification of remote sensing data for change detection. J. Shandong Jianzhu Univ. **58**(3), 225–238 (2008)
20. Wang, X.L., Bao, Y.H.: Study on the methods of land use dynamic change research. Prog. Geogr. **18**(1), 83–89 (1999)
21. Wu, J.G.: Landscape Ecology: Pattern, Process, Scale and Hierarchy, 2nd edn. Higher Education Press, Beijing (2007)

Wetland Monitoring Application in Panjin City Based on Remote Sensing Data of Wide-Band Imaging Spectrometer of Tiangong-2

Jing Xing[1,2] and Dan Meng[1,2](✉)

[1] College of Resources Environment and Tourism, Capital Normal University, Beijing 100048, China
mengdan@cnu.edu.cn
[2] Beijing Laboratory of Water Resource Security, Beijing 100048, China

Abstract. The Liaohe Delta wetland is important in China for its abundant resources. In addition, the current situation of Panjin city wetland is significant for protecting and restoring the Liaohe wetland ecology. The suitability of Tiangong-2 remote sensing imagery and non-remote sensing data to coastal wetland monitoring was studied in this paper. Based on spectral features of various areas, a method of partition classification is proposed. Information about Panjin wetland resources, including four types of natural wetland and three types of artificial wetland was extracted by spectral analysis combining feature shape, patch area and texture characteristics, and classification accuracy was evaluated with Landsat data. The overall accuracy of the partition classification method reached 89.89%, and the kappa coefficient was 0.8748, The feasibility of remote sensing extraction of coastal wetland information in Panjin city based on Tiangong-2 data is validated. The research show that the wetland area is 3184.89 km^2, of which the natural wetland area is 1129.98 km^2, accounting for 40.53% of the total area; and the artificial wetland area is 1893.91 km^2, accounting for 59.47% of the total area. Among these, paddy fields occupy the largest area, accounting for 55.29% of the total wetland area. In order of decreasing area, the others are reed swamps, mudflat wetlands, *Suaeda salsa* swamps, aquaculture farms, reservoirs/ponds, and river wetland.

Keywords: Tiangong-2 · Wetland monitoring · Remote sensing
Partition extraction · Panjin city

1 Introduction

The Liaohe Delta coastal wetland is located at the junction of the Liaohe and the Daliaohe Estuary. It is one of four major estuary deltas in China and is also China's main oil and grain production base [1]. Due to geographical location, the Liaohe Delta wetland was affected by rivers and oceans. With the acceleration of urbanization and reclamation activities, some coastal wetlands in this area were seriously degraded [2, 3]. By analyzing the distribution of coastal wetlands, we can understand the causes, mechanisms and evolution of coastal wetlands, and provide a scientific basis for sustainable development of coastal wetland resources. The Liaohe Delta area has become a

Y. Gu et al. (Eds.): *Proceedings of the Tiangong-2 Remote Sensing Application Conference*, LNEE 541, pp. 300–311, 2019.
https://doi.org/10.1007/978-981-13-3501-3_28

hotspot research area in recent years, due to the concentrated distribution of various types of coastal wetlands, and strong human disturbance. Some scholars have studied the status and distribution characteristics of wetland resources in the Liaohe Delta with the help of remote sensing and geographic information system (GIS) technology, and established the Liaohe Delta wetland resource classification system [4]. In 1997, Huang Guilin used 3S technology to classify Liaohe Delta wetland types and areas, as well as the changes to and causes of the Shuangtaizi River Nature Reserve and wetland types [5]. Landsat thematic mapper (TM) satellite imagery from 1986, 1995, 2000 and 2006 was adopted to monitor changes to wetlands in Panjin city through visualization interpretation, computer automatic identification, and field verification, and the causes of the changes were analyzed [6]. However, use of Tiangong-2 data to reveal the distribution of wetland in the Liaohe Delta is still rare. The Tiangong-2 wide-band imaging spectrometer is a new generation optical remote sensor with wide-band and wide-field coverage, as well as image-spectrum merging. This is the first opportunity for large-scale measurement of visible light, short-wave infrared, and thermal infrared multi-spectra on a single instrument, as well as field full-push scanning imaging with integrated functions. The wide-band imaging spectrometer has 14 channels in the visible and near-infrared, and the spectral range is 0.4 μm to 1.0 μm. It has multiple bands, narrow spectral range and continuous banding, which are conducive to accurate interpretation of wetland information.

In this study, the suitability of remote sensing data from wide-band imaging spectrometer of Tiangong-2 for monitoring coastal wetland have been explored. Tiangong-2 remote sensing data and GIS technology were used to analyze Panjin wetland types, area and distribution. This information will be beneficial to ensure the background data of the wetland ecosystem and its biodiversity in the Liaohe Delta, and will lay a foundation for the protection and rational use of wetland resources.

2 Study Area and Data

2.1 Study Area

The Liaohe Delta is located in southwestern Liaoning Province, at the top of Liaodong Bay in the Bohai Sea. It is an alluvial marine plain formed by the Liaohe, Daliaohe, and Dalinghe rivers, including parts of Panjin and Yingkou cities. The geographical position is 121°25′ to 112°55′ E, and 40°40′ to 41°25′ N, in the concentrated wetland area southern part of northeastern China with extensive development of agriculture, petroleum industry, shallow sea breeding industry, and reed planting industry. Panjin City lies on the alluvial marine plain that is the main core of the Liaohe Delta. Landform include tidal plain and tidal flat. Panjin has a warm temperate continental semi-humid monsoon climate, and regional climate differentiation is not obvious. The annual average temperature is 8.3 °C to 8.4 °C, the frost-free period is 167 d to 174 d, the annual average rainfall is 611.6 mm to 640.0 mm, and the annual average evaporation is 1 669.6 mm. The annual sunshine hours are 2786.0 h [7]. There are more than 20 rivers in Panjin, which gather in the Shuangtaizi Estuary and the Daliaohe River before emptying into the sea. Panjin city has a complete system of administrative divisions and statistical data, so we selected this city in this study as an area of focus to analyze the applicability of Tiangong-2 remote sensing data in coastal wetland monitoring (Fig. 1).

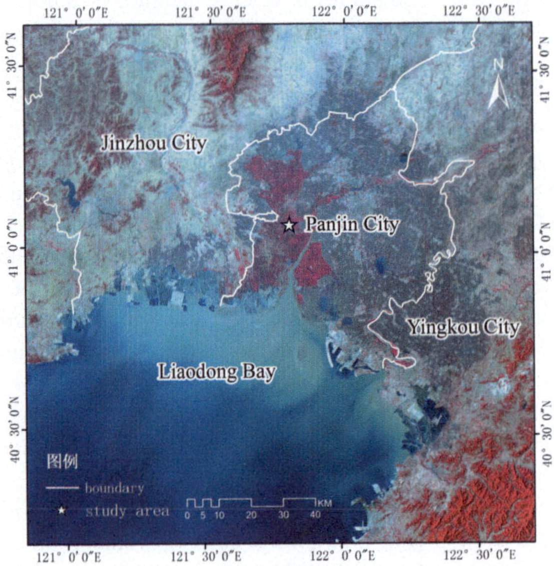

Fig. 1. Location of the study area (Combination of 5/8/10 channel visible)

2.2 Data

The Tiangong-2 Space Laboratory was launched on September 15, 2016, and was in orbit on September 23, 2016. Data services were provided after June 2017. Tiangong-2 is equipped with a new space application load device. Instruments include a wide-band imaging spectrometer, three-dimensional imaging microwave altimeters and ultraviolet edge imaging spectrometers for earth observation and space geoscience. The wide-band imaging spectrometer is the first in the world to be packaged as a single instrument. It integrates visible, short-wave infrared and thermal infrared multi-spectral large field of view full-push scan imaging. During orbit, it is mainly used for medium (higher) ground resolution; large-scale ground object monitoring, suitable for inland lakes, land and atmospheric exploration, as well as observation of ocean and coastal water color and temperature [8]. This study uses visible and near-infrared imagery from the Tiangong-2 wide-band imaging spectrometer. Because river wetlands, reservoirs and aquaculture farms are water bodies that change with the seasons, wet season data were used. Furthermore, the chlorophyll content in the summer vegetation growth period is high, which will be conducive to identifying marsh wetlands. Cloud cover should be less than 10% to ensure the quality of remote sensing imagery. Therefore, the remote sensing imagery data on June 1st, 2018 was selected. System radiation and geometric corrected image data were downloaded from the official manned space applications data sharing and distribution website (http://www.msadc.cn/), with a spatial resolution of 100 m. Lands at 8 operational land imager (OLI) remote sensing image of Panjin on May 5th, 2018 were used to evaluated the accuracy of Tiangong-2 data for wetland extraction. Image download from OLI official website (http://glovis.usgs.gov/) [9].

3 Methods

3.1 Remote Sensing Image Preprocessing

Atmospheric correction. The images acquired by sensors on satellites need to be processed before use. To obtain the true reflectivity of the surface, radiation correction is needed to reduce the effects of sensors, atmospheric conditions, and solar radiation. The radiation correction in this study was done with ENVI 5.1 software. Lelvel-2 grade Tiangong-2 data was used in the study. There are the radiance products after system radiation correction and geometric correction. The DN value is radiance amplified by a certain magnification; the magnification parameter is included in the companion XML file. Then, atmospheric correction was done using the fast line-site atmospheric analysis of hypercubes (FIAASH) atmospheric correction method. Because there is no basic parameter of the Tiangong-2 in the ENVI software, it is necessary to edit the header file including the center wavelength and the band width corresponding to the 14 bands (Table 1). In addition, the spectral response function of the Tiangong-2 wide-band imaging spectrometer is required to complete the parameter setting of FIAASH atmospheric correction.

Table 1. Visible and near-infrared spectra from the wide-band imaging spectrometer (Unit: nm).

Band name	Spectrum range	Central wavelength	Bandwidth	Band name	Spectrum range	Central wavelength	Bandwidth
V1	970 ~ 990	980	20	V8	655 ~ 675	665	20
V2	930 ~ 950	940	20	V9	610 ~ 630	620	20
V3	895 ~ 915	905	20	V10	555 ~ 575	565	20
V4	845 ~ 885	865	40	V11	510 ~ 530	520	20
V5	810 ~ 830	820	20	V12	480 ~ 500	490	20
V6	740 ~ 760	750	20	V13	433 ~ 453	443	20
V7	677.5 ~ 687.5	682.5	10	V14	403 ~ 423	413	20

To check the effects of atmospheric correction, spectral profiles of the same pixel, sampled from vegetation, before and after atmospheric correction are shown in Fig. 2. Figure 2a is the original DN value before atmospheric correction, and Fig. 2b is the reflectivity magnified 10,000 times after atmospheric correction. Figure 2 shows that atmospheric correction removed the influence of part of the atmosphere.

Geometric correction. Tiangong-2 data was geometrically corrected by Landsat OLI imagery with the right geometric information for the same time period. Quadratic polynomial and bilinear interpolation were selected as the parameters, and the above

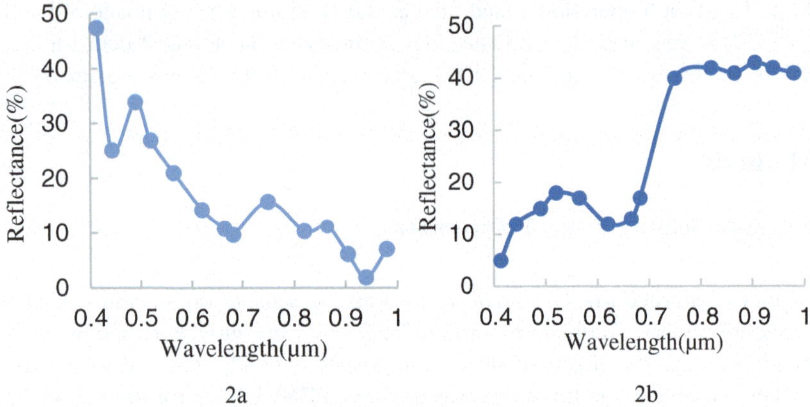

Fig. 2. Vegetation spectrum curve comparison before (2a) and after atmospheric correction (2b)

image was corrected for geometric accuracy. The root mean square error was controlled within 0.5 pixels. Due to bias between images, the corrected image projection and coordinates need to be converted to a common system, which ensures accuracy verification of wetland classification and allows a comparison. Then, the images were cut and enhanced using the boundaries of the study area.

3.2 Determination of Wetland Classification System

Current wetland classifications according to Ramsar Convention and the National Technical Regulations for Wetland Resources Survey are widely used [10]. Human production activities in the delta zone of this study area are severe. In the wetland classification system, wetland formation dynamics are the dominant factor, and the hydrological, ecological and plant dominant communities of the wetland are comprehensively considered. A wetland classification system of two levels was adopted in the study. In the first level, wetland was classified as natural or artificial. Natural wetlands indicate that biotic and abiotic elements in the landscape were not affected or interfered with by human activities. The relationship between energy flow and material flow in the landscape is still a natural ecosystem, or human beings just transform and manage the natural ecosystem slightly, without changing its system or landscape type, for instance, by artificially managing a reed field. Artificial wetlands are landscapes that were strongly disturbed by human activities. In the artificial wetland, many landscape elements have been artificially changed, and humans have formed new and different ecosystems. Humans need to invest a lot of energy into the system and take out products such as paddy fields and reservoirs. In the second level, wetlands were classified by hydrological status (water accumulation) and landscape type. In combination with the actual situation of wetlands in Panjin city and the marks that can be interpreted by Tiangong-2 remote sensing images, a wetland classification system suitable for the study area was established (Table 2, Fig. 3).

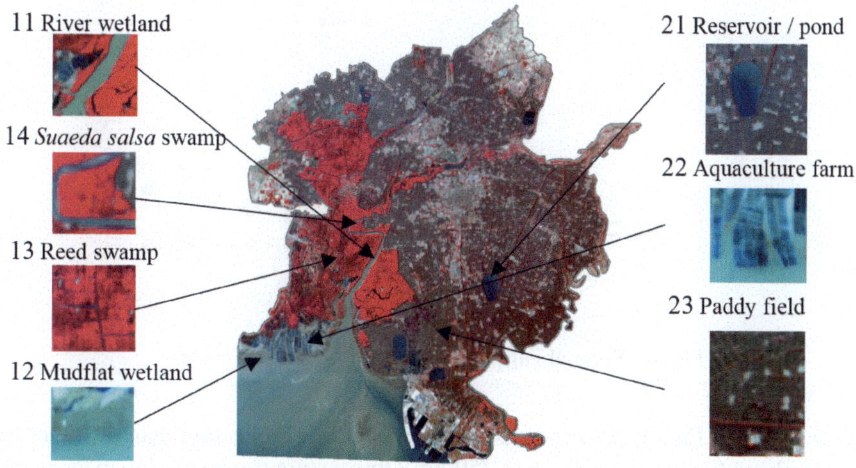

Fig. 3. Wetland classification system in Panjin city (Combination of 5/8/10 channel visible)

Table 2. Wetland classification system in Panjin city.

Primary classification	Code	Secondary classification	Explanation
Natural wetland	11	River wetland	Permanent, seasonal or intermittent rivers
	12	Mudflat wetland	An erosional zone of water between the coastal high tide and low tide
	13	Reed swamp	The dominant species in the community are the Mesozoic meadows of reeds and ponded swamps
	14	Suaeda salsa swamp	The dominant species in the community is Suaeda salsa, mainly distributed in floodplain and jiangxin continent
Artificial wetland	21	Aquaculture farm	Artificial wetlands for aquaculture purposes
	22	Paddy field	Plain paddy fields
	23	Reservoir/pond	Artificially artificial water storage areas

3.3 Wetland Information Extraction

According to the established wetland classification system and the characteristics of the study area, spectral characteristics of ground objects were analyzed. Spectral differences of water bodies, vegetation, and mudflats are obvious (Fig. 4), making them easier to classified. Using partition extraction [11], the study area is initially classified into water bodies, vegetation, and mudflats, then these three types are further classified separately.

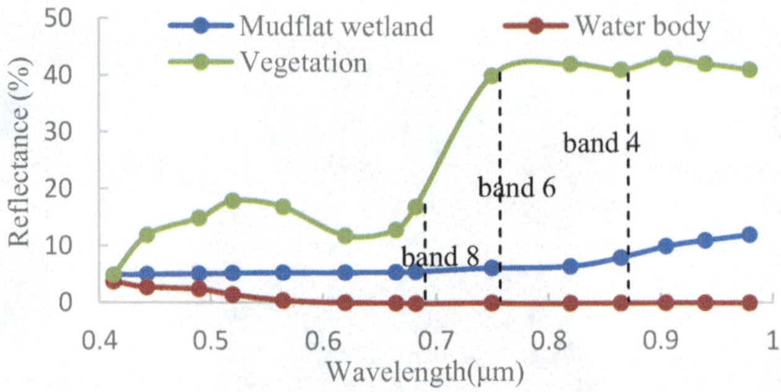

Fig. 4. Spectral curves of three kinds of ground objects

According to Fig. 4, water bodies have a much lower reflection than the other two classes in band 6 (central wavelength 750 nm); this feature can be used to extract water bodies in the study area [12]. Vegetation has a distinct absorption in the red band and strong reflectance in the near-infrared band. The vegetation index is designed by combining the reflectance of the red band and the near-infrared band. The Normalized Different Vegetation index (NDVI) is widely used, and its formula NDVI = (NIR-R)/(NIR + R) is calculated by using band 4 in the near-infrared bands and the red band 8 of the Tiangong-2 data. NDVI can effectively separate plant growing areas, mudflat, and buildings. Based on the threshold of band 6 and NDVI, the study area was classified into three major areas. A detailed classification of the three types was done separately according to the technical flowchart in Fig. 5.

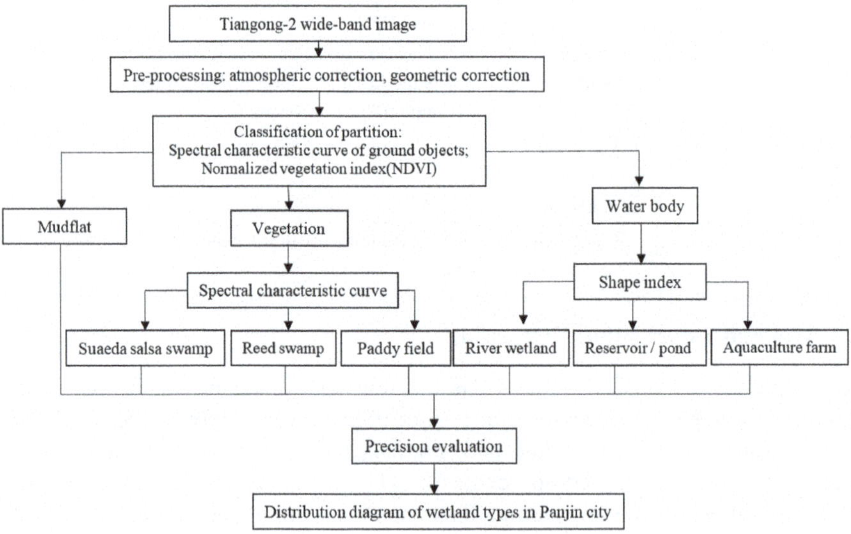

Fig. 5. Technology roadmap

Water body information extraction. Rivers, aquaculture farms, and reservoirs in the study area are the subjects of water classification. The spectral differences of water bodies are not significant, but their shapes are. For example, small rivers are curved strips and large lakes have regular geometric shapes. According to these features, the secondary classification of water can be carried out. The shape index [13] is an effective expressions of patch-shape features. The shape index indicates the degree of regularity of the patch. The more irregular the shape, the smaller the shape index is. The patch area of a water body can also effectively identify the type. For example, the patch area of some offshore waters is significantly larger than that of a pond. Therefore, confusing water body can be distinguished based on both shape index and patch area. Aquaculture farms can be extracted according to regular textures. Some rivers are discontinuous and cannot be identified by the shape index. In this regard, with the help of GIS tools, the preliminary classification of water body raster data is converted into vector data, and these water bodies are distinguished by visualization interpretation to improve the classification accuracy.

Extraction of vegetation information. The vegetation information includes coastal wetlands (including reeds, *Suaeda salsa*, and paddy fields) and some green spaces in the city center. The spectral characteristics of typical vegetation can be used for secondary classification. The spectral characteristic curves of *Suaeda salsa* and reeds are shown in Fig. 6. The two types of ground objects can be identified by using the peak value of reeds in the green band (band 10) and the valley value of *Suaeda salsa* in this band, as opposed to their peak value in the red band (band 8). Due to water information, the NDVI can be combined with the low reflectance feature of band 6 for paddy field extraction.

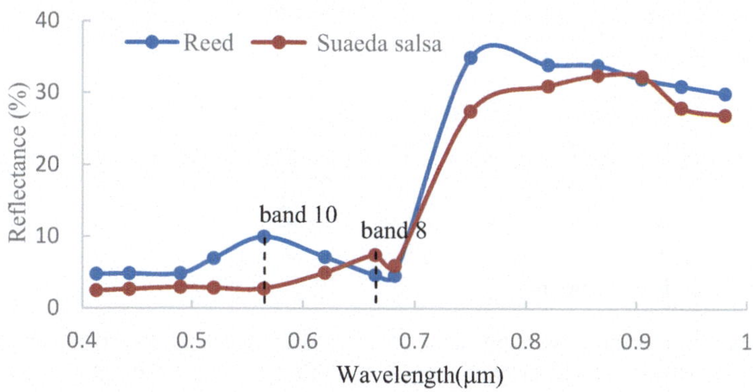

Fig. 6. Spectral curves of reed and Suaeda salsa

Beach information extraction. Since the spectral characteristics of mudflats and buildings are very similar, when the mudflat are extracted above a certain threshold of NDVI, the extracted information mixed with a lot of information of the building, and a small amount of water information. Mudflats in the study area are mainly distributed near the estuary, and have obvious spatial distribution characteristics. Combining these features, manual visual interpretation is performed to improve classification accuracy.

4 Classification Results and Accuracy Verification

4.1 Area and Spatial Distribution of Various Types of Wetlands

In 2018, wetlands occupied 3184.89 km^2 in the study area. As described in Table 3, paddy fields occupy the largest area, accounting for 55.29% of the total. In order of decreasing area, the others are reed swamp, mudflat wetland, *Suaeda salsa* swamp, aquaculture farm, reservoir/pond and river wetland.

Figure 7 also shows that Panjin city has a wide range of wetlands, accounting for about 80% of the entire study area. The regional resource development in Panjin city is centered on paddy field, reed fields, and aquaculture farms. Paddy fields are the most widely distributed, largely in the eastern part of the research area. Reed swamps are the main type of the wetland on the north shore of Liaodong Bay, distributed west of the Shuangtaizi River, with a small amount on the east side. The development of reed fields in this region is in line with ecological rules and has economic value. *Suaeda salsa* swamp are mainly distributed on both sides of river in the center of the study area. Mudflat wetlands are mainly distributed near the mouth of the Liaohe River. There are aquaculture farms along the coast, mainly because the estuary is suitable for this purpose. River wetlands mainly concentrated along the Shuangtaizi River in the center and the Daliao River in the southeastern part of the study area.

Table 3. Area and proportion of various types of wetlands in Panjin city in 2018.

	Natural wetland					Artificial wetland				Total
	River wetland	Mudflat wetland	Reed swamp	*Suaeda salsa* swamp	Subtotal	Aquaculture farm	Paddy field	Reservoir/pond	Subtotal	
Area (km^2)	51.94	147.10	971.08	120.86	1290.98	80.72	1760.86	52.33	1893.91	3184.89
Proportion (%)	1.63	4.62	30.49	3.79	40.53	2.54	55.29	1.64	59.47	100

4.2 Classification Accuracy

Because Landsat remote sensing data were widely used in the landuse/landcover classification, as well as with higher spatial resolution, wetland data imaged by Landsat 8 on May 5th, 2018 were used to evaluated the accuracy of Tiangong-2 data for wetland extraction, using the confusion matrix method [14]. The confusion matrix was established to calculate overall classification accuracy. In addition, the Kappa coefficient of classification accuracy was calculated for various types of wetlands. The size of the Kappa coefficient indicates the consistency of classification results for different wetland types and the spatial distribution of real objects. When the Kappa coefficient is less than

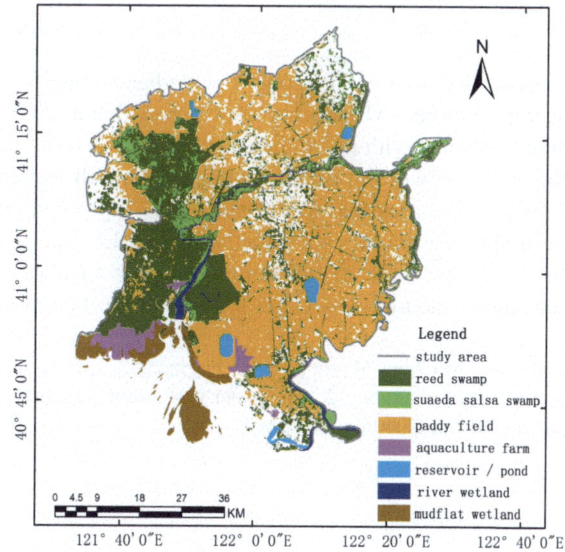

Fig. 7. Distribution of wetland types in Panjin city in 2018

0.4, the consistency degree is not ideal. When the Kappa coefficient is between 0.4 and 0.75, the degree of agreement between the two is ordinary. When the Kappa coefficient is greater than 0.75, it indicates a good consistency between the two [15].

Table 4 indicates that Tiangong-2 data are highly consistent with Landsat classification data. The accuracies of producer and user are mostly over 90%, indicating that the classification results are good. Among these, the extraction precision for paddy field is very high.

Table 4. Classification accuracy of all types of wetlands.

	River wetland	Mudflat wetland	Reed swamp	*Suaeda salsa* swamp	Aquaculture farm	Paddy field	Reservoir/pond	Total	User accuracy (%)
River wetland	649	0	0	0	0	0	20	699	97.01
Mudflat wetland	0	892	0	0	83	12	0	987	90.37
Reed swamp	0	31	2534	289	0	0	0	2916	88.79
Suaeda salsa swamp	0	0	186	1246	0	0	0	1432	87.01
Aquaculture farm	0	0	0	0	1894	45	0	1939	97.68
Paddy field	10	0	0	0	294	3588	0	3892	92.19
Reservoir/pond	64	0	0	0	0	0	678	742	91.37
Total	723	923	2720	1597	2271	3645	689	12577	
Producer accuracy(%)	89.76	96.64	93.16	81.17	83.40	98.44	97.13		
Overall accuracy = 89.89%, Kappa coefficient = 0.8748									

5 Conclusion

In this study, the application of the Tiangong-2 wide-band imaging spectrometer to wetland monitoring was explored. Monitoring results show that in 2018, Panjin city has a wide distribution of wetlands, with paddy fields mainly in the east, and large areas of reed swamp distributed in the western of the study area. Overall wetland classification accuracy using Tiangong-2 data reached 89.88%, and the kappa coefficient was 0.8748. Compared with the traditional classification method using Landsat data, the partition classification method improves the classification accuracy to a certain extent, and achieves remote sensing extraction of wetland information in Panjin City.

Acknowledgement. Thanks to China Manned Space Engineering for providing space science and application data products of Tiangong-2. This paper was jointly supported by National Key R&D Program of China (2017YFC0406004).

References

1. Liu, T., Liu X., Du J., Song K.: Pattern and change of Coastal Wetlands in the Liaohe river delta for 5 periods. Wetland Sci. **15**(4), 622–628 (2017)
2. Wang, Yu., Li, Z., Jia, Q.: Soil Freeze-thaw characteristics of Phragmites Australis marshes in Panjin. Wetland Sci. **14**(3), 295–301 (2016)
3. Tian, Q., Tian, H.: Study on historical process of the land subsidence in Panjin wetlands. Wetland Sci. **4**(5), 607–610 (2016)
4. Liu, H.Y., Lu, X.G., Liu, Z.Q.: Study on wetland resources and regional sustainable development in Liaohe Delta. Sci. Geogr. Sin. **20**(6), 545–551 (2000)
5. Huang, G., Zhang, J., Li, Y.: Classification and status analysis of Liaohe delta wetland. For. Resour. Manag. **4**, 51–56 (2000)
6. Junling, W., Ying, M.: Analysis of remote sensing monitoring of wetland change in Panjin city. Ecol. Environ. **8**, 67–69 (2009)
7. Guilin, H., Ping, H.: Study on wetland landscape evolution in Liaohe Delta–A case study of Panjin city. For. Resour. Manag. **3**, 82–89 (2011)
8. Official website for data sharing and release of manned space applications. http://www.msadc.cn/rwjstgeh/. Accessed 8 Nov 2018
9. Official website for United States Geological Survey. http://glovis.usgs.gov/. Accessed 9 Nov 2018
10. State Forestry Bureau: National Technical Regulations for Wetland Resources Surve, pp. 3–4 (2008)
11. Mingxiang, N., Gengxing, Z., Zunying, L.: Extracting of remote sensing information of wetland in Nansihu area based on multi-subarea and multi-layer techniques. Geogr. Geo-Inf. Sci. **20**(2), 45–48 (2004)
12. Lin, C., Gong, Z., Zhao, W.: The extraction of wetland hydrophytes types based on medium resolution TM data. Acta Ecol. Sin. **30**(23), 6460–6469 (2010)
13. Reif, M., Froh, R.C., Lane, C.R., Autrey, B.: Mapping isolated wetland in a Karst landscape: GIS and remote sensing methods. GIScience Remote Sens. **46**, 187–211 (2009)

14. Fangfang, L., Yonghong, J.: Study on method of extracting wetland and its changed area based on Landsat TM images. Sci. Surv. Mapp. **33**(2), 147–149 (2008)
15. Min, W., Xinhua, G.: The land use classification based on Landsat 8 remote sensing image—A case study of Anqu demonstration community in Honyuan County of Sichuan Province. Pratacultural Sci. **32**(5), 694–701 (2015)

Analysis of Flow Regime in the Turbidity Maximum Zone of Yangtze Estuary Based on Texture Features of Tiangong-2 Remote Sensing Images

Lizhi Teng[1], Heqin Cheng[1,2(✉)], and Yuanying Qiao[1]

[1] State Key Laboratory of Estuarine and Coastal Research, East China Normal University, Zhongshan N. Road 3663, Shanghai 200062, China
51163904014@stu.ecnu.edu.cn, hqch@sklec.ecnu.edu.cn
[2] Institute of Eco-Chongming (IEC), 3663 N. Zhongshan N. Road 3663, Shanghai 200062, China

Abstract. Two regions in the South Passage of the turbidity maximum zone (TMZ) of the Yangtze Estuary were selected for an analysis of the flow regime by remote sensing images. The remote sensing images of these regions obtained by the wide-band imaging spectrometer in Tiangong-2 in the maximum flood during spring tide in the dry season and in flood season were processed by a Gabor filter to get the texture images on the ocean surface. The flow direction was extracted by visual interpretation, and the semi-quantitative flow regime information was obtained by calculation of the texture entropy in the images. The results show that the texture information in the remote sensing images processed by this method is abundant. The streamline is clear, which can clearly reflect the changes in the flow regime in different seasons. The texture features of these images reflect the variation in the flow regime around Nanhuizui after the construction of the reclamation project in 2014. Remote sensing images of Tiangong-2 are feasible in the semi-quantitative interpretation of the flow regime in the TMZ with high suspended sediment concentration (SSC) in the estuary.

Keywords: Tiangong-2 · Texture features · Flow regime · Turbidity maximum zone · Yangtze Estuary

1 Introduction

Estuarine flow regime is complicated. Compared with physical models, numerical simulations, analytical simulations, and other traditional methods, satellite remote sensing inversion can provide synchronous continuous macroscopic real-time data of the estuarine flow pattern. Dagestad et al. [1] used synthetic aperture radar (SAR) data from 2007 to 2011 to analyze the flow distribution on the ocean surface. Ding et al. [2] used optical remote sensing images combined with GIS technology to study the hydrodynamics and sediment transport in the Hanjiang Estuary. Yun et al. [3]

Y. Gu et al. (Eds.): *Proceedings of the Tiangong-2 Remote Sensing Application Conference*, LNEE 541, pp. 312–322, 2019.
https://doi.org/10.1007/978-981-13-3501-3_29

qualitatively described the flow distribution and sediment distribution on the surface of the Yangtze Estuary by Landsat satellite images. Fluid information obtained from optical remote sensing images is more economical and feasible. However, it is a qualitative description, whereas development towards a more quantitative and semi-quantitative description is necessary.

Texture is one of the inherent characteristics of an image, which is shown by the variation in color or gray in the images. It can provide spatial distribution information of structure in the images, e.g., depth and surface of the object [4]. The remote sensing images of the estuarine area normally display a significantly linear texture, which is the flow trace left by the surface sediment in the river when it moves along with the flow [5]. Therefore, the flow line of surface water can be analyzed by interpreting and identifying the texture structure of the remote sensing image of surface water with high SSC in an estuary. Until now, very few tools have been effectively applied to this issue.

Gabor filter is widely used in texture analysis because it is similar to a biological vision system, easy to tune direction and radial frequency bandwidth [6–8]. A Gabor filter is a useful tool for interpreting texture features of remote sensing images to obtain flow regime information in an estuary [5].

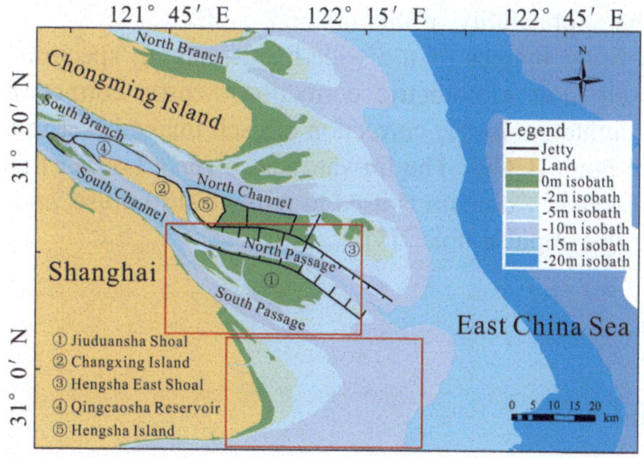

Fig. 1. Sketch map of Yangtze River Estuary.

The Yangtze Estuary is an example of case II water [9]. The SSC in the TMZ of the Yangtze Estuary is high, and the movement of sediment is active. The water color changes significantly in this area. The texture information interpreted by the remote sensing image is more abundant, which can fully reflect the movement of suspended sediment in the Yangtze Estuary with the water. In recent years, a series of major projects, such as the reclamation of Pudong International Airport, reclamation of Nanhui East shoal, and regulation of Deep Waterway Project, have greatly changed the hydrodynamic environment of the Yangtze Estuary [10]. Therefore, taking two typical regions in the South Passage of the Yangtze Estuary as an example (the regions in the red box in Fig. 1), the Gabor filter is used to process the remote sensing images obtained by the wide-band imaging spectrometer in Tiangong-2 to obtain the macroscopic texture image, and texture features are statistically analyzed to obtain the semi-quantitative flow information. The feasibility of this method is assessed by comparing the flow regime in-formation and historical data obtained from remote sensing images of the Landsat TM inversion under the same method, so that the remote sensing images obtained by the wide-band imaging spectrometer in Tiangong-2 can offer a reference for the study of hydrodynamic and geomorphological changes brought by engineering construction in the estuary and coastal areas.

2 Data Sources and Preprocessing

To reflect the characteristics of the flow regime in this study area in different seasons, the visible-near-infrared remote sensing images of the Yangtze Estuary obtained by the wide-band imaging spectrometer in Tiangong-2 on January 22, 2017 (maximum flood during spring tide in dry season) and July 26, 2017 (maximum flood during spring tide in flood season) were selected. Compared with the visible-near-infrared remote sensing satellite image of the Yangtze Estuary obtained by the near-infrared band TM4 of Landsat TM on December 3, 2013 (maximum flood during spring tide in dry season), the variation in the flow regime around Nanhui after the new reclamation in 2014 is analyzed to test the applicability of this method.

The remote sensing images of the wide-band imaging spectrometer used in this study are processed after radiometric correction, sensor correction, and geometry correction. The accurate geometric correction of each image is carried out according to the ground control point (GCP). The forward and inverse Fourier transform (FFT) filter, based on the frequency domain is used to remove strip noise, and the median filter is used to remove the impulse noise [5].

3 Methods

3.1 Gabor Filter

Use of the Gabor filter is one of the common methods in texture extraction [4], which uses the multi-scale and multi-direction characteristics of the Gabor wavelet to analyze images and extract texture features. Filtering texture information is obtained through controlling the center frequency and bandwidth, based on the theory that texture is a narrow-band signal, and each filter only allows the texture corresponding to its frequency to pass through, while other texture energy is suppressed [11].

The commonly used even-symmetric 2-D Gabor filters can be expressed as

$$\mathrm{h(x, y)} = \frac{1}{2\pi\sigma_\mu\sigma_v} exp\left[-\frac{1}{2}\left(\frac{\mu^2}{\sigma_\mu} + \frac{v^2}{\sigma_v}\right)\cos(\omega\mu)\right] \tag{1}$$

$(\mu = \mathrm{x}\cos\theta + y\sin\theta\ ,\ \mathrm{v} = -\mathrm{x}\sin\theta + y\cos\theta)$

In this expression, h(x, y) is the pixel value of point (x, y) on the filtered image; θ is the direction of the Gabor filter; μ is the direction of the axis perpendicular to θ; σ_μ, σ_v are the standard deviations of Gaussian envelope, used to control the size of the filter template; and ω controls the frequency of the filter [12].

In this study, this method is implemented in MATLAB. The best filter image is obtained by changing the direction angle of the filter.

3.2 Interpretation and Calculation of Flow Direction

The linear texture of estuarine water can be regarded as the flow path trace of surface sediment, i.e., the instantaneous streamline. The flow direction of each position at that time can be described as the tangent of the texture according to the Euler method. The remote sensing images filtered by Gabor are imported into ArcGIS, and the streamline is drawn along the linear texture with high luminance value. The direction is calculated by the layer calculator in ArcGIS according to the coordinates of the first, middle, and last points shown in the attribute table. The calculation principle is as follows:

$$\alpha = 90^\circ - \tan^{-1}[(y_2 - y_1)/(x_2 - x_1)] \times 180/\pi \tag{2}$$

$$\beta = 270^\circ - \tan^{-1}[(y_2 - y_1)/(x_2 - x_1)] \times 180/\pi \tag{3}$$

In this expression, α indicates the direction of the ebb streamline; β indicates the direction of the flood streamline; (x_1, y_1) are the geographical coordinates of the current point; and (x_2, y_2) are the geographical coordinates of the next point.

3.3 Semi-quantitative Analysis of Flow Velocity Based on Texture Entropy

The entropy of information reflects the amount of information in the image, which is an important feature in texture analysis. The larger the entropy is, the more texture is in the image. In turbulent flow, the current velocity is high, the thin and dense texture in the image is widespread, and the texture entropy is large. In laminar flow, there is lower current velocity, less texture in the image, and lower texture entropy. Therefore, it could be inferred that the magnitude of the current velocity is positively correlated with the entropy value of the texture image. Because of the complexity of the flow regime in the estuary, texture features are random, and the texture entropy of the flow regime is analyzed by a statistical analysis method based on a gray-level co-occurrence matrix [13].

The gray-level co-occurrence matrix is a spatial correlation matrix of gray tone, which is composed of the joint probability density P (*i*, *j*, d, θ) of the gray level Z (*i*, *j*). The correlation coefficients between a pixel and its specific neighborhood are reflected by the probability of transition from one gray level to another between two pixels in a specific direction (θ) and a specific distance (d). This matrix reflects the comprehensive information of the image in the interval, direction, amplitude of change, and speed. Therefore, a texture filter such as mean, variance, synergy, or information entropy

(ENT) can be carried out in ENVI accordingly. The calculation principle is as follows [14]:

$$\mathrm{ENT} = -\sum_{0<j\leq n}^{0\leq i\leq m} P(i,j)\log P(i,j) \tag{4}$$

In this expression,$P(i,j)$ is the value of the i row and the j column of gray-level co-occurrence matrix, and the m and n are the total number of rows and columns of the gray-level co-occurrence matrix.

4 Processing Results and Discussion

According to the quality of reflection of SSC in the pre-processed images, the V7 channel (0.6775–0.6875 μm) was chosen to analyze the texture of remote sensing images. Taking the remote sensing image of the Yangtze Estuary obtained by the wide-band imaging spectrometer in July 26, 2017 as an example, the texture image in the directions of 2π/3, 3π/4, 4π/5, 5π/6, 6π/7, and 7π/8 was obtained by changing the direction angle θ of the Gabor filter in MATLAB (see Fig. 2).

The filtered images in the 2π/3 and 7π/8 directions show less texture information because of the cancellation of the sine and cosine terms in these directions. Meanwhile, the filtered images in the 5π/6 direction is more suitable for interpreting the streamline, because of the clear texture and significant information of the remote

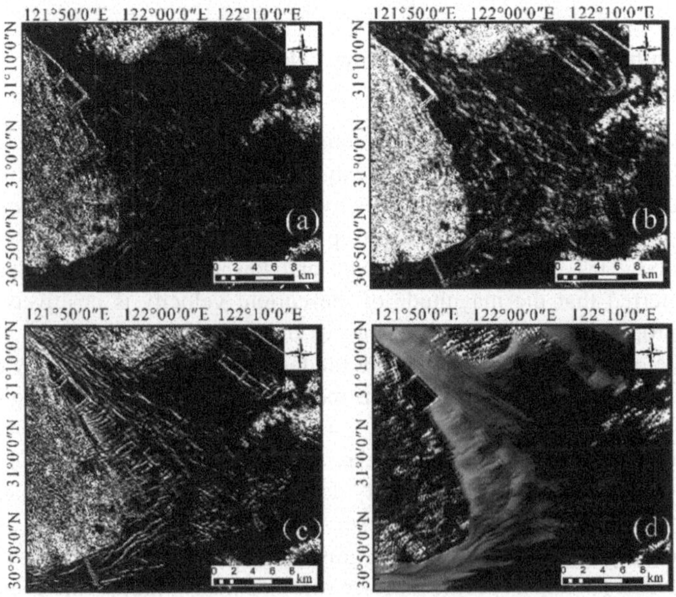

Fig. 2. Images processed by Gabor filters in the (a) 2π/3, (b) 3π/4, (c) 5π/6, and (d) 7π/8 directions. The source remote sensing image of the South Passage of the Yangtze Estuary was obtained by the wide-band imaging spectrometer in Tiangong-2 on 07/26/2017.

sensing image. Therefore, we choose this direction to process the remote sensing images of the typical regions.

The linear texture of the remote sensing image processed by the Gabor filter is essentially the flow path trace left by the surface sediment when it moves with the current. Therefore, the higher the SSC, the more abundant the texture information is. In the Yangtze Estuary, because of the large quantities of sediment transported from upstream, the SSC in the flood season is significantly higher than that in the dry season [15], which makes the texture information obtained in flood season more abundant in two observation areas (see Figs. 3 and 5).

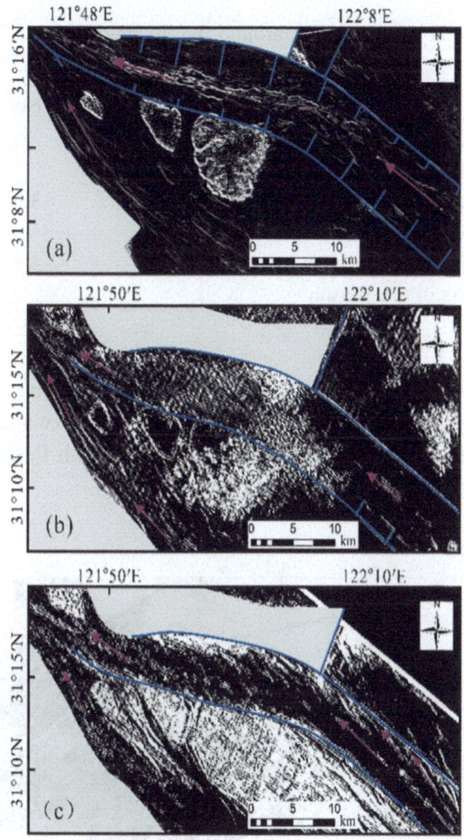

Fig. 3. Flow regime around the diversion point of the South and North passage (a) in 2013, maximum flood during spring tide in dry season, (b) in 2017, maximum flood during spring tide in flood season, (c) in 2017, maximum flood during spring tide in dry season

4.1 Flow Regime in Upper Section of South Passage

With the construction of the deep waterway channel, the flow regime in the upper section of the South Passage has significantly changed. The flood tide moves northward from 310° to 330° along the shoreline and meets the flood tide from the North Passage at the front of the submerged breakwater. Due to the beam action of the double levees and spur dikes in the deepwater channel, the streamlines in the center of the North Passage channel are concentrated, and the flow direction is about 309°, which is consistent with the trend of the deepwater channel. The results are similar in different seasons, and they are consistent with the results analyzed by Landsat TM (see Table 1 for details). In the maximum flood during spring tide in flood season (Fig. 3b), a flood current flows in the deep waterway channel over the northern levee with a direction of 325°, which is caused by the angle between the direction of the channel and the flow direction [16, 17].

Table 1. The flood flow direction around the diversion point of the South and North passage

	South passage (°)	North passage (°)
Landsat TM, maximum flood during spring tide in dry season	325	286
Tiangong-2, maximum flood during spring tide in flood season	330	309
Tiangong-2, maximum flood during spring tide in dry season	327	300

Figure 4 is the image of texture entropy distribution obtained through texture analysis in this study area. Around the submerged breakwater, the flood current velocity in the South Passage is higher than that in the North Passage, which is similar to the previous numerical simulation results [18, 19].

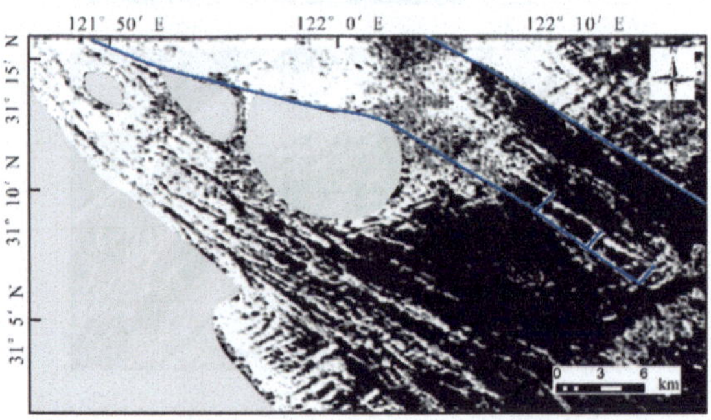

Fig. 4. The image of texture entropy distribution around the diversion point of the South and North Passage (07/26/2017)

4.2 Flow Regime in Nanhuizui

With the reclamation of the Nanhui East shoal, the nearshore water in Nanhui is affected by the land boundary. The flood current is divided into the northern and southern branches near Nanhuizui.

Affected by the reclamation in 2014, there is no clear texture in the reclamation area where the flow pattern is disordered, while the flow regime in the central area of the channel is less affected by the project (see Table 2 for details), which is similar to the result from the analysis of the remote sensing image obtained by Landsat in 2013, as the construction of the reclamation of the Nanhui East shoal straightens the shoreline of Nanhui and the flow path [20]. In the flood season, the north branch of the flood tide enters the south passage of the Yangtze Estuary at about 310° (Fig. 5b), and it enters the Yangtze Estuary in a direction of 320° in the dry season (Fig. 5c). At the same time, the flow regime of south of the diversion point is less affected by the reclamation project, the southern branch of the flood tide enters the Hangzhou Bay at 230° in the flood season (Fig. 5b) and in the direction of 250°. In the dry season (Fig. 5c), t which is consistent with the tidal current direction of Hangzhou Bay [21]. The reclamation project has not changed the location of the diversion point.

Table 2. The flood flow direction around Nanhuizui

	The north edge of Nanhuizui (°)	The south edge of Nanhuizui (°)
Landsat TM (maximum flood during spring tide in dry season)	310	275
Tiangong-2 (maximum flood during spring tide in flood season)	320	250
Tiangong-2 (maximum flood during spring tide in dry season)	310	230

In the distribution image of the texture entropy (Fig. 6), the flow in the South Passage decreases from the diversion point of the South and North Passage to the entrance gradually, and the flood flow velocity in the southern edge of Nanhuizui is generally larger, which is basically consistent with the interpretation of the Landsat TM image and the results of previous numerical simulations [5].

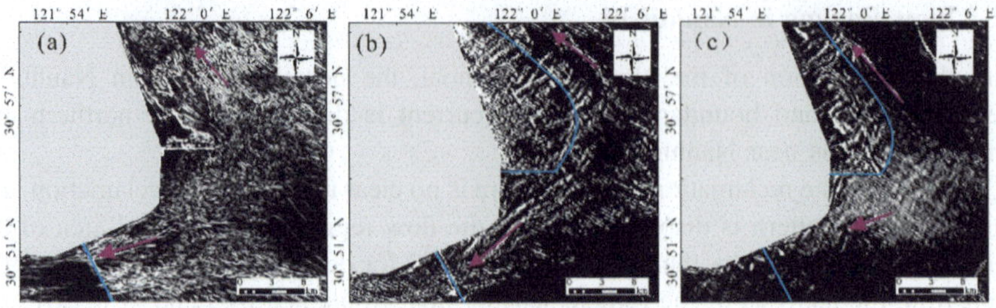

Fig. 5. Flow regime around Nanhuizui (a) in 2013 maximum flood during spring tide in dry season, (b) in 2017 maximum flood during spring tide in flood season, (c) in 2017 maximum flood during spring tide in dry season

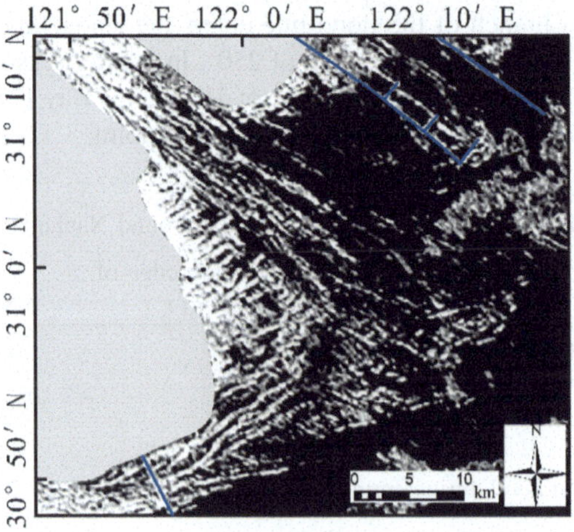

Fig. 6. The image of texture entropy distribution around Nanhuizui (07/26/2017)

5 Conclusion and Suggestion

Taking the South Passage of the Yangtze Estuary as an example, the flow regime around the diversion point of the South and North passage and Nanhuizui are semi-quantitatively interpreted, based on the texture features of the water surface in the images obtained by the wide-band imaging spectrometer of Tiangong-2.

The remote sensing images processed by a Gabor filter are informative, and streamlines obtained from these images are clear. The analysis results are consistent with actual measurement data.

The texture features can reflect the changes of flow regime in different season. Determined by the characteristics of the Yangtze Estuary where the SSC is higher in the flood season than the dry season, the texture information obtained in the flood season is more abundant than that in the dry season.

After the construction of the reclamation project in 2014, the flow regime in the upper section of the South Passage has not changed much. There is no clear texture in the reclamation area where the flow regime is disordered. The central area of the channel is less affected by the project, which is similar to the results of the analysis of historical remote sensing images obtained by Landsat.

This demonstrates that it is feasible to apply remote sensing images of the wide-band imaging spectrometer in Tiangong-2 to the semi-quantitative interpretation of the flow regime in the TMZ of an estuary with high SSC. It is suggested that the observation frequency and resolution of the imager should be improved in the future, and the application of this method in the TMZ should be further developed by combination with observation data from the three-dimensional imaging microwave altimeter. This will make a significant contribution to the construction and maintenance of major projects and the monitoring of strategic resources and ecological environments, such as water and soil.

Acknowledgements. We thank China Manned Space Engineering for providing the wide-band imaging spectrometer data products of Tiangong-2, and the National Nature Science Foundation of China-The Netherlands Organization for Scientific Research-Engineering and Physical Sciences Research Council (NSFC-NWO-EPSRC) (51761135023), and the China Geological Survey (DD20160246) for support.

References

1. Fernández-Prieto, D., Sabia, R.: Remote Sensing Advances for Earth System Science. Springer, Heidelberg (2013)
2. Ding, X.Y., XU, X.X.: A remote sensing analysis of characteristics of suspended sediments movement in Hangjiang estuary. Remote. Sens. Land Resour. **19**(3), 71–73 (2007)
3. Yun, C.X.: Recent Developments of the Changjiang Estuary. China Ocean Press, Beijing (2004). Chinese
4. Li, Y., Meng, X.P.: Image texture feature detection based on Gabor filter. J. Chang. Univ. Technol. (Natural Science Edition), **29**(1), 78–81(2008). Chinese
5. Qiao, Y.Y., Cheng, H.Q., Song, Z.K., et al.: Research on information semi-quantitation of estuarine flow pattern based on TM remote sensing image. Bull. Soil Water Conserv. **34**(5), 118–123 (2014). Chinese
6. Ulaby, F.T., Kouyate, F., Brisco, B., et al.: Texture features for browsing and retrieval of image data. IEEE Trans. PAMI **18**(8), 837–842 (1996). Chinese
7. Lei, Y.S., Wang, S.M., Qin, R.: Edge information extraction algorithm of CT cerebrovascular medical image based on imaginary part of Gabor filter. J. Tianjin Univ. **40**(7), 833–838 (2007). Chinese
8. Clausi, D.A., Jernigan, M.E.: Designing Gabor filters for optimal texture separability. Pattern Recogn. **33**(11), 1835–1849 (2000)
9. Ma, H.: The optical characteristics of the three elements in high turbidity water of the Yangtze River Estuary and its influence on the inversion of TSM, East China Normal University, Shanghai (2015). Chinese
10. Xie, H.L., Dai, Z.J., Zuo, S.H.: Morph dynamic processes of the south passage of the Yangtze Estuary (1959–2013). Ocean. Eng. **33**(5):51–59 (2015). Chinese
11. Yv, J., Guan, Z.Q.: Research on transformation method of target texture feature description under different imaging conditions. Bull. Surv. Mapp. (s1), 375–379 (2012). Chinese

12. Chen, X.G., Feng, J.F.: Fast Gabor filtering. Acta Autom. Sin. **33**(5), 456–461 (2007). Chinese
13. Li, S., Ge, Y., Li, D.Y.: Visualizing presentation of the attribute uncertainty in classified remotely sensed imagery. Remote. Sens. Land Resour. **18**(2):20–25 (2006). Chinese
14. Wang, H.M., Shi, P.: Methods to extract images texture features. J. Commun. Univ. China Sci. Technol. **13**(1), 49–52 (2006). Chinese
15. Shen, H.T., Pan, D.A.: Turbidity Maximum in the Changjiang Estuary. China Ocean Press, Beijing (2001). Chinese
16. Chen, W., Li, J.F., Li, W.H.: Recent suspended sediment transport in bifurcation area of north and south passage of the Yangtze Estuary. Resour. Environ. Yangtze Basin **22**(7), 865 (2013). Chinese
17. Fan, Z.Y.: The effect of deep waterway project on current and salinity in Changjiang Estuary. East China Normal University, Shanghai (2011). Chinese
18. Chen, W., Gu, J., Qin, X., et al.: Numerical analysis of the sediment deposition in the upper reach of the deepwater navigation channel in the Changjiang River Estuary. Chin. J. Hydrodyn. **27**(2), 199 (2012). Chinese
19. Pan, L.Z.: Impacts of deep waterway project on morphological change within the north passage of Changjiang Estuary, East China Normal University, Shanghai, China (2012). Chinese
20. Cao, Y., Zhu, J.Z.: Discuss on hydrodynamic influence of the Nanhui reclamation in Yangzi Estuary. National Congress on Hydronamics, Shanghai (2003). Chinese
21. Huang, G., Chen, S., Zhang, G.A.: Characteristics of water and sediment movement in Nanhui side beach and its influence on reclamation project Yangtze River. **38**(1), 60–63 (2007). Chinese

Preliminary Study of Oceanic Eddies Detected by Tiangong-2 Interferometric Imaging Radar Altimeter

Yuhang Wang[1,2,3], Weiya Kong[1,2,3], and Jinsong Chong[1,2(✉)]

[1] National Key Laboratory of Science and Technology on Microwave Imaging, Beijing 100190, China
iecas_wang@126.com, iecas_chong@163.com
[2] Institute of Electronics, Chinese Academy of Sciences, Beijing 100190, China
[3] School of Electronics, Electrical and Communication Engineering, University of Chinese Academy of Sciences, Beijing 100190, China

Abstract. Oceanic eddy, which plays an important role in ocean thermal cycling, is a significant branch of oceanic scientific research. Mesoscale eddies can be observed by multiple sensors. However, sub-mesoscale or small scale eddies cannot be detected by conventional radar altimeters. The Interferometric Imaging Radar Altimeter (InIRA) on Tiangong-2 provides a large amount of images for the observation and investigation of sub-mesoscale or small scale eddies. 100 InIRA images obtained during 2016–2018 are used to investigate possible mechanisms of the appearance of eddies in InIRA images. Result shows that under a moderate wind speed of 3–5 m/s, film mechanism is the main form of oceanic eddies and under a considerable near-water wind speed of 5–12 m/s, and the wave/current interaction mechanism comes into action. The interferometric data of InIRA are processed to detect eddies after a series of procedures. It is shown that the phase change is corresponding to ocean surface height anomaly induced by eddy. Moderate Resolution Imaging Spectroradiometer (MODIS) sea surface temperature (SST) and chlorophyll-a (CHL) data are used to verify eddies identified from InIRA images. Result shows that the center of eddy with lowest SST corresponds to the highest CHL concentration.

Keywords: Tiangong-2 · Oceanic eddies
Interferometric imaging radar altimeter · Interferometric phase

1 Introduction

Oceanic eddy plays a significant role in the interactions among physics, biology and biogeochemistry of the ocean. The movement of eddies will agitate seawater and expand the scope of biological organisms, thus affecting the distribution of organics and the transportation of heat and salt in the ocean [1]. Therefore, oceanic eddies can be observed by multiple sensors, such as radar altimeter, optical/infrared sensor and synthetic aperture radar (SAR). Mesoscale eddies with horizontal scales more than 100 km can be observed by traditional radar altimeters using sea surface height anomalies [2]. However, sub-mesoscale or small scale eddies with diameters ranging

Y. Gu et al. (Eds.): *Proceedings of the Tiangong-2 Remote Sensing Application Conference*, LNEE 541, pp. 323–332, 2019.
https://doi.org/10.1007/978-981-13-3501-3_30

1 km to 100 km cannot be detected, they can only be observed by high-resolution optical sensors or SAR. SAR is capable of acquiring data all-day and all-weather, while optical sensors are vulnerable to clouds and weather. Conventional SAR can only provide two-dimensional surface features, leaving out sea surface anomalies of eddies unmeasured. Interferometric synthetic aperture radar (InSAR) has ability to measure ocean surface topography with high-resolution and its wide swath is an advantage to observe oceanic eddies.

InSAR has been used to measure topography in Shuttle Radar Topography Mission (SRTM) [3]. Topography change is detected by InSAR altimetry via the phase of two SAR complex images in the same region [4]. Due to the stochastic movement, the two SAR complex images of ocean surface should be acquired simultaneously. Besides, ocean surface topography altimetry demands centimeter-level precision, so the InSAR parameters need to be adapted for the strict requirement, and the near-nadir incidence is the most unique feature in the existing InSAR systems [5].

The Interferometric imaging radar altimeter (InIRA) is an experimental payload on Tiangong-2, which was launched on September 15th 2016 [5]. The incidence of InIRA is 1°–8°, so the swath width achieve more than 40 km. The incidence of conventional SAR is 20°–60°, determining that the received signals are mostly backscattered from ocean surface [6]. However, under 1°–8° incidence, the reflected signal will dominate, which makes the data of InIRA is different from other InSAR systems.

The following observational data are used in the present study: InIRA single look complex (SLC) images, Moderate Resolution Imaging Spectroradiometer (MODIS) sea surface temperature (SST) and chlorophyll-a (CHL) concentration data. The SLC images are used for eddy imaging mechanism analyzation and eddy detection. The SST and CHL concentration data are used to verify eddies identified from the InIRA SLC images. To our knowledge, this is the first time that eddies has been studied by simultaneously acquired InIRA, SST and CHL data.

The paper is organized as follows: In Sect. 2, we analyzed eddies imaging mechanism using 100 InIRA images and divided eddies into two types. Section 3 is devoted to the processing of InIRA data and realizing the detection of eddies. In Sect. 4, we present SST and CHL data to verify the eddy identified from one InIRA image and discuss its characteristics. In Sect. 5, we come to some conclusions and give several suggestions on how to detect different scale eddies in the future.

2 Mechanisms of Eddy Structure Visualization in InIRA Images

In order to study the specific features of eddy structures in InIRA images, we analyzed altogether 100 images of different water areas over the period from 2016.9 to 2018.7. The corresponding wind data is acquired from European Centre for Medium-Range Weather Forecasts (ECMWF). Eddy structures appearing in InIRA images can be divided into two different mechanisms.

At moderate wind speeds of 3–5 m/s, eddies usually appear in InIRA images due to the surface films over the sea surface. Such surface films alter sea surface tension and gravity-capillary waves by smoothing ripples and cause the diminishing of normalized

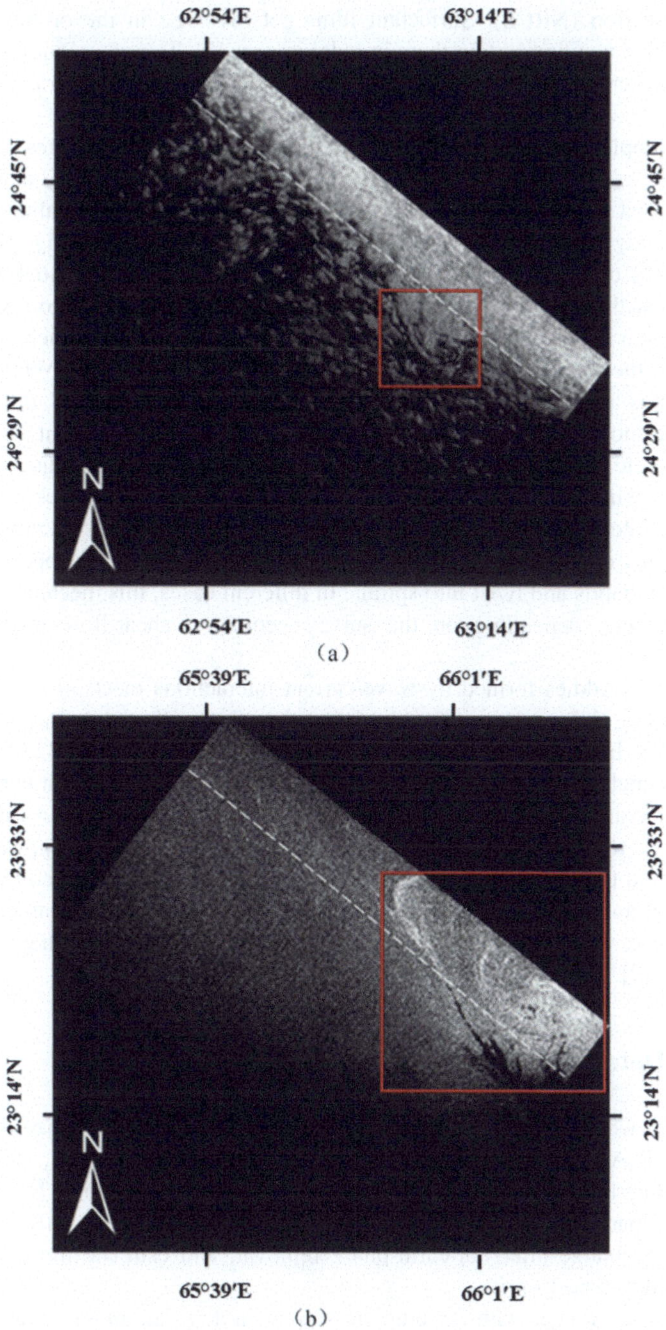

Fig. 1. Manifestation of eddies formed by film mechanism in InIRA images obtained on (a) November 11, 2017 at 00:02:23 UTC and (b) November 8, 2017 at 01:19:36 UTC. The white line is an intensity boundary around 4° incidence.

radar cross section (NRCS). Surfactant films get involved in the orbital motion of eddies, in such a way imprinting them in radar images [7]. Therefore, eddies formed by surface films generally manifest themselves as dark, thin and concentric slicks on the sea surface.

Some examples of eddies formed by film mechanism in InIRA images are shown in Fig. 1. Figure 1a was obtained on November 11, 2017 at 00:02:23 UTC in the north of the Arabian Sea. The red box highlights a small scale eddy with diameter of about 25 km. The eddy has a spiral arm and manifest as dark thin slicks. In Fig. 1b, the same water area was captured by InIRA on November 8, 2017 at 01:19:36 UTC. A mesoscale eddy with diameter of about 55 km is discernible in the red box. As shown in Fig. 1, an intensity boundary can be found around 4° incidence, which is shown as white line. In the near range, the films is brighter than adjacent area. While in the far range, the films are shown as darker features. When incidence is 1–4°, due to a small surface slope induced by ocean eddy, the radar signals reflect straight back into the receiver. Therefore, films under the incidence 1–4° is brighter than adjacent area.

At a high wind speed of 5–12 m/s, eddies appear in InIRA images as a result of wave/current interactions in the zones of current shear and manifest themselves in the form of regions of increased NRCS. Eddies of this type generally appear as several bright, narrow bands and twist into spirals. In different cases, this mechanism works in different manners, depending on the survey geometry, shear-flow intensity, wind velocity, etc. [8].

Examples of eddies formed by wave/current interactions mechanism are provided in Fig. 2. In Fig. 2a, a mesoscale eddy detected in the northeastern Bay of Bengal is presented. This InIRA image was obtained on October 7, 2017 at 06:32:05 UTC. The eddy with diameter of about 58 km has an eye center and several bright narrow bands. In Fig. 2b, a couple of eddies in eastern of Hokkaido is visualized in InIRA image obtained on October 9, 2017 at 06:19:37 UTC. The diameter of the upward eddy is about 9 km and the other is about 10 km. The upward eddy was formed by west–east Ocean current and the downward one was generated by east-west Ocean current. Due to the couple of eddies' well-formed patterns winding into a spiral, they were called spiral eddies [9].

3 Eddy Detection

In this paper, we will use InIRA SLC data from Tiangong-2 to realize detection of eddy. The InIRA SLC data processing usual include three main steps, including registration, flat-earth phase removal and system parameters calibration [10]. Since the calibration of parameters of InIRA has not been proceeded yet, we only focus on the relative phase change after flat-earth-phase removal, and estimate the existence and location of eddy based on it.

The intensity image with an eddy is shown in Fig. 3a. In the near range, the intensity is stronger than the surrounding area. The registered phase of two InIRA complex images is shown in Fig. 3, and the whole ocean area is covered by the flat-earth phase.

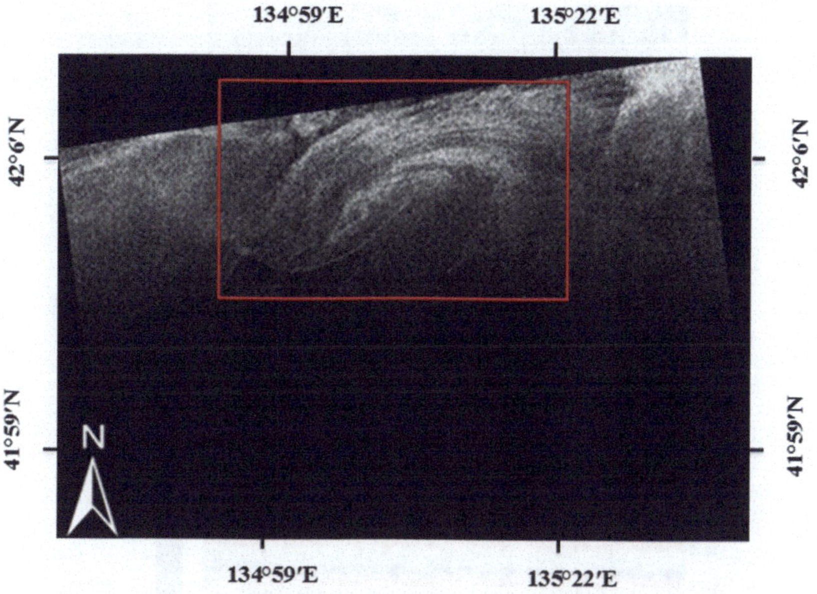

(a)

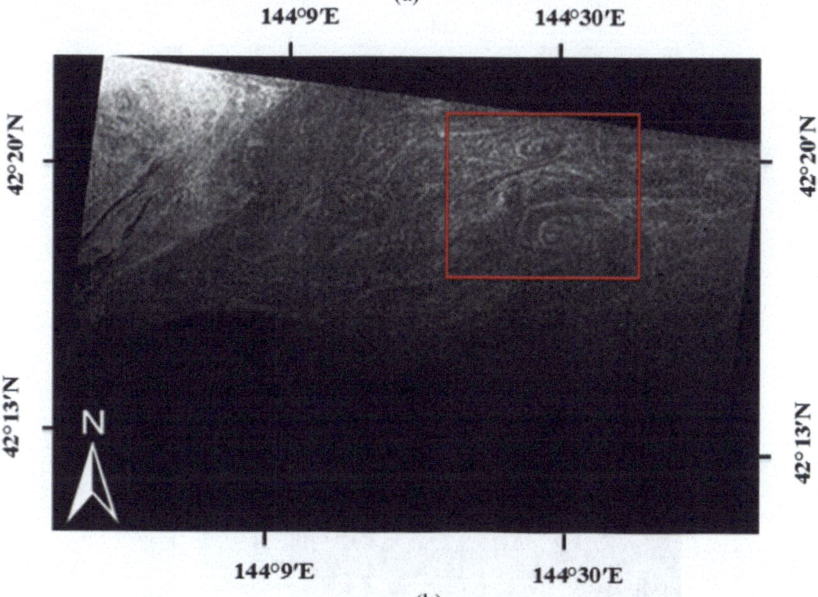

(b)

Fig. 2. Manifestation of eddies formed by wave/current interactions mechanism in InIRA images obtained on (a) October 7, 2017 at 06:32:05 UTC and (b) October 9, 2017 at 06:19:37 UTC.

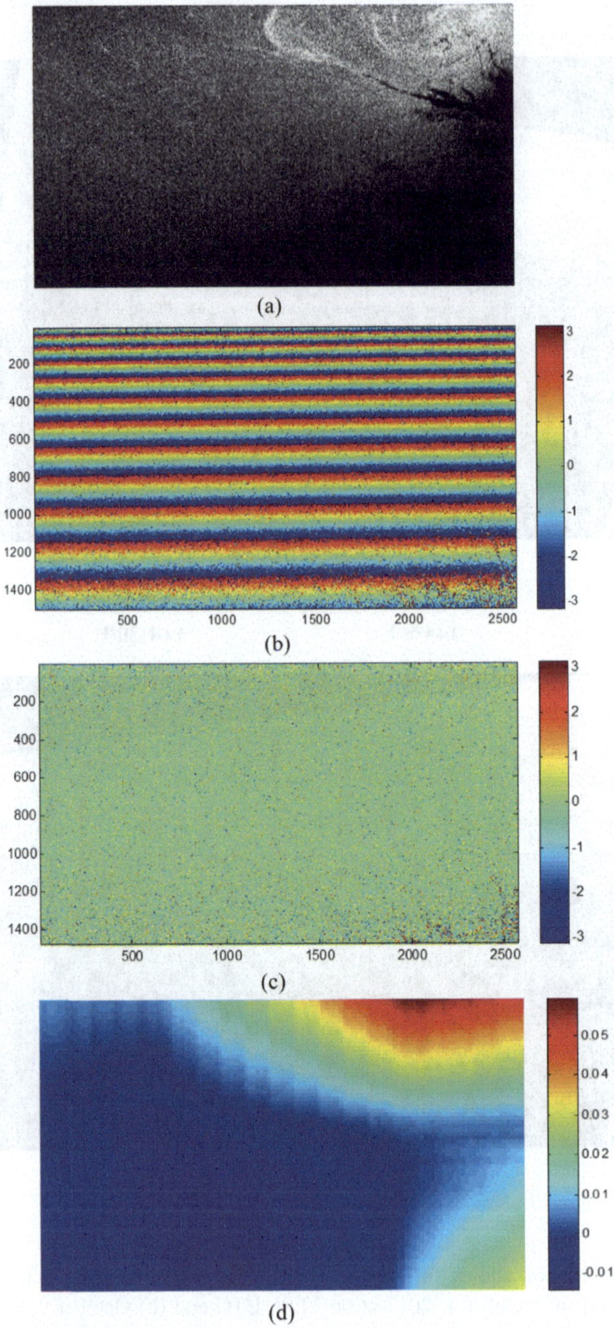

Fig. 3. (a) InIRA Intensity image with an eddy; (b) registered phase of two InIRA complex images; (c) interferometric phase after removing the flat-earth phase; (d) multi-looked interferometric phase.

To acquire the relative phase, the flat-earth phase need to be removed. The flat-earth phase can be calculated by [11]

$$\varphi = \frac{2\pi B \sin(\theta - \alpha)}{\lambda} = \varphi_{fringe} + \varphi_0 \tag{1}$$

where B is the baseline length, θ is incidence angle, α is baseline tilt angle and $\varphi_0 = 2\pi n$ is the wrapped phase. φ_{fringe} is the phase fringe obtained from actual data, as shown in Fig. 3b. Since the ocean surface level height change in local area is minor, its corresponding phase is very small.

A multi of azimuth line of the sea surface in the phase map was selected and averaged. After the flat-earth phase is unwrapped, the phase curve is obtained. Then the phase curve fitting is performed with Eq. (1) as the fitting formula. The combination of parameters $(B, \alpha, \theta, \varphi_0)$ is changed constantly, and the optimal parameter combination is obtained on the basis of the minimum RMSE. The parameters are calibrated by Eq. (2). Therefore, the nominal flat-earth phase of the overall image is calculated and eliminated [5].

$$\overset{\wedge}{(B, \alpha, \theta, \varphi_0)} = \arg\min[\varphi_{fringe} - \varphi(B, \alpha, \theta, \varphi_0)] \tag{2}$$

The interferometric phase after removing the flat-earth phase is shown in Fig. 3c. The phases have small changes and tend to be near zero among the ocean area. This is because under the InIRA system parameters, the interferometric phase corresponding to the 10 cm sea level change is only 0.01–0.001 rad from the near end to the far end of the swath, while the phase range is $[-\pi, \pi]$ in Fig. 3c. Since the dynamic range of the interferometric phase is too large and the phase noise is obvious, even if there is a certain fluctuation of the sea surface, the elevation phase is difficult to be highlighted on the interferometric phase map at this spatial resolution.

The interferometric phase are multi-looked to reduce the phase noise, as shown in Fig. 3d. The range of phase reduces obviously and the phases associate to ocean surface topography finally emerges on the 5 km grid. By comparing with Fig. 3a, it can be seen that the ocean surface height anomaly may be induced by eddy. We will provide verification for this result in the next section.

4 Discussion

To verify the eddy recognized from InIRA SLC data, MODIS SST and ocean color data were used to confirm the existence and location of eddy. The MODIS SST data and CHL concentration data are acquired from Ocean Data System (ODPS) and capable to display sea surface signatures of eddies.

A cyclonic eddy results in local downwelling, while an anticyclonic leads to local upwelling. The core of cyclonic eddy appears cool feature, and the center of anticyclonic eddy is warmer than the surrounding area. In Fig. 4a, an eddy was found on an InIRA image on November 8, 2017 at 01:19:36 UTC. Figure 4b shows the same eddy

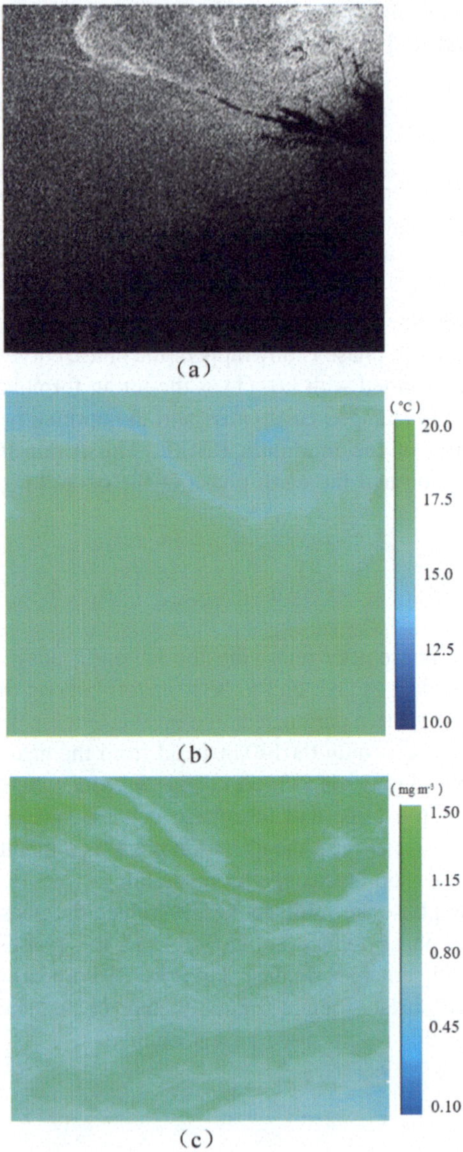

Fig. 4. (a) InIRA image acquired on November 8, 2017 at 01:19:36 UTC. (b) MODIS SST image obtained November 8, 2017 at 03:55:00 UTC. (c) MODIS CHL image acquired November 8, 2017 at 03:55:00 UTC

detected by MODIS on November 8, 2017 at 03:55:00 UTC. In Fig. 4b, we can see the cool eddy in the upside of the SST map. It can be seen that the core of the eddy and the spiral arms emanated from the core are about 2–3° cooler than the surrounding waters.

The downwelling and upwelling caused by eddies can cause the exchange of nutrients and plankton, which results in the higher CHL concentration of cyclonic eddies area and lower CHL concentration of anticyclonic eddies than the adjacent water [12]. As shown in Fig. 4c, the eddy was detected by MODIS on November 8, 2017 at 03:55:00 UTC. Due to the surfactant films get involved in the orbital motion of the eddy, the CHL concentration in the core of the eddy and the spiral arms is about 0.5–1 mg m^{-3} higher than the adjacent waters. From Fig. 4b and c, we can see that the center of the eddy with lower SST and with higher CHL concentration coincide.

5 Summary

More than 100 InIRA images obtained during 2016.9–2018.7 of different area are analyzed, the most important results gained are listed as follows.

Eddy structures appear in InIRA images can be divided into two different mechanisms. Under a moderate wind speeds of 3–5 m/s, film mechanism is the main form of oceanic eddies and about 65% of all eddies detected in InIRA images are visualized. Under a considerable near-water wind speed of 5–12 m/s, the wave/current interactions mechanism comes into action and visualizes about 35% of eddy structures in InIRA images.

By registering the InIRA SLC data and removing flat-earth phase, the result shows that the interferometric phase change is corresponding to the ocean surface height anomaly induced by eddy. The MODIS SST and CHL concentration data verify the eddy identified from InIRA image. In addition, the center of the eddy with lowest SST corresponds to the highest CHL concentration. Due to the calibration of parameters of InIRA has not been proceeded yet, we can only realize the preliminary detection of eddies. The data of conventional radar altimeters, such as Topex and Poseidon, can be used to validate mesoscale eddies after the parameter calibration in the future study.

The small scale and sub-mesoscale eddies are a significant portion of ocean, while they cannot be detected by conventional radar altimeters. Therefore, the present research on small scale and sub-mesoscale eddies are limited to case studies. Tiangong-2 InIRA provide a large amount of images for the observation and investigation of small scale and sub-mesoscale eddies. However, very small scale and large scale eddies also cannot be detected by InIRA due to its limited resolution or swath width. In addition, the SWOT (Surface Water and Ocean Topography) satellite whose primary payload is a wide-swath InSAR altimeter, is scheduled to be launched in 2021, allowing the investigation of sub-mesoscale and small scale eddies on a global scale.

Acknowledgement. Thanks to China Manned Space Engineering for providing InIRA data products of Tiangong-2. We also thank NASA for providing MODIS SST and ocean color data products.

References

1. Karimova, S., Gade, M.: Improved statistics of sub-mesoscale eddies in the Baltic Sea retrieved from SAR imagery. Int. J. Remote Sens. **37**(10), 2394–2414 (2016)
2. Alpers, W., Brandt, P., Lazar, A., et al.: A small-scale oceanic eddy off the coast of West Africa studied by multi-sensor satellite and surface drifter data. Remote Sens. Environ. **129**, 132–143 (2013)
3. Rodriguez, E., Morris, C.S., Belz, E.J.: A global assessment of SRTM performance. Photogramm. Eng. Remote Sens. **72**(3), 249–260 (2006)
4. Ulander, L.M.H., Askne, J.: Repeat-pass SAR interferometry over forested terrain. IEEE Trans. Geosci. Remote Sens. **33**(2), 331–340 (1995)
5. Kong, W., Chong, J., Tan, H.: Performance analysis of ocean surface topography altimetry by Ku-Band Near-Nadir interferometric SAR. Remote Sens. **9**(9), 933 (2017)
6. Jackson, C.: Synthetic Aperture Radar Marine User's Manual, 1st edn. U.S. Department of Commerce, Washington, USA (2004)
7. Karimova, S.: Spiral eddies in the Baltic, Black and Caspian seas as seen by satellite radar data. Adv. Space Res. **50**(8), 1107–1124 (2012)
8. Karimova, S.S., Lavrova, O.Y., Solov'ev, D.M.: Observation of eddy structures in the Baltic Sea with the use of radiolocation and radiometric satellite data. Izv. Atmos. Ocean. Phys. **48** (9), 1006–1013 (2012)
9. Yamaguchi, S., Kawamura, H.: SAR-imaged spiral eddies in Mutsu Bay and their dynamic and kinematic models. J. Oceanogr. **65**(4), 525–539 (2009)
10. Rodriguez, E., Martin, J.M.: Theory and design of interferometric synthetic aperture radars. IEEE Proc. F **139**(2), 147–159 (1992)
11. Jin, G., Xu, Q., Zhang, H.: Synthetic Aperture Radar Interferometry, 1st edn. National Defense Industry Press, Beijing, China (2014)
12. Gaube, P., Mcgillicuddy, D.J., Chelton, D.B., et al.: Regional variations in the influence of mesoscale eddies on near-surface chlorophyll. J. Geophys. Res. Ocean. **119**(12), 8195–8220 (2015)

Estimation of Lake Level Using Tiangong-2 InIRA Data

Jingjuan Liao[1(✉)] and Hui Xue[1,2]

[1] Key Laboratory of Digital Earth Science, Institute of Remote Sensing and Digital Earth, Chinese Academy of Sciences, Beijing 100094, China
liaojj@radi.ac.cn

[2] University of Chinese Academy of Sciences, Beijing 100049, China

Abstract. Tiangong-2 space laboratory, launched on 15 September, 2016, carries an Interferometric Imaging Radar Altimeter (InIRA). InIRA can conduct interferometry of ocean and land with high precision and wide swath. In this study, nine Qinghai-Tibet Plateau lakes are selected to analyze the preliminary performance of lake level estimation using the InIRA data. Based on bias and standard deviation of estimated lake level, the lake levels obtained from InIRA Level1 data (InIRA lake levels) are compared with the lake levels directly derived from Cryosat-2 Synthetic Aperture Radar interferometry (SARin) Level1b (L1b) data, and the accurate lake levels obtained from the retracking Cryosat-2 SARin L1b data. The results show that the bias of absolute InIRA lake levels is 4.105 m, and the mean standard deviation of InIRA point lake levels is 1.063 m. It indicates that InIRA Level1 data is more stable than Cryosat-2 SARin L1b data and relative lake levels with high precision can be expected to estimate using the further processed InIRA data.

Keywords: Tiangong-2 · InIRA · Cryosat-2 · Lake level estimation

1 Introduction

Lake levels are important indicators for regional and global environmental changes. In-situ gauges can measure high precision lake level data, but are expensive to maintain and challenging to operate in remote areas. Satellite radar altimetry has now been used successfully for more than two decades to measure lake levels as an addition to gauge measurements. The technique has, due to the large footprint, primarily been used for large-sized lakes, and measured lake levels with an accuracy of up to centimeter level [1–6]. At present, since the on-board radar altimeters operate in along-track measurement mode, it is difficult to measure smaller lakes due to their limited coverages and revisit cycles [7–9]. The Cryosat-2, launched in 2010 and with high along-track resolution, makes it possible to accuratcly monitor lake level through retracking processing [10–12]. Kleinherenbrink et al. [13, 14] have investigated the performance of Cryosat-2 data in SARin mode and found a good agreement with Jason-2 measurement. Tourian et al. [15] used Cryosat-2 measurements to study the shrinkage of Lake Urmia in Iran. Villadsen et al. [16] estimated river levels at virtual stations for the

Y. Gu et al. (Eds.): *Proceedings of the Tiangong-2 Remote Sensing Application Conference*, LNEE 541, pp. 333–343, 2019.
https://doi.org/10.1007/978-981-13-3501-3_31

Brahmaputra and Ganges Rivers. However, the water level difference often occurs due to the orbital drift and low revisit cycles when lake levels were estimated using Cryosat-2 data. It is necessary to use new altimetry data to monitor high-precision lake levels.

Tiangong-2 space laboratory, launched on 15 September, 2016, was equipped with an interferometric imaging radar altimeter (InIRA). InIRA can conduct interferometry of ocean and land with high precision and wide swath [17], and makes it possible to measure the water levels of smaller lakes. In this study, nine Qinghai-Tibet Plateau lakes are selected to analyze the preliminary performance of lake level estimation using the InIRA data, and the lake level derived from the Cryosat-2 Synthetic Aperture Radar Interferometry (SARin) data are used to verify the lake levels derived from the InIRA data in order to assess the estimation accuracy of lake level using the InIRA data.

2 Study Lakes and Data

2.1 Study Lakes

Nine Qinghai-Tibet Plateau lakes are selected as the study objects, and scattered in northern, southern, central and eastern parts of Qinghai-Tibet Plateau (Fig. 1). Among them, Chibuzhang Co has the largest area (541 km^2), Yulin Lake has the smallest area (13 km^2), and the average area is about 232 km^2.

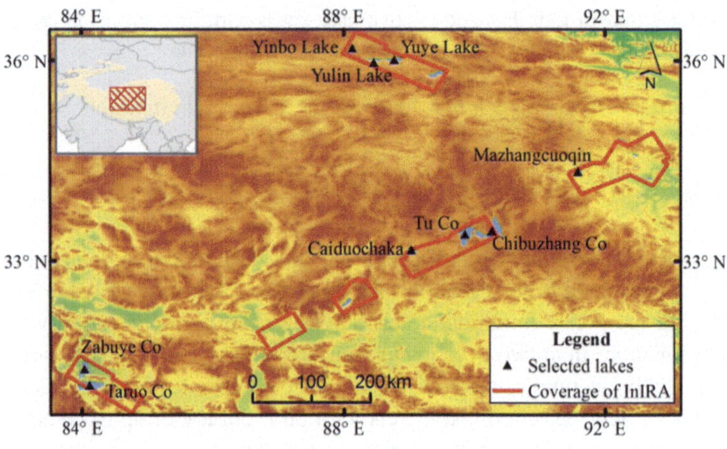

Fig. 1. Distribution of the lakes and coverage of InIRA data (background is the DEM of Shuttle Radar Topography Mission (SRTM))

2.2 Data Used in This Study

Tiangong-2 InIRA Data

Tiangong-2 InIRA data used in this study was acquired on September 23, 2016, and its operating frequency is Ku band (13.58 GHz), the interferometric baseline is 2.3 m, the swath width is more than 30 km, the observation angle is less than 8°, and the pixel resolution is 30 m in range. The data uses WGS-84 as the reference ellipsoid, and has been performed various correction, including instrument error, radial orbit error, and propagation delay error. Table 1 shows the statistics of InIRA level 1 data for the lakes in this study, and indicates that InIRA has a more observations than other spaceborne altimeters. Figure 2 shows the InIRA data covering these lakes, and some lakes, such as Taruo Co, Chibyzhang Co, and Caiduochaka, are not completely covered by the InIRA data. Figure 3 shows the coverage of InIRA data over these lakes.

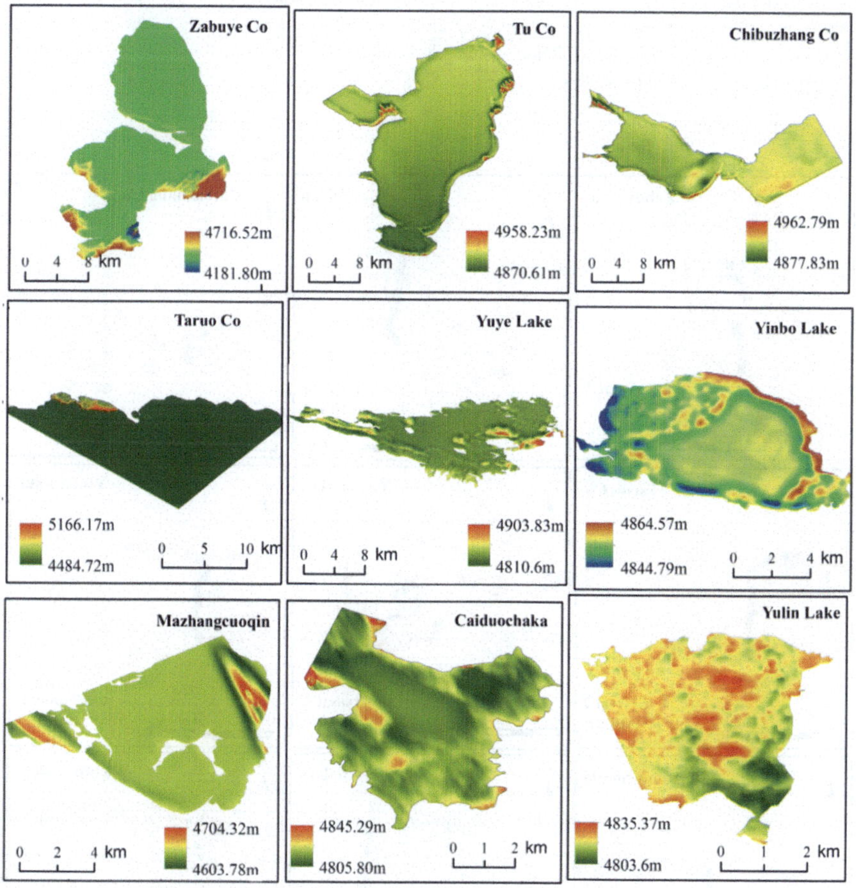

Fig. 2. The InIRA data over these lakes used in this study

Cryosat-2 Data

The operating frequency of Cryosat-2 is Ku band (13.58 GHz), and its interferometric baseline is 1.168 m. The SARin mode of Cryosat-2 has an along-track resolution of about 300 m, and the footprint is up to 15 km in the cross-track direction [18]. The Cryosat-2 data used in this study is Level 1b product of SARin mode, which contains

Table 1. Statistics of InIRA Level1 data and Cryosat-2 data in these lakes in this study

Lake Name	Area (km^2)	Standard deviation of InIRA[a] (m)	Number of observation points from InIRA[a]	Number of observation points from Cryosat-2	Observation date of Cryosat-2 (Year/month/day)	Time difference between two satellite observation (Days)
Zabuye Co	259	11.223	312953	80	20161014	21
Tu Co	448	2.381	547721	64	20160930	7
Chibuzhang Co	541	10.043	383560	16	20160903	20
Taruo Co	485	6.143	199571	53	20161014	21
Yuye Lake	147	2.325	193071	21	20161002	9
Yinbo Lake	50	1.429	64979	14	20160905	18
Mazhangcuoqin	68	16.581	61741	17	20160926	3
Caiduochaka	75	3.003	11240	26	20161002	9
Yulin Lake	13	3.607	9872	8	20161006	13

[a]The statistics are not processed by outliers.

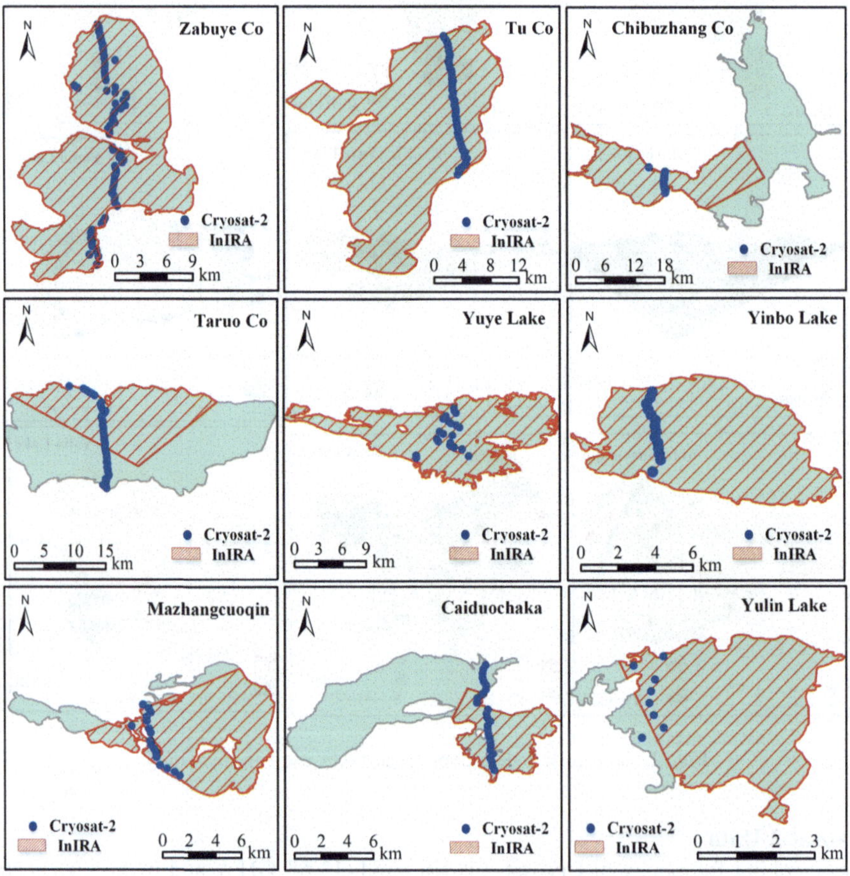

Fig. 3. Cryosat-2 crossing locations and InIRA data coverage for each lake

the orbit information and echoes received by the altimeter, also called waveforms. Level 1b data product is un-retracking data, so the waveforms were retracked using the Multiple Waveform Persistent Peak (MWaPP) retracker [11], which has proven to provide valid results for lake levels using retracking Cryosat-2 data. In order to synchronize with the InIRA observations, only the Cryosat-2 data for 30 days before and after September 23, 2016 is selected. Table 1 lists the details of Cryosat-2 data, and shows the number of observation points is at least 8 and the maximum is 80. Time difference of data acquisition between Cryosat-2 and InIRA is at least 7 days and up to 21 days. Figure 3 shows the observation locations of Cryosat-2 and InIRA for these lakes.

Mask Data of the Lakes

In this study, the mask data of the lakes were obtained from the 2014 Lake Dataset of Qinghai-Tibet Plateau published by Wan et al. [19] and 5 Landsat-8 Operational Land Imager (OLI) images acquired around September 23, 2016. Table 2 shows the details of the Landsat-8 OLI images used in this study.

Table 2. The Landsat-8 OLI images used in this study

ID	Data ID	Acquisition date	Time difference from InIRA observation (Days)	Covered lakes
1	LC81410352016299LGN00	20161025	32	Yinbo Lake
2	LC81400352016276LGN00	20161002	7	Yulin Lake, Yuye Lake
3	LC81420382016274LGN00	20160930	7	Zabuye Co, Taruo Co
4	LC81390372016253LGN00	20160909	14	Caiduochaka, Tu Co Cuibuzhang Co,
5	LC81380362016294LGN00	20161020	27	Mazhangcuoqin

3 Method

In this section we present the method to extract the lake level, including delineation of lake extent and estimation of lake level using Cryosat-2 and InIRA data. The mask data of the lakes were firstly used to delineate the lake extent in ArcGIS. Then, the Cryosat-2 and InIRA data within each lake were extracted using lake extent. Finally, the Cryosat-2 and InIRA data within each lake were used to calculate the lake level. Sections 3.1 and 3.2 describe how to calculate lake level using Cryosat-2 and InIRA data, respectively. In order to compare the lake levels derived from Cryosat-2 and InIRA data, we only use the InIRA data within 300 m centered on the Cryosat-2 measurements to calculate the lake level extracted from the InIRA data.

3.1 Estimation of Accurate Lake Level Using Cryosat-2 Data

Previous studies have shown that the accurate lake levels can be estimated using the retracking Cryosat-2 data [6, 10, 12]. Due to the lack of the in-situ lake levels on the Qinghai-Tibet Plateau, we used the lake levels extracted by Cryosat-2 data as a reference to verify the lake levels extracted by the InIRA data. The method of the lake level estimation using Cryosat-2 data is as following:

(1) Calculate the retracking water level H of each crossing point of the lake using Cryosat-2 L1b data, The formula is:

$$H = a - R_{wd} + R_{gc} + R_{retrack} - N \tag{1}$$

Where a is the distance of the satellite from the reference ellipsoid. R_{wd} is range from the satellite to the lake. R_{gc} indicates error correction, including correction of ionosphere, dry and wet troposphere, solid tide, ocean tide and pole tide. $R_{retrack}$ is retracking correction obtained by MWaPP method [11]. N is the height of EGM2008 geoid corresponding to the reference ellipsoid, and the size of the calculated grid is $1' \times 1'$. Except for $R_{retrack}$ and N, other parameters can be obtained from L1b data.

(2) Calculate accurate water levels from the water levels obtained by Eq. (1) using the Static Error Mixture Model [10].
In addition, we calculate the un-retracking water level (CSL 1b) using Cryosat-2 L1b data, and this water level is calculated in the same way as the accurate water level except that $R_{retrack}$ is not corrected in Eq. (1).

3.2 Estimation of Lake Level Using InIRA Data

In this study, the lake level estimated by the InIRA data includes the overall water level and along-track water level, indicating the water level in the lake surface and along the track. The estimation steps are as follows:

(1) Calculate the point water level H_{tg} from the InIRA data, and the corrections including pole tide, solid tide, ocean tide and geoid were applied to all InIRA data within each lake. The correction parameters are the same as in Sect. 3.1. The formula is:

$$H_{tg} = InIRA - (PoleCor + SolidCor + OceanCor) - N \tag{2}$$

Where $InIRA$ is the InIRA data, and $PoleCor$, $SolidCor$ and $OceanCor$ are the corrections including pole tide, solid tide, and ocean tide. N is the same as in Eq. (1).

(2) All point water levels were used to calculate the overall and along-track water level from the InIRA data using the Static Error Mixture Model.

4 Results and Discussion

4.1 Comparison of Absolute Water Level

For each lake, we calculated the along-track and overall water levels derived from the InIRA data, and the un-retracking and accurate water level derived from the Cryosat-2 L1b data, respectively. The amount of InIRA and Cryosat-2 measurements used to calculate various water levels is shown in Table 1. Table 3 shows the difference between accurate water level from the Cryosat-2 retacking data and along-track and overall water levels from the InIRA data, un-retracking water level from the Cryosat-2 L1b data, respectively. RMSE in Table 3 are the root mean square errors between accurate Cryosat-2 water level and the InIRA water levels, un-retacking Cryosat-2 water level of nine lakes, and they can indicate the estimation biases of various water levels. The table shows that the RMSE of the InIRA overall water level is lower of 0.298 m than that of the InIRA along-track water level, so the InIRA overall water level can reflect the water level of the lakes. However, the RMSE of the InIRA overall water level is 4.105 m with a deviation of meter level, but this value is significantly lower than the un-retracking Cryosat-2 water level. The previous study presented that the high-precision water level could be obtained using retracking Cryosat-2 data, and the measurement accuracy could reach decimeter level for the lakes in Tibetan Plateau [14], indicating that the accuracy of the water level estimated by the InIRA data could also be increased if the InIRA level 1 data is further processed. Therefore, the further study can be conducted in the future when more InIRA data were acquired.

Table 3. Comparison of InIRA along-track and overall water levels, un-retracking Cryosat-2 water levels and accurate Cryosat-2 water levels

No.	Lake level	Difference minimum (m)	Difference maximum (m)	RMSE (m)
1	InIRA along-track water level	1.140	10.452	4.403
2	InIRA overall water level	0.797	8.952	4.105
3	Un-retracking Cryosat-2 water level	0.937	44.306	17.546

Figure 4(a) shows the InIRA along-track and overall water levels, un-retracking and accurate Cryosat-2 water level of nine lakes, and Fig. 4(b) shows the difference between the InIRA along-trackwater level, overall water levels, un-retracking Cryosat-2 water level and accurate Cryosat-2 water level. They show that the InIRA overall water levels over nine lakes have consistent changes with the InIRA along-track water levels. The differences between the un-retracking Cryosat-2 water level and the accurate Cryosat-2 water level are large, especially for these lakes such as Chibuzhang Co, Tu Co, Taruo Co and Yulin Lake. The reason is that the Cryosat-2 data in the four lakes have more multi-peak waveforms in the near shore areas of these lakes, and it is difficult to obtain accurate water level from the Crysat-2 data if these multi-peak waveforms are not retracked [6, 10, 12], so we can only estimate lower precision water level derived from the un-retracking Cryosat-2 data.

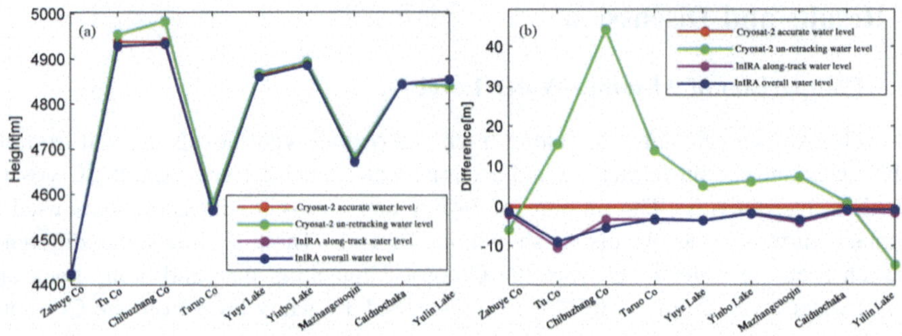

Fig. 4. Comparison of the InIRA along-track and overall water levels, un-retracking and accurate Cryosat-2 water level of nine lakes in the Qinghai-Tibet Plateau

4.2 Comparison of Standard Deviation

Table 4 shows the comparison of standard deviations between the InIRA point water level, Cryosat-2 un-retracking water level and Cryosat-2 accurate water level. The standard deviation of each lake is calculated by removing the outliers, and the outliers of the InIRA and Cryosat-2 data are removed respectively using the double error and Grubbs criterion [20, 21]. The smaller the standard deviation is, the more stable the measurement is, and the easier it is to obtain the accurate lake level. From the table, InIRA point water levels have the mean standard deviation of 1.063 m, and the mean standard deviation is much higher than that of the Cryosat-2 accurate water level, but significantly less than that the Cryosat-2 un-retracking water level.

Table 4. Comparison of standard deviations of InIRA point water level, Cryosat-2 un-retracking water level and Cryosat-2 accurate water level

No.	Lake level	Minimum standard deviation (m)	Maximum standard deviation (m)	Mean Standard deviation (m)
1	InIRA point water level	0.628	1.859	1.063
2	Cryosat-2 un-retracking water level	0.921	4.958	2.657
3	Cryosat-2 accurate water level	0.114	0.207	0.178

Figure 5 shows the standard deviations between the InIRA point water level, Cryosat-2 un-retracking water level and Cryosat-2 accurate water level in the nine lakes. From the figure, except for Mazhangcuoqin Lake, the standard deviations of the InIRA point water levels in other lakes are significantly less than that of the Cryosat-2 un-retracking water levels, and the standard deviations of the Cryosat-2 un-retracking

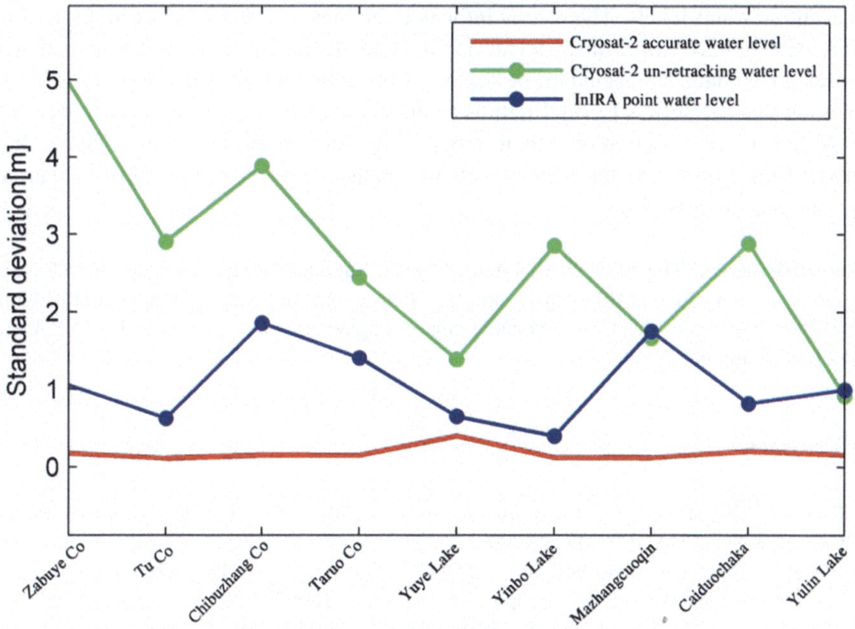

Fig. 5. Comparison of the standard deviations of the InIRA point water level, Cryosat-2 un-retracking water level and Cryosat-2 accurate water level of nine lakes

water levels in Yuye and Yinbo Lakes are close to the Cryosat accurate water levels. The results indicate that the InIRA Level 1 data is more stable than the Cryosat-2 L1b data, and the InIRA data could estimate high-precision relative lake level through the further correction processing because the InIRA has more observations than other altimaters, so it is very important for monitoring the water levels of smaller lakes. For the lake level estimation using altimeter data, the number of observations is critical to the accuracy of the water level extraction. Cryosat-2 can measure the larger lakes with better water level estimation accuracy due to more observations, while for smaller lakes, the measurement accuracy is not satisfactory due to fewer observations. However, InIRA can obtain more observations even in smaller lakes due to its three-dimensional imaging capability, thus ensuring the accuracy of water level estimation. From Table 1, for the smaller lakes, such as Yinbo Lake and Yulin Lake, the InIRA observations are more than Cryosat-2 observations. These indicate that the InIRA has much higher stability than Cryosat-2, and the InIRA has great potential for estimating the water level of smaller lakes.

5 Conclusion

In this study, nine lakes in the Qinghai-Tibet Plateau were selected to analyze the performance of water level estimation using the Tiangong-2 InIRA data. This study was conducted by comparing the InIRA water levels with the Cryosat-2 un-retracking and accurate water levels. The results indicated the bias of water level estimation of the InIRA data is 4.105 m with a deviation of meter level, but it is better than that of Cryosat-2 L1b data. The standard deviation of the InIRA point water level is 1.063 m, indicating that the InIRA Level 1 data is more stable than the Cryosat-2 L1b data. The InIRA has more observations than other altimeters, even in smaller lakes. It is important for monitoring the water levels in smaller lakes when more InIRA data are acquired and processed.

Acknowledgments. This research was funded by the National Key Research and Development Program of China (2016YFB0501501), and the National Natural Science Foundation of China (41871256). We thank the China Manned Space Engineering for providing the InIRA data products of Tiangong-2.

References

1. Crétaux, J.F., Birkett, C.: Lake studies from satellite radar altimetry. Comptes Rendus Geosci. **338**(14–15), 1098–1112 (2006)
2. Birkett, C.M.: The contribution of TOPEX/POSEIDON to the global monitoring of climatically sensitive lakes. J. Geophys. Res. Atmos. **100**(C12), 25179–25204 (1995)
3. Morris, C.S., Gill, S.K.: Evaluation of the TOPEX/POSEIDON altimeter system over the Great Lakes. J. Geophys. Res. Ocean. **99**(C12), 24527–24539 (1994)
4. Gao, L., Liao, J., Shen, G.: Monitoring lake-level changes in the Qinghai-Tibetan Plateau using radar altimeter data (2002–2012). J. Appl. Remote Sens. **7**(1), 073470 (2013)
5. Liao, J., Gao, L., Wang, X.: Numerical simulation and forecasting of water level for qinghai lake using multi-altimeter data between 2002 and 2012. IEEE J. Sel. Top. Appl. Earth Obs. Remote. Sens. **7**(7), 609–622 (2014)
6. Jiang, L., Nielsen, K., Andersen, O.B., Bauer-Gottwein, P.: Monitoring recent lake level variations on the Tibetan Plateau using CryoSat-2 SARIn mode data. J. Hydrol. **544**, 109–124 (2017)
7. Crétaux, J.F., Abarca-Del-Río, R., Bergé-Nguyen, M., et al.: Lake volume monitoring from space. Surv. Geophys. **37**(2), 269–305 (2016)
8. Xu, Y., Gao, L., Zhang, Y.: New generation altimetry satellite SWOT and its reference to China's swath altimetry satellite. Remote. Sens. Technol. Appl. **32**(1), 84–94 (2017). (In Chinese)
9. Sui, X., Zhang, R., Wan, X.: Research on microwave interferometric altimetry systems and performances. Spacecr. Eng. **25**(2), 13–18 (2016). (In Chinese)
10. Nielsen, K., Stenseng, L., Andersen, O.B., Villadsen, H., Knudsen, P.: Validation of CryoSat-2 SAR mode based lake levels. Remote Sens. Environ. **171**, 162–170 (2015)
11. Villadsen, H., Deng, X., Andersen, O.B., Stenseng, L., Nielsen, K., Knudsen, P.: Improved inland water levels from SAR altimetry using novel empirical and physical retrackers. J. Hydrol. **537**, 234–247 (2016)
12. Göttl, F., Dettmering, D., Müller, F., et al.: Lake level estimation based on CryoSat-2 SAR altimetry and multi-looked waveform classification. Remote. Sens. **8**(11), 885–901 (2016)
13. Kleinherenbrink, M., Ditmar, P., Lindenbergh, R.: Retracking cryosat data in the SARin mode and robust lake level extraction. Remote Sens. Environ. **152**, 38–50 (2014)
14. Kleinherenbrink, M., Lindenbergh, R.C., Ditmar, P.G.: Monitoring of lake level changes on the Tibetan Plateau and Tian Shan by retracking Cryosat SARIn waveforms. J. Hydrol. **521**, 119–131 (2015)

15. Tourian, M., Elmi, O., Chen, Q., Devaraju, B., Roohi, S., Sneeuw, N.: A spaceborne multisensor approach to monitor the desiccation of Lake Urmia in Iran. Remote Sens. Environ. **156**, 349–360 (2015)
16. Villadsen, H., Andersen, O.B., Stenseng, L., Nielsen, K., Knudsen, P.: CryoSat-2 altimetry for river level monitoring—evaluation in the Ganges-Brahmaputra river basin. Remote Sens. Environ. **168**, 80–89 (2015)
17. Bao, Q., Lin, M., Zhang, Y., et al.: Wind speed inversion for imaging microwave altimeter. J. Remote. Sens. **21**(6), 835–841 (2017). (In Chinese)
18. European Space Agency (ESA). CryoSat-2 Product Handbook[OL]. https://earth.esa.int/web/guest/missions/esa-operational-eo-missions/cryosat
19. Wan, W., Xiao, P., Feng, X., et al.: Monitoring lake changes of Qinghai-Tibet Plateau over the past 30 years using satellite remote sensing data. Chin. Sci. Bull. **8**(8), 701–714 (2014)
20. Gao, L.: Monitoring the changes in lake level and glacier elevation in the Qinghai-Tibet Plateau using satellite altimetry data. Ph.D. Dissertation of University of Chinese Academy of Sciences (2014). (In Chinese)
21. He, P.: Some methods of deleting inordinate values from measuring data. Aviat. Metrol. Meas. Technol. **15**(1), 19–22 (1995). (In Chinese)

Index

Y. Gu et al. (Eds.): *Proceedings of the Tiangong-2 Remote Sensing Application Conference*, LNEE 541, pp. 345–346, 2019.
https://doi.org/10.1007/978-981-13-3501-3

The manufacturer's authorised representative in the EU is Springer Nature Customer Service Centre GmbH, Europaplatz 3, 69115 Heidelberg, Germany. If you have any concerns regarding our products, please contact ProductSafety@springernature.com

Printed and bound by CPI Group (UK) Ltd, Croydon, CR0 4YY
25/09/2026
02222900-0015